Theoretical and Practical Teaching Strategies for K-12 Science Education in the Digital Age

Jason Trumble
University of Central Arkansas, USA

Sumreen Asim
Indiana University Southeast, USA

Joshua Ellis
Florida International University, USA

David Slykhuis
Valdosta State University, USA

A volume in the Advances in Early Childhood and K-12 Education (AECKE) Book Series

IGI Global
PUBLISHER of TIMELY KNOWLEDGE

Published in the United States of America by
IGI Global
Information Science Reference (an imprint of IGI Global)
701 E. Chocolate Avenue
Hershey PA, USA 17033
Tel: 717-533-8845
Fax: 717-533-8661
E-mail: cust@igi-global.com
Web site: http://www.igi-global.com

Library of Congress Cataloging-in-Publication Data

Names: Trumble, Jason, DATE- editor. | Asim, Sumreen, DATE- editor. | Ellis, Joshua, DATE- editor. | Slykhuis, David, DATE- editor.
Title: Theoretical and practical teaching strategies for K-12 science education in the digital age / Jason Trumble, Sumreen Asim, Joshua Ellis, and David Slykhuis, editors.
Description: Hershey, PA : Information Science Reference, [2022] | Includes bibliographical references. | Summary: "The purpose of this book is to communicate how to increase science educators' understanding of the needs of digital age learners, develop theoretical and practical teaching strategies that align with science content, and integrate technologies for learning seamlessly"-- Provided by publisher.
Identifiers: LCCN 2022031076 (print) | LCCN 2022031077 (ebook) | ISBN 9781668455852 (hardcover) | ISBN 9781668455890 (paperback) | ISBN 9781668455869 (ebook)
Subjects: LCSH: Science--Study and teaching (Elementary)--United States. | Science--Study and teaching (Secondary)--United States. | Science--Study and teaching (Elementary)--Technological innovations--United States. | Science--Study and teaching (Secondary)--Technological innovations--United States.
Classification: LCC Q181 .T446 2022 (print) | LCC Q181 (ebook) | DDC 507.1/273--dc23/eng20220919
LC record available at https://lccn.loc.gov/2022031076
LC ebook record available at https://lccn.loc.gov/2022031077

This book is published in the IGI Global book series Advances in Early Childhood and K-12 Education (AECKE) (ISSN: 2329-5929; eISSN: 2329-5937)

British Cataloguing in Publication Data
A Cataloguing in Publication record for this book is available from the British Library.

Advances in Early Childhood and K-12 Education (AECKE) Book Series

Jared Keengwe
University of North Dakota, USA

ISSN:2329-5929
EISSN:2329-5937

Mission

Early childhood and K-12 education is always evolving as new methods and tools are developed through which to shape the minds of today's youth. Globally, educational approaches vary allowing for new discussions on the best methods to not only educate, but also measure and analyze the learning process as well as an individual's intellectual development. New research in these fields is necessary to improve the current state of education and ensure that future generations are presented with quality learning opportunities.

The **Advances in Early Childhood and K-12 Education (AECKE)** series aims to present the latest research on trends, pedagogies, tools, and methodologies regarding all facets of early childhood and K-12 education.

Coverage

- Poverty and Education
- Bullying in the Classroom
- Urban K-12 Education
- K-12 Education
- Head Start and Pre-K Programs
- Curriculum Development
- Literacy Development
- Standardized Testing
- Performance Assessment
- Learning Outcomes

IGI Global is currently accepting manuscripts for publication within this series. To submit a proposal for a volume in this series, please contact our Acquisition Editors at Acquisitions@igi-global.com or visit: http://www.igi-global.com/publish/.

The Advances in Early Childhood and K-12 Education (AECKE) Book Series (ISSN 2329-5929) is published by IGI Global, 701 E. Chocolate Avenue, Hershey, PA 17033-1240, USA, www.igi-global.com. This series is composed of titles available for purchase individually; each title is edited to be contextually exclusive from any other title within the series. For pricing and ordering information please visit http://www.igi-global.com/book-series/advances-early-childhood-education/76699. Postmaster: Send all address changes to above address.

Titles in this Series

For a list of additional titles in this series, please visit: www.igi-global.com/book-series/advances-early-childhood-education/76699

Practical Strategies to Reduce Childhood Trauma and Mitigate Exposure to the School-to-Prison Pipeline
Belinda M. Alexander-Ashley (Independent Researcher, USA)
Information Science Reference • © 2023 • 350pp • H/C (ISBN: 9781668457139) • US $215.00

Preparing Pre-Service Teachers to Integrate Technology in K-12 Classrooms Standards and Best Practices
C. Lorraine Webb (Texas A&M University, San Antonio, USA) and Amanda L. Lindner (Texas A&M University, San Antonio, USA)
Information Science Reference • © 2022 • 340pp • H/C (ISBN: 9781668454787) • US $215.00

Cutting-Edge Language and Literacy Tools for Students on the Autism Spectrum
Katharine P. Beals (Drexel University, USA)
Information Science Reference • © 2022 • 298pp • H/C (ISBN: 9781799894421) • US $215.00

Handbook of Research on Family Literacy Practices and Home-School Connections
Kathy R. Fox (University of North Carolina Wilmington, USA) and Laura E. Szech (University of North Carolina Wilmington, USA)
Information Science Reference • © 2022 • 353pp • H/C (ISBN: 9781668445693) • US $270.00

Rethinking Inclusion and Transformation in Special Education
Maria Efstratopoulou (United Arab Emirates University, UAE)
Information Science Reference • © 2022 • 348pp • H/C (ISBN: 9781668446805) • US $215.00

Best Practices for Trauma-Informed School Counseling
Angela M. Powell (Lone Star College, USA)
Information Science Reference • © 2022 • 331pp • H/C (ISBN: 9781799897859) • US $215.00

Disciplinary Literacy as a Support for Culturally and Linguistically Responsive Teaching and Learning
Leslie Haas (Xavier University of Louisiana, USA) and Jill T. Tussey (Buena Vista University, USA)
Information Science Reference • © 2022 • 391pp • H/C (ISBN: 9781668442159) • US $215.00

Handbook of Research on Using Motor Games in Teaching and Learning Strategy
Pedro Gil-Madrona (University of Castilla-La Mancha, Spain)
Information Science Reference • © 2022 • 511pp • H/C (ISBN: 9781799896210) • US $270.00

IGI Global
PUBLISHER of TIMELY KNOWLEDGE

701 East Chocolate Avenue, Hershey, PA 17033, USA
Tel: 717-533-8845 x100 • Fax: 717-533-8661
E-Mail: cust@igi-global.com • www.igi-global.com

Editorial Advisory Board

Table of Contents

Detailed Table of Contents

Technology Integration in Practice

This chapter presents three cases of high school science teachers collaborating with personnel at three different universities to design and implement innovative, curriculum-aligned geospatial lessons involving hands-on data collection, following the authors' Socio-Environmental Science Investigation (SESI) model. Each case is presented in the context of professional development activities conducted at each university, the curriculum of the collaborating teacher, the students' engagement with the instructional sequence, and the teacher's reflections following implementation. All teacher professional development, collaborative curriculum development, and classroom implementation were conducted using ArcGIS Online and its affiliated tools. Implications and conclusions for other teachers' and teacher educators' development of similar, geospatially-integrated instruction are discussed.

This chapter describes the development, characteristics, and applicability of a web-based interactive notebook, the Science Practices Innovation Notebook (SPIN), for use by high school students and teachers during science investigations. SPIN integrates data practices, computational thinking (CT), and self-regulated learning (SRL) principles to support high school students as they engage in science investigations. SPIN uses CT tactics and underlying SRL principles to optimize student progression through five core data practices. SPIN also incorporates loop mechanisms that foster communication, prompting, and feedback among individual students, their teachers, and peers. SPIN gives teachers data about student progress, communication, and learning analytics that assists in providing high quality feedback to students. This chapter will describe how the theoretical underpinnings of SPIN reinforce each other, how collaborative interactions between teachers and researchers led to SPIN development and design, and how SPIN implements support for student data practices in an online setting.

Preface

Science and technology have had a long and intertwined history. Scientific discoveries allow for the creation of new technologies, and new technologies allow science to expand and grow into new areas. In a similar fashion, science education has had a long and intertwined history with technology. Science teachers have been capitalizing on technological advances from as far back as the advent of photography and "moving pictures," bringing opportunities to students in the classroom that were not available to them before.

We are now in a place where science education and science teachers are inundated with options for bringing technology into their classrooms (Asim et al, 2022). At the same time, science teachers are faced with an ever-growing cascade of new initiatives to add engineering, computer science, inquiry, problem-based learning, STEM, 21st century skills, and more to their teaching repertoire. If that doesn't make the life of a science teacher difficult enough, there is also now an expectation that, at any moment, whatever is happening in a classroom can be translated into an equivalent online experience. Can science educators respond to these pressures and continue to prepare a scientifically literate society with the thoughtful inclusion of relevant technologies?

This book aims to give hope to the science educator that all of this is possible! Technology can and should be integrated into science teaching and learning. Technology can help students of all levels to achieve and learn. Science and technology can be combined to help students learn computational thinking skills and engineering design. Combined with innovative pedagogy, science and technology can invigorate and engage students in learning science in ways that were not possible even a few years ago. In the pages that follow, you will find examples of these approaches that include practical applications for science educators.

The book opens with three chapters focused on how to bring science education to all learners, regardless of their ability level, social, cultural, or religious backgrounds, or if they are in a physical classroom or online. Boda (DISc) provides a framework, Design-based Inclusive Science (DISc), with an example of how to put the approach into practice. While this framework was developed with the idea of ensuring that disabled students and students from marginalized populations would receive an optimal science education experience, it can be considered to be like a Universal Design for Learning for science education, where good teaching for these populations is good teaching for everyone.

While Boda was focused on students who have traditionally struggled in science classrooms, the second chapter by Johnson-Leslie is focused on meeting the needs of the gifted and talented population through the implementation of makerspaces. With the testing culture that has been prevalent throughout the education system in recent decades, these students have not received as much focus as they will typically 'pass' with limited teacher intervention. The last chapter in this section focuses on meeting the needs of all science learners while throwing in the added dimension of teaching online. This chapter by Hatoum provides examples of how to keep all students motivated in an online environment and includes ways to simulate and do hands-on lab activities.

The second section of the book focuses on the more recent pressures that have been added to science education, incorporating engineering and computational thinking. Canbazoğlu presents a series of examples of how to use a simple and relatively inexpensive piece of equipment, Arduino, to incorporate engineering design into STEM activities. Several classroom examples are given to illustrate how this can be accomplished. Jacques's chapter presents ideas on how to use the engineering design process, and problem-based learning meets the challenges of the Next Generation Science Standards. They contend that computational thinking is a natural bridge between the engineering design process and problem-based learning and provide a series of steps to implement this in your classroom with a lesson example.

The third section of this book focuses on specific examples of incorporating different technologies into science education. Popejoy presents three different cases for using ArcGIS to have students collect authentic data on authentic problems in their local area. Burton focuses on harnessing the ubiquity of online access to create web-based interactive science notebooks. One advantage of this method is for allowing for more student autonomy in the pace of their learning. Lastly, Vieyra illustrates how to harness the power of the microcomputer that each student is carrying in their pocket. The chapter illustrates two examples of using smartphones to enhance science instruction using multiple examples that connect the classroom to the real world.

The last section of the book provides some more general principles for the effective integration of technology into the science classroom. The first two chapters in this section examine two issues that have become common since the turn of the century, incorporating 21st century skills, and finding a way to keep up with the ever-increasing onslaught of information and ideas that are available online. Ali presents a crosswalk between 21st Century skills and the Technological, Pedagogical, and Content Knowledge (TPACK) framework. They find that science teachers tend to be content-focused first, and while science teaching experience leads to an increase in content knowledge, it is negatively correlated with technology knowledge. Tsybulsky examines how the principles of digital curation can be applied to the science classroom to sort through the abundance of technology and online resources while helping students to better handle the new rigors of social media and online communities.

The last three chapters of the book provide principles and steps for going from thinking about using technology in a science classroom to effectively putting it into practice. Krahenbuhl provides a set of four principles for integrating technology and provides examples of each of these in practice in an elementary school setting. Schwortz presents a review of the literature on how technology has been used in science education and builds on these examples to give vignettes and guidance on how to put this research into practice. Finally, Unruh, Giles, and Willis close the book with an exploration of how the thoughtful use of vicarious experiences in a science education methods course can promote science teacher candidates' understanding of how content, pedagogy, and technology intersect in the classroom.

Whether you are a pre-service science teacher, a practicing K-12 science teacher, or a university science educator, we hope that you will find this book useful. We have strived in each chapter to communicate how to increase science educators' understanding of the needs of digital-age learners, develop theoretical and practical teaching strategies that align with science content, and integrate technologies for learning with fidelity. The long-standing interconnectedness of science and technology is not going away - in fact, we believe that it will be ever-increasing. Hopefully, this compendium will help you to navigate this ever-changing landscape.

Jason Trumble
University of Central Arkansas, USA

Sumreen Asim
Indiana University Southeast, USA

Joshua Ellis
Florida International University, USA

David Slykhuis
Valdosta State University, USA

REFERENCE

Asim, S., Ellis, J., Slykhuis, D., & Trumble, J. (2022). Three domains for technology integration in science teacher education. *Contemporary Issues in Technology & Teacher Education*, *22*(2). https://citejournal.org/proofing/three-domains-for-technology-integration-in-science-teacher-education/

Section 1
Science Education for All Learners

Chapter 1
Disability-as-Possibility:
Leveraging Technology for Design-Based Inclusive Science (DISc)

Phillip Boda
University of Illinois Chicago, USA

ABSTRACT

This chapter presents a theory of design-based inclusive science (DISc) to help science teachers and researchers navigate how to explicitly design for all body-mind-spirits in the room. First, it outlines the background of design-based research, and then it describes the affordances and limitations among these approaches. The chapter then describes the lacking inquiries on disabled students in science education around justice and describes the importance of teachers to see themselves, and be seen, as designers.

INTRODUCTION

Design research (DR), and its subsequent progeny Design-based Research (DBR), has been primarily described as a researcher-focused endeavor that incorporates feedback inputs to inform design impact and interface features. Turning the power of design and its possibility for science teachers in the digital age on its head recognizes that students' lived realities in and out of the classroom can be leveraged to inform how to dream of and desire new ways of creating learning environments using technology in K-12 spaces among teachers and students. Broaching this task with teachers, though, demands the witnessing of those who have been excluded by design in many of these educational research endeavors, even if they focus on racial justice (Boda et al., 2022). To move toward this goal means to also engage with students labeled with disabilities as often being positioned peripherally within design-based approaches specific to disciplinary endeavors, and even among inquiries that examine forms of oppression across race and disability in education (Kulkarni et al., 2021). This reality demands that if we leverage technology to design for more just learning, we must develop an acuity among science teachers about what this looks like on the ground, complementing what has been developed in academia.

DOI: 10.4018/978-1-6684-5585-2.ch001

This chapter presents a Theory of Design-based Inclusive Science (DISc) to help science teachers (and researchers) navigate how to explicitly design for all Body-Mind-Spirits in the room to feel like they are in the act of Becoming, that they Belong, and so they feel Beloved (Boda, In press). First, this approach outlines Design-based Research and then describes the affordances and limitations among these approaches in relation to social design experiment work (Gutiérrez et al., 2017). The chapter then describes lacking research inquiries with Disabled students in the field of science education around justice and describes the importance of teachers seeing themselves, and being seen, as designers. Finally, the DISc approach describes a set of projects where marginalized students were central and proximal to the designs created by science teachers and a design-based research team. Through this interweaving of narrative, literature, and notions of possibility advocated more broadly (Warren, 2021), I invoke an argument about how science teachers can start to see themselves as participant designers of their own classroom learning by engaging with a praxis of understanding the importance of design-based principles imbued with sociocultural features that have been taken up in other fields like English (Fowler-Amato & Warrington, 2017), but are no less important for science teaching for more just futures.

Design (-based) Research in Science Education Leveraging Technology

Research into teaching and learning environments has been primarily seen as concerned with creating disciplinary artifacts designed with specific purposes, methods, and goals – albeit sometimes the assumptions of these designs are not well unpacked in relation to the normative grammars of schooling or analyzed for their sociocultural affiliations (Ladson-Billings, 2018). Drawing from Learning Sciences research, Design-based Research (DBR; Design-based Research Collective, 2003) is seen on one hand as an inquiry into the design process itself, or, rather, viewing 'design *as* research' (Svihla & Reeves, 2016), but it can also be seen as the iterative impact of negotiations that occur among research partners, teachers, and students to fulfill specific disciplinary goals, and achieve greater equity among standardized outcomes. More recently, DBR has also been utilized to bring awareness and challenge socially unjust conditions in ways that support multiply marginalized youth to envision a world that is not yet actualized in the day-to-day designs they encounter within schools (cf. Gutiérrez & Jurow, 2016; Shea & Sandoval, 2020; Vakil, 2020). This latter theorization of Design-based Research as a space for invention and transformation is the focus on this chapter's claim: If science teachers seek to leverage technology toward justice, their designs must be inclusive and support all needs.

In these nuanced explanations of design research traditions in education and beyond, often the explicit defining of benchmarks that would constitute success plays a crucial role in both the iterative nature of research across disciplinary commitments, as well as how and through what measures such approaches seek to meet their intended goals. However, critical scholars in DBR argue these goals should not remain static, nor should they envelope rigid expectations of a 'one-size-fits-all' mentality (Gutiérrez et al., 2017), lest they fall into a process merely supporting the standardized status quo. Alternatively, taking up this more justice-oriented approach toward DBR asks, instead, if we are to encourage teachers to view digital-aged youth as innovators and creators of the future worlds in which they will experience power and space, our science instruction demands that we first help further the imaginative possibilities of what science inquiries could be guided by, how they are conducted, and whom they serve. Taking form as a pragmatic curriculum design approach guided by these ideas, science teachers would ask students about their interests, out-of-school Lives, and Dreams for the future. These would be subsequently tied into the adaptations to provided curriculum or newly designed activities to engage these digital-aged youth. This process of engaging with students who are in K-12 classrooms is not new, although with the onset of greater inclusive contexts where general and special education students learn together and the pertinence for science/STEM teachers to see themselves, and be seen, as designers is paramount. Teachers, and especially disciplinary expert teachers, must first see that their classrooms are designed a certain way from history; but, when they engage with their classrooms as designed spaces, they can imagine more inclusive systems.

Science education scholars have over the past twenty years reported many research projects intended to serve historically marginalized populations in STEM degrees and careers more broadly but often struggle to move beyond standardized goals that are designed to meet all students' needs toward scalable and lasting transformations (cf. Boda & Riley Miller, 2022; Kim et al., 2007; Krajcik et al., 2008; Mehalik et al., 2008; Semali et al., 2015). This recognition is not a lament of lack, per se, but rather an engagement with the reader about how the history of science education research is, like science disciplines more broadly, ever-evolving and therein in need of revision in terms of our goals and purposes for K-12 science teaching and learning. In turn, there is, unfortunately, a paucity of science education research at the multiple nexus of marginalization integrating disability which comes at the expense of students most vulnerable to exclusion from participation in science (Boda, 2018). While some exploratory and empirical studies observe what marginalized students experience in science classrooms, they do little to design specific innovations, or design for transformative futures, focused on disrupting systemic inequities among the pedagogy, curriculum, and assessment used – such as explicitly integrating students' Dreams and aspirations for content learning (Warren, 2021). To challenge this gap in the literature, this chapter explicates the design decisions made to develop more inclusive science learning focused on how technology is used in justice-oriented pursuits toward more concrete actualizations of science for all, beyond issues of access via accommodations and adaptations.

What Has Been Explored About Students Labeled With Disabilities in Science Education?

Disabled students are consistently marginalized even within the justice-oriented science education community (Boda, 2019), as well as often misinterpreted by researchers attempting to analyze social justice work that may be named 'Disability Justice' (Boda, 2022b). I argue, we need to radically re-center the purpose and promise of technology when designing science learning environments to support those most vulnerable to the negative effects related to the imposition of standards-focused aims: students labeled with disability(ies). Past scholarship has explicitly emphasized technology in education for 'employability' (Williams-Diehm et al., 2018), which on its own doesn't engage these youth with how they understand science content in their lives, nor does it think about them as having the agency and intelligibility to help teachers envision a space for their Body-Mind-Spirits as Belonging in K-12 science/STEM classrooms. In other words, when we think about designing with Disabled and multiply marginalized students toward participation in science, we must not merely stay at the level of accommodations and 'quick-fixes' to what is often perceived as a hindrance to 'efficient' science/STEM instruction. This 'amelioration of academic disparity' does little to work toward justice. Instead, what if we viewed *disability-as-possibility*, as argued more broadly from Black education scholars (Warren, 2021). This would require science teachers to engage with Disabled students beyond content and academics: To see disability beyond 'fixing' and Disabled youth as leaders (Berne et al., 2018).

Furthermore, adopting only a pragmatic approach to inclusive science/STEM that only seeks to support students through accommodations leaves the central goal for 'inclusion' of students labeled with disabilities in the individual and lacks notions of systemic ableism. These approaches date back to work in science education by Salend (1998) for students labeled with disabilities around developing life skills rather than seeking to disrupt ableist and racist systems of oppression that deny voice/choice from those students (Reid & Knight, 2006). This neglect of engaging Disabled students' humanities and personalities denies the power and promise of inclusive education beyond deficit narration and positioning (cf. Fitch et al., 2021; Nusbaum, 2013; Reeves et al., 2022; Thomas, 2013). This neglect also denies a rigorous scientific inquiry that sees these students as sites of possibility, replacing this opportunity with a technician's approach to diverse Body-Mind-Spirits where the goal is to 'fix' the student to assimilate into the rigid conformist-based, 'normal' classroom. This upholds a static commitment of teachers as de-professionalized technical executors of content, rather than *designers* of their science classroom spaces that afford their students opportunities to engage with reform-minded practices (NGSS Lead States, 2013). When thinking about Disabled students at multiple identity nexuses of marginalization, this type of pragmatic solution without a deep consideration of systemic designs erases disability as a culture and an identity that deserves respect to support these students' Lives-Hopes-Dreams. In simpler terms, when giving science teachers various tools to 'support' Disabled students without discussing why tools – 'accommodations' – are needed in the first place, does justice for no one. An example of this is when teachers receive an Individualized Education Plan for students, or a 504, and it says they need extra time on tests and quizzes. Instead of just adding time for these individual students, what if a science teacher focuses more on developing deep questions about the content being learned and asked students to apply them in a portfolio and presentation format? Forming deeper questions, in this context, disrupts the status quo of often-used assessments in science and aligns with the most recent NGSS policy.

Indeed, reviews of research concerning students with disabilities in science education contexts mirror this technocratic commitment with little attention paid to justice-oriented goals more broadly (cf. Comarú et al., 2021; Kaldenberg et al., 2015; Tosun, 2022; Villanueva et al., 2012). However, the maintenance of these standardized goals for those multiply marginalized youth labeled with disabilities is not without contention. Sullivan and Thorius (2010) and Connor (2012) have pushed back for education theorists to conceptualize what students labeled with disabilities mean for the larger field of justice-oriented education research. Pugach et al. (2021) align with the above scholars, noting the absence of disability in much of teacher education:

... disability remains an intractable problem in relationship to social justice, requiring a need to engage far greater complexity—and plain hard work—to transcend its precarious status quo. Assuring disability a consistent, unambiguous seat at the social justice table necessitates robust, innovative understandings of the full dimensionality of social justice to achieve the deep transformation of teacher education, predicated on collective ownership. Located at the interstices of identity and instruction, disability may offer a unique site where transformational questions of identity and instruction can be challenged simultaneously and recursively. As such, disability may hold the potential to serve as a pivot upon which teacher educators oriented toward social justice begin not only to address these ambiguities, but simultaneously help right long-standing wrongs with regard to disability, thus strengthening social justice as a common endeavor. (p. 247)

My work in science education research with new teachers and graduate students has shown that viewing disability along these axes of power and difference remains a struggle for all education stakeholders (Boda, 2019). Moreover, in my collaborative work with Disability Studies in Education (DSE) scholars (Boda et al., 2022; Kulkarni et al., 2021), we have been part of a wave in (science) education research toward justice that has started taking up inclusive interrogations of disability beyond deficit frames of what these students 'lack,' along with contributors to Koomen et al. (2018), in ways that define disability beyond in need of accommodations and, instead, imagine that students labeled with disabilities merely don't fit into the designs of historical classrooms where the premise and promise of education was solely about social stratification. What is needed is a re-engagement with science teachers to explore the ways that technology can be designed specifically to move beyond disciplinary commitments that exclude by design. In sum, this chapter follows what has been asked by Ware (2001): How can we infuse justice into the design goals we have for *all* students learning science using technology? And *do we dare do it*? The answers to these questions guide this chapter's narrative and purpose.

TEACHERS AS DESIGNERS OF LEARNING

The onset of accountability reforms in education have led to the need for teachers to take on more responsibility for the students they are charged to serve – no matter their own prior training, personal background, or cultural affiliation with their students. Whether you agree or disagree with the policies and practices that educational reforms bring – i.e., *No child left behind* and *Race to the top* – what is evident in this larger sphere of educational reform across the country and the world is that (science) teachers are expected to be forever learners, which was, surprisingly, something I've anecdotally found when speaking with veteran teachers with over 30+ years. This does not have to mean the same thing for all teachers, though; some teachers do not adopt this stance because it might not feel necessary – others adopt it out of necessity: That as student demographics change, so must approaches to build connections. For science teachers this means that when they engage their students, they need to know about their Lives and use that knowledge to design instruction in ways that are relevant to disciplinary expectations (NGSS Lead States, 2013), and align with justice-oriented lenses to help students feel like they have the right to voice their needs and future Hopes for their education (Calabrese-Barton & Tan, 2019).

As older generations of teachers engage with youth as digital consumers and producers, this means that these educators, while they might not enjoy or thrive in the space of technology, are required to respond in kind to the ways in which it is used by youth and how those types of interfaces could be productive in science/STEM classrooms (Emdin et al., 2021). Taking this perspective toward the digital-aged evolution of teaching means that teachers must not wait for top-down professional development to guide their learning. Instead, as has been advocated in English education (Fowler-Amato & Warrington, 2017), science teachers need to be seen as robust designers of their science classroom's curriculum, pedagogy, and assessment. This does not mean that this looks the same for all teachers but it does recognize and make evident that science teachers are powerful stakeholders in designing for equitable science (Kali et al., 2015). The question remains: How do science teachers do this when not able to work with researchers?

Designing With Marginalized Students as Innovation Experts

Recently, Vakil and McKinney de Royston (2022) put forth that youth should be seen as 'philosophers of technology' in ways that move toward a radical pragmatics of how we can support student learning beyond content acquisition among informal contexts. Drawing on this work as a frame of reference and theorization, youth *in schools* should also be seen as *Innovation Experts* who are active agents of their learning where they can apply the content they learn to their real lives outside school. That is, marginalized youth can teach the educators charged to serve them novel ways in which technology could be infused within science/STEM classrooms if given the space to showcase their brilliance. In the broader education research field, participatory design is often a term used to describe this process of how one lead person or persons, often with more authority or perceived knowledge base, engages their population by conducting a research study around an agreed-upon topic of inquiry. Participatory *action* research is like this but is also concerned with how the lead agent themselves change during this process rather than just the artifact/design impact on their population. Teachers are often asked to be disciplinary experts and pedagogical designers, and this pressure to help students succeed is thereafter compounded by the context in which they teach. For teachers serving primarily marginalized groups across race, gender, and ability, teaching a discipline like science in a social-constructivist manner while also designing for specific accommodations for students labeled with disabilities can be seen as, or could become, an insurmountable task. This approach is not the only way to pursue inclusion.

What I am asking the reader to consider is what if the social constructivist stance toward science education did not need the explicit *addition* of accommodations, specifically in relation to the literature presented above on the consistent prevalence of this type of work done in science education. Rather, what would it mean if we helped science/STEM teachers become designers of their own learning environment designs in their own classrooms *with their students* using technology? In other words, starting with students' perspectives of their science teaching and learning context would be the primary locus where information could be collected to improve teachers' instruction and then used in coordination with their students to iteratively refine incremental changes made through this co-generative dialogic process (see Pang et al., 2021 as a guide for researchers and Emdin's 2016 chapter on 'Co-generative Dialogues' for a teacher primer). This also means that science teachers need to first address the barriers that require the addition of accommodations so that there are no longer barriers, specifically to support students including those labeled with a disability or multiply marginalized, as well as those positioned and treated as 'lacking.' In this way, science teachers would focus less on supporting a single group of students, and radically change their teaching such that the pedagogies they use to support science learning are not rife with exclusionary assumptions (Keates & Clarkson, 2003).

Stapleton (2021) argues for a new approach to these types of projects called *Teacher Participatory Action Research* (TPAR). In this approach, teachers are seen as one of the most important designers in classroom contexts, but their approach to design is always mediated by the political terrain of their locale and community. Insomuch as teachers are designers of this type of instruction, students are also innovators of how to apply technology to their lives and communities. Like the availability of robust resources and innovative designs often privileged to affluent, white, able-bodied/mind, males in selective schools, Stapleton puts forth that while not all designs are plausible in all contexts, the nature of how teachers play a role in designing novel classroom instruction should not be underestimated. Venturing into this conceptualization further, what this means is that recognizing the affordances and constraints of a particular schooling context is the first step in understanding how to change our instructional choices.

Figure 1. Three Tenets and Guiding Questions of the Design-based Inclusive Science Approach

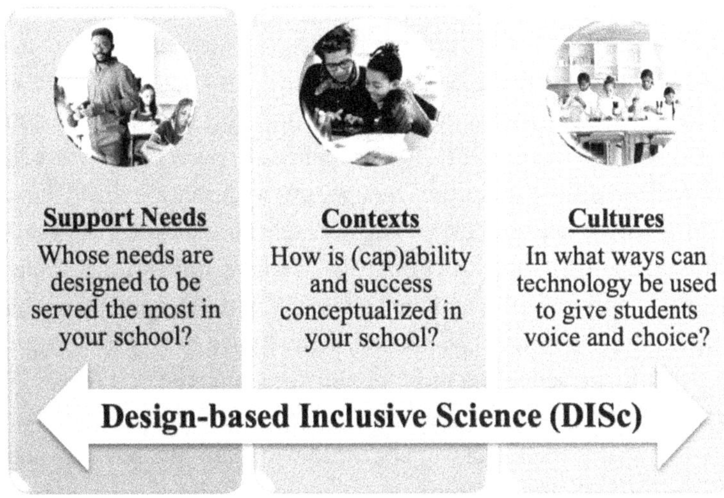

So, for example, in the DBR study by Krajcik et al. (2008), teachers' input was highly valued, but the goals of applying a more critical benchmark focused on students' locales and cultures were secondary, and inevitably did not get engaged with at all. However, in Gutiérrez et al. (2017), their review across social design experiments tells a picture of not focusing on content as a decontextualized component of students' lives and instead seeking to infuse the students' humanities in the project itself. This shift takes the form of witnessing teachers as often being the replicators of the status quo rather than designers (Lee, 2021), which I argue can be disrupted.

When the Rubber Hits the Road: Two Case Studies of the DISc Approach

A Design-based Inclusive Science (DISc) approach toward justice values when students' *support needs, contexts,* and *cultures* are seen as integral and interwoven components to design for inclusive science among teachers. These three tenets manifest when teachers open space for students' brilliance to shine. Figure 1 provides these three tenets and some guiding questions.

Drawing on these tenets means that science teachers must first engage with a social-ecological inventory of their classroom and school before they engage the broader context of the students' communities, locales, and cultures. This can be done partially by using a self-report checklist like the one published by Villa and Thousand (2016), or it can involve a Community of Practice group with other teachers and administration scheduled through peer observation and feedback cycles to focus on *inclusion as a process* but never a destination (Duncan et al., 2021). As stated above, these initial steps should leverage the innovation expertise of students as a guide. To reiterate, self-assessing and peer collaboration is the first step to building toward inclusion.

Next, science teachers need to use that information to engage with parents, their students, teaching peers, and administration. These discussions and actions, again, are always going to be functionally dependent on the political terrain a teacher faces (Stapleton, 2021), so this is not to say that every inclusive approach will be the same or reach the same extent of inclusivity. Do not despair! This should not dismay a new or veteran science teacher from trying to improve their practices, especially with digital-aged youth; again, teachers are forever learners. Rather, when teachers bear witness to the realities of their own schooling system and collegial atmosphere, they have a choice to make: What *can* I do? By enveloping other teachers and their students in these pursuits, science teachers start the process of changing their classrooms by using small iterative moments of inspiration – they start to think about their classrooms as spaces for them to design transformative opportunities and start to view their students' Lives-Hopes-Dreams as sites of possibility (Warren, 2021), not challenges to overcome. When this shift takes place, from 'what do I do now?' to 'what if I did this?', engaging with marginalized students in their locales and communities becomes more important because these teachers start to view their classroom as a site of innovation, thereafter, designing an instructional space where students are at the center.

By unpacking students' support needs, and then engaging with the context in which these needs have become manifested as needing 'additions,' science teachers have begun the process of understanding their classroom, their school, and schooling more broadly as a practice of culture. This is extremely important. Working on the first two steps without understanding how schooling embodies and emboldens certain cultural practices negates the history of societal impacts on the purpose and promise of formal education systems (Cole, 2005). Schools are institutions, and all teachers need to recognize their agency, or lack thereof, when thinking about their classrooms as sites where broader societal aims are infused and valued (Emdin, 2016; Lee, 2021). This last component, Culture, uses the first two as guiding strands toward developing an understanding of students as humans that exist outside of schools, with their unique humanities that involve their ways of knowing, being, and feeling. It also addresses the most recent requests by justice-oriented scholars to re-make the ways in which rights are constituted in classrooms through the valuing of students' interests, motivations, and worldviews in such a way as to think about science learning as a deeply contextualized relationship (Calabrese-Barton & Tan, 2019). In what follows, I present two cases of where these components were leveraged in K-12 science.

Teaching Science in 'the Hood' to 'Those Kids'

I have previously written about how disability and whiteness operate inter-subjectively in urban schools (i.e., the positionings that define someone's personhood and abilities by virtue of them signaling a specific culture/ability through their way of knowing, being, and feeling). This signaling framed how Black boys in my classroom, and previously in other teachers' classrooms, experienced *Exclusion from participation in science* (2018). This work, while academic in its nature, was guided by a critical approach I took toward my own teaching of science in 'the hood' to 'those kids.' This paraphrase draws from Emdin's (2016) seminal work *For white folks who teach in the hood... and the rest of y'all too*, where he argues that well-meaning white folks, as well as people of color who affiliate themselves with whiteness, end up teaching Black and Brown youth in 'the hood,' also bringing with them stereotypes of Black and Brown youth as gangsters, deviant, and, I argue, inherently disabled in relation to comparative indices of success ascribed by whiteness, coloniality, and ableism (see Boda et al., 2022 for a research unpacking). In this context, I was charged to teach a group of students who had 'failed high school science multiple times.' Given my own commitments to support all students, I sought a path to success beyond this initial rhetoric and outside the positioning of 'these students' as inherently deficit.

Initially, my students were, to put it lightly, resistant – for any teacher new to a school and community, this is par for the course, anywhere. However, it was not me as a person or as a teacher against whom they were pushing back. My students later explained that when they walked into the class and saw it was a science class with some of the similar cues of traditional science instruction – rows of seats, a projector of slides, and a stereotypic white-passing male instructor – they expected what they previously experienced to: Lectures, notes, homework, and tests. I started the year rocky by exploring the topic of Neuroscience and soon found out that the questions students had about the world more broadly were both fascinating and brilliant. This is where I started to take inventory of what support needs my students would require if I were to help them learn science using technology. Many of my students were avid informal users of technology, like the unrestricted ways researchers have advocated for in relation to being 'philosophers of technology' (Vakil & McKinney de Royston, 2022). However, when I dug deeper into what my students needed to be successful, what I found were rampantly held assumptions about what digital-aged youth do with technology in schools and how to support them. In this initial inquiry with my students, I found that what I needed to take away was risk.

Time and time again, Black and Brown youth with and without labels of disability must prove that they are competent, oftentimes in ways that are completely foreign to their ways of knowing, being, and feeling outside of school. Recognizing this phenomenon, I envisioned ways to think about how students would not have to take such public risks of proving their competence and, instead, sought to position them as brilliant experts of their lives and the world they have experienced. This led me to use technology to help the students conduct their own research study that involved the collection of data from their administration and my fellow teachers about their science content knowledge, answering questions they had themselves. Again, this is complicated and did not happen overnight, or without several bumps. I worked for weeks to select the right administrative people and teachers that would be of interest for the students to interview, be available to visit my class, and want to participate in our work. Through two iterations of this learning cycle that removed the barrier of risk and positioned my students as experts/competent, and then worked directly within the confines of what was possible in my school, what emerged were extremely powerful questions and presentations of science content, aligned with standards, showcasing that not only could these students learn, but they could teach the adults in the room.

In Figure 2, I show the ending of this first iteration where students, instead of merely exploring and answering their science questions about the world to no consequence or priority of the broader schooling contexts, my Black and Brown youth with and without disabilities asked their most pertinent disciplinary questions to the principal of the school, audio recorded the principal's responses with a peer making short anecdotes for efficiency in identifying the principal's science content knowledge, researched the scientific explanations of these specific science content phenomena, and then presented these juxtapositions between the response provided by the administration and the scientific community, educating their authority figures.

These presentations were rife with students challenging, and at many points overruling, the authority figures that often saw them as deviant, defiant, and disabled (Boda, 2018). It was fascinating to see the dynamic change when the students heard and recorded their 'superiors' answers to simple science content questions that, while they speculated were wrong at the time, later on, they got the chance to – indeed – call them out for their inaccuracy but, also, educate these figures in the school community about the correct answer and showcase their brilliance.

This process then expanded to really think more deeply about how these students, who I was told were 'those kids' and just 'needed the credit,' could be supported. By embodying an act of counter-storying, these students were presented to my teaching peers as not only having bright and engaged questions about the world, but also that they could also find the scientific answers to those questions, even when met with an alternative claim by an authority figure in their life: Adults in their school. By engaging with this process of first identifying what my students needed to feel like they could be seen as competent

Figure 2. Presenting: Why does the moon look different and how do we know the answer?

and did not have to prove that competence (their support needs), and then enveloping the affordances and constraints of the school I was charged to serve them in (contexts), I was able to create an innovative and inclusive science learning environment by design. In other words, my students were always the innovation experts that helped *me* find a way to teach *them*, and in doing so also helped me engage with technology in ways that wasn't driven by my goals alone, or by standardized reform, but their Dreams. As I transitioned out of the classroom and into the research space, these lessons guided more designs.

SCIENCE IN THE CITY

During my period at Stanford working with Dr. Bryan Brown and his *Science in the City* fellows (Brown, 2021), we leveraged a Research-Practice Partnership model with local schools and elementary teachers to support transformative science instruction with these communities. While in the previous example where I was the lead teacher for the work, in this circumstance I was working proximal to students, with their teachers and their community members, to design instruction that would leverage technological tools in novel ways to teach them science. Like the example above in a high school science class, the designs built for this work centralized the local contexts in which students lived – to help them see *science in their local community contexts.* By virtue of engaging first the teachers of these students, this shifted approach toward Design-based Inclusive Science (DISc) is thus described here as not central but proximal. This distinction is important for science teachers, their colleagues, and to researchers: To name a design project as 'centering' students, these youths' voices – Lives-Hopes-Dreams – need to be the primary focus of attention; however, what most research in science education has built design off of

Table 1. Virtual reality 360 lesson 'Flaming Hot Cheetos Food Chain' standards alignment

Type	Learning Artifact Content	Standards
Text & Audio of Text	Slide 6 • *Energy comes from the food we eat* • *Eating the stored energy in animal meat is less efficient than stored energy in food from plants* • *All energy can eventually be found to come from the sun* • *Food sources that get their energy directly from the sun are more efficient for humans to eat*	CA NGSS DCI Grade 5 **LS2.A: Interdependent Relationships in Ecosystems** The food of almost any kind of animal can be traced back to plants. Organisms are related in food webs in which some animals eat plants for food and other animals eat the animals that eat plants. Some organisms, such as fungi and bacteria, break down dead organisms (both plants or plants parts and animals) and therefore operate as "decomposers." (5-LS2-1)
Graphic & Text & Auditory	Slide 8 • *Food sources that get their energy directly from the sun are called 'producers'* • *If an animal gets its energy from a 'producer' they are called 'consumers'* • *If a consumer gets its energy from another consumer, they are labeled as a 'secondary,' 'tertiary,' or 'final' consumer* • *If a food source gets their energy from a dead consumer, they are called 'decomposers'* • *When we describe the relationship between producers, consumers, and decomposers we call this a 'food chain'*	
Virtual Reality & Text	Slide 19 • *2:10 - What you eat affects how healthy you are* • *2:18 – Dialysis is something unhealthy people must do because of their bad eating habits*	

are more proximal attendances of students' instructional experiences through the eyes of their teachers. This is not to say one is more valuable than another, but instead to merely make the distinction clear. We must never forget whose voices are present, and whose are excluded by design.

Over this multi-year project, we interviewed students, co-generatively designed scopes and sequences of science lessons working with these teachers, tested pilots of these designs, and sought to define the impact of our designs to support more positive attitudes toward science with these youth. The start of these projects was, however, centralized around identifying how we could help the teachers, and in what ways we could support their students to learn science such that they see the relevancy of the content and not just memorize it for a test. This involved a yearly set of meetings to describe the current population each set of teachers were charged to serve in their classes, which changed each year, and so these initial check-ins were the litmus from which we gauged the extent we could manifest broader transformative goals. Again, these designs did not emerge merely by our imposition and collaboration with teachers; we actively sought to make notes about students' interaction with the technology, as well as how they spoke to one another during these experiences. Using Virtual Reality (VR) 360 technology created an engaging space for these Black and Brown youth with and without labels of disability to think why science might be important and provided a space for their teachers to see their brilliance.

These initial support needs meetings weren't labeled as such but were interwoven with the ways that we as a research team sought to explore the impact of VR 360 beyond the current technology as a novel experiential moment of exposure. Within the context of these schools, the teachers still needed to teach science to their elementary grade level students! Recognizing that context, and then having iterative meetings to adjust the designs we made, were benchmarks to not only think about how to support students learning science, but also think about what is, and may not be, possible to implement in that context with these teachers. Going through this second step of the theory of Design-based Inclusive Science (DISc) brought forth countless moments of productive conflict and critique: Teachers have cer-

Figure 3. Elementary students in the Bay Area, CA using VR 360 technology

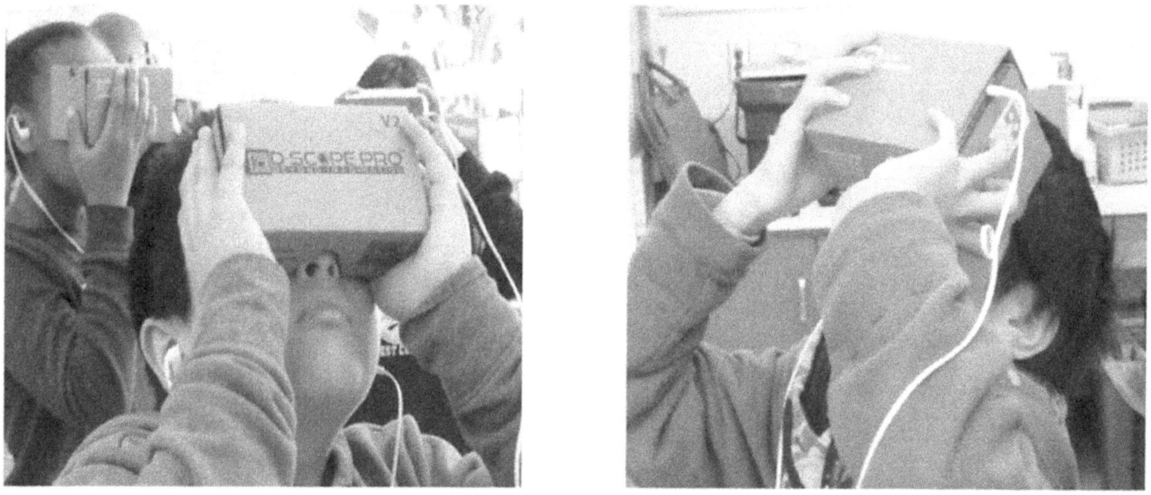

tain ways they want their classrooms to operate, and we as researchers also wanted to push the type of instruction happening in those contexts beyond the traditional implementation of science teaching that occurred therein. But this was not a negative component to our work; this was an *integral* lens needed to design for a more inclusive learning environment realistically and pragmatically. In Table 1, the reader is provided an alignment between engagement type, content, and standards designed in these activities.

Across the multiple years where we implemented these VR 360 experiences, both students and teachers were learning about design, science pedagogy, and the ways that success could be reframed. In Figure 3, I provide a still shot of students using the VR 360 technology.

In this still shot, students are directly engaged with the technology to learn science. As I presented above in the first example, technology became a conduit tool to support inclusive science in ways that took away risk and pointed the importance of technology to facilitating a different pedagogical and assessment structure rather than changing the curriculum itself. In this project, the curriculum itself was changed with the pedagogy and assessments being designed directly into the technological platform. Both versions of using technology to support more inclusive science teaching view students as powerful sites of possibility (Warren, 2021), but the employment of technology was very different. Here, the example of using technology for digital-aged youth aligns more with Vakil and McKinney de Royston's (2022) radical proposition that technology can be used by students to engage with sociopolitical realities. However, it departs from the ethereal context of informal learning to design such a transformative engagement with local context and students' communities such that disciplinary goals are not left agnostic to the importance of technology to afford an opportunity for teachers to grow as instructional designers.

From this set of projects, working with the same elementary teachers over multiple years, there started to be a shift in the ways that these participating teachers viewed their own teaching, and we attributed that paradigm shift in part to our inclusion of their voices in the design process, as well as the students' experiences in their classrooms when learning science with us. The context was, yes, then learned through technology, but the goals of the project were much broader than that. We saw teachers think about, and start to Dream of, transformative ways that technology could be used to teach curriculum, support the design of pedagogy, and differentiate the way to assess their students' learning. An example helps here:

Our projects explicitly wanted to remove the barrier for Multilingual students to engage with English-only content. Thus, the designed experience infused their home language into instruction through a dual-language approach to value a translanguaging dialogue between content, student, and teacher that is becoming a prominent field of inquiry in science education more broadly (cf. Fine, 2022; Pierson et al., 2021; Suárez, 2020). Assessments also provided a space for this dialectic maneuvering between language fluencies in that we provided the option for students to use a voice-to-text function to answer short-answer questions. These design-based infusions were not additions for some; they were available and permitted to be leveraged by any student. Here is where the culture of the science classroom took on a novel approach for youths' use of technology, and manifested support needs, context, and culture in developing Design-based Inclusive Science.

DISCUSSION OF LESSONS LEARNED

The power of centering students' Lives-Hopes-Dreams in the creation of science/STEM curriculum is not the status quo right now in science education research. However, what I have elaborated on here are two examples of how to enact this process as a teacher and working within research groups from local universities. These examples also showcase the ways in which that centering of students' voices and lived experiences doesn't look the same or get infused in the same way when thinking about inclusive science education for Black and Brown youth with and without labels of disability. Here, again, we must remember that inclusive education is not an end point, nor is it a single benchmark. Indeed, to reiterate with great emphasis, there is also no inclusive education without the inclusion of students labeled with disabilities; to name such a practice as inclusive and yet not include Disabled youth in the population being served is an injustice to the inclusive education movement (Fitch et al., 2021; Nusbaum, 2013; Reeves et al., 2022; Thomas, 2013). We must stop naming such work inclusive unless it, in fact, includes by design. And in this way, what is brought forth by the work I present here is a template, yet fluid, set of examples where both teachers and researchers can use in the future to think about how to design for relationality and sociopolitical justice for all students (Calabrese-Barton & Tan, 2019).

Implications for Digital Youth Empowerment Through Inclusive Science

The power of this Theory of Design-based Inclusive Science (DISc), however, does not end just with curriculum design, or how to think about pedagogy and assessment types as ways to think about inclusion. Circling back to Vakil and McKinney de Royston's argument on students as 'philosophers of technology' (2022), I put forth through these exemplars and the schema provided around the guiding questions that this type of inclusive science design should not remain distinct to informal educational spaces. If we as researchers toward justice wish to engage with the radical possibility of transforming our schools, we must do our work in schools! This, indeed, comes with limitations and oftentimes can lead to what many scholars may think are 'less just' manifestations of inclusion, justice, and sociocultural realities. And yet, I stake the claim that such work is powerful, like my colleague Srikala Naraian when she describes how teachers don't ever reach inclusive education: They are always in the process of negotiating what they know about their students, what they believe they can do in their schooling context, and the extent to which they feel comfortable 'rocking the(ir) boat,' so to speak (2010; 2017; 2021). And as we broach a new wave of inquiry into these sub-fields of science education to support digital-aged youth, what this chapter provides is a starting point, not an ending point, to engage with Design-based Inclusive Science (DISc), and our work to build transformative futures for all.

REFERENCES

Berne, P., Morales, A. L., Langstaff, D., & Invalid, S. (2018). Ten principles of disability justice. *WSQ. Women's Studies Quarterly*, *46*(1), 227–230. doi:10.1353/wsq.2018.0003

Boda, P. A. (2018). Exclusion from participation in science: Confessions from an ally on the other side of the fence. In M. Koomen, S. Kahn, C. Atchison, & T. Wild (Eds.), *Toward inclusion of all learners through science teacher education* (pp. 301–311). Sense/Brill Publishing. doi:10.1163/9789004368422_033

Boda, P. A. (2019). Conceptualizing the margins in science education: The limits of multicultural analyses. *Cultural Studies of Science Education*, *14*(2), 493–514. doi:10.100711422-019-09926-x

Boda, P. A. (2022). On the methodological and epistemological power of epistemic (de) centering as a reflexive praxis of resistance toward disability justice. *Qualitative Research Journal.*

Boda, P. A. (In press). How do we disrupt the labeling of disability and difference as deficit in urban schools? In S. R. Steinberg & E. Adjapong (Eds.), 19 Urban questions: Teaching in the city (3rd ed.). New York, NY: Peter Lang.

Boda, P. A., Nusbaum, E., & Kulkarni, S. (2022). From 'what is' toward 'what if' through intersectionality: Problematizing ableist erasures and coloniality in racially-just research. *International Journal of Research & Method in Education*, *45*(4), 1–14. doi:10.1080/1743727X.2022.2054981

Boda, P. A., & Riley Miller, A. (2022). Educational technologies for multicultural science learning: Comparing reform and critical design studies. In M. Atwater (Ed.), *International handbook of research in multicultural science education.* Springer. doi:10.1007/978-3-030-83122-6_3

Brown, B. A. (2021). *Science in the city: Culturally relevant STEM education.* Harvard Education Press.

Calabrese Barton, A., & Tan, E. (2019). Designing for rightful presence in STEM: The role of making present practices. *Journal of the Learning Sciences*, *28*(4-5), 616–658. doi:10.1080/10508406.2019.1591411

Cole, M. (2005). Cross-cultural and historical perspectives on the developmental consequences of education. *Human Development*, *48*(4), 195–216. doi:10.1159/000086855

Comarú, M. W., Lopes, R. M., Braga, L. A. M., Batista Mota, F., & Galvão, C. (2021). A bibliometric and descriptive analysis of inclusive education in science education. *Studies in Science Education*, *57*(2), 241–263. doi:10.1080/03057267.2021.1897930

Connor, D. (2012). Does Dis/ability Now Sit at the Table (s) of Social Justice and Multicultural Education? A Descriptive Survey of Three Recent Anthologies. *Disability Studies Quarterly*, *32*(3). Advance online publication. doi:10.18061/dsq.v32i3.1770

Design-Based Research Collective. (2003). Design-based research: An emerging paradigm for educational inquiry. *Educational Researcher*, *32*(1), 5–8. doi:10.3102/0013189X032001005

Duncan, J., Punch, R., & Croce, N. (2021). Supporting primary and secondary teachers to deliver inclusive education. *Australian Journal of Teacher Education (Online)*, *46*(4), 92–107. doi:10.14221/ajte.2021v46n4.6

Emdin, C. (2016). *For White folks who teach in the hood... and the rest of y'all too: Reality pedagogy and urban education*. Beacon Press.

Emdin, C., Adjapong, E., & Levy, I. P. (2021, July). On Science Genius and Cultural Agnosia: Reality Pedagogy and/as Hip-Hop Rooted Cultural Teaching in STEM Education. *The Educational Forum*, *85*(4), 391–405. doi:10.1080/00131725.2021.1957636

Fine, C. G. M. (2022). Translanguaging Interpretive Power in Formative Assessment Co-Design: A Catalyst for Science Teacher Agentive Shifts. *Journal of Language, Identity, and Education*, *21*(3), 191–211. doi:10.1080/15348458.2022.2058858

Fitch, E. F., Hulgin, K. M., & Coomer, M. N. (2021). How "special needs" vouchers deceive the public and silence the right to inclusive education. *International Journal of Qualitative Studies in Education*, 1–21. doi:10.1080/09518398.2021.1957174

Fowler-Amato, M., & Warrington, A. (2017). Teachers as designers: Social design experiments as vehicles for developing antideficit English education. *Literacy Research: Theory, Method, and Practice*, *66*(1), 358–372. doi:10.1177/2381336917719054

Gutiérrez, K. D., Cortes, K., Cortez, A., DiGiacomo, D., Higgs, J., Johnson, P., Ramón Lizárraga, J., Mendoza, E., Tien, J., & Vakil, S. (2017). Replacing representation with imagination: Finding ingenuity in everyday practices. *Review of Research in Education*, *41*(1), 30–60. doi:10.3102/0091732X16687523

Gutiérrez, K. D., & Jurow, A. S. (2016). Social design experiments: Toward equity by design. *Journal of the Learning Sciences*, *25*(4), 565–598. doi:10.1080/10508406.2016.1204548

Kaldenberg, E. R., Watt, S. J., & Therrien, W. J. (2015). Reading instruction in science for students with learning disabilities: A meta-analysis. *Learning Disability Quarterly*, *38*(3), 160–173. doi:10.1177/0731948714550204

Kali, Y., McKenney, S., & Sagy, O. (2015). Teachers as designers of technology enhanced learning. *Instructional Science*, *43*(2), 173–179. doi:10.100711251-014-9343-4

Keates, S., & Clarkson, J. (2003). Design exclusion. *Inclusive Design*, 88-102.

Kim, M. C., Hannafin, M. J., & Bryan, L. A. (2007). Technology-enhanced inquiry tools in science education: An emerging pedagogical framework for classroom practice. *Science Education*, *91*(6), 1010–1030. doi:10.1002ce.20219

Koomen, M., Kahn, S., Atchison, C. L., & Wild, T. A. (Eds.). (2018). *Towards inclusion of all learners through science teacher education*. Brill Publishers. doi:10.1163/9789004368422

Krajcik, J., McNeill, K. L., & Reiser, B. J. (2008). Learning-goals-driven design model: Developing curriculum materials that align with national standards and incorporate project-based pedagogy. *Science Education*, *92*(1), 1–32. doi:10.1002ce.20240

Kulkarni, S., Nusbaum, E., & Boda, P. A. (2021). DisCrit at the margins of teacher education: Informing curriculum, visibilization, and disciplinary integration. *Race, Ethnicity and Education*, *24*(5), 654–670. doi:10.1080/13613324.2021.1918404

Ladson-Billings, G. (2018). Thc social funding of race: The role of schooling. *Peabody Journal of Education*, *93*(1), 90–105. doi:10.1080/0161956X.2017.1403182

Lee, S. Y. (2021). Rethinking teacher agency: Cybernetics, action research, and the process-oriented rationality. *Journal of Curriculum Studies*, *53*(6), 821–840. doi:10.1080/00220272.2020.1858451

Mehalik, M. M., Doppelt, Y., & Schuun, C. D. (2008). Middle-school science through design-based learning versus scripted inquiry: Better overall science concept learning and equity gap reduction. *Journal of Engineering Education*, *97*(1), 71–85. doi:10.1002/j.2168-9830.2008.tb00955.x

Naraian, S. (2010). General, special and… inclusive: Refiguring professional identities in a collaboratively taught classroom. *Teaching and Teacher Education*, *26*(8), 1677–1686. doi:10.1016/j.tate.2010.06.020

Naraian, S. (2017). *Teaching for inclusion: Eight principles for effective and equitable practice*. Teachers College Press.

Naraian, S. (2021). Making inclusion matter: Critical disability studies and teacher education. *Journal of Curriculum Studies*, *53*(3), 298–313. doi:10.1080/00220272.2021.1882579

NGSS Lead States. (2013). *Next Generation Science Standards: For States, By States*. The National Academies Press.

Nusbaum, E. A. (2013). Vulnerable to exclusion: The place for segregated education within conceptions of inclusion. *International Journal of Inclusive Education*, *17*(12), 1295–1311. doi:10.1080/13603116.2013.826292

Pang, V. O., Alvarado, J. L., Preciado, J. R., & Schleicher, A. R. (2021). Culturally relevant education: Think local within a holistic orientation. *Multicultural Perspectives*, *23*(1), 3–16. doi:10.1080/15210960.2021.1877546

Pierson, A. E., Clark, D. B., & Brady, C. E. (2021). Scientific modeling and translanguaging: A multilingual and multimodal approach to support science learning and engagement. *Science Education*, *105*(4), 776–813. doi:10.1002ce.21622

Pugach, M. C., Matewos, A. M., & Gomez-Najarro, J. (2021). Disability and the meaning of social justice in teacher education research: A precarious guest at the table? *Journal of Teacher Education*, *72*(2), 237–250. doi:10.1177/0022487120929623

Reeves, P., Ng, S. L., Harris, M., & Phelan, S. K. (2022). The exclusionary effects of inclusion today:(re) production of disability in inclusive education settings. *Disability & Society*, *37*(4), 612–637. doi:10.1080/09687599.2020.1828042

Reid, D. K., & Knight, M. G. (2006). Disability justifies exclusion of minority students: A critical history grounded in disability studies. *Educational Researcher*, *35*(6), 18–23. doi:10.3102/0013189X035006018

Salend, S. J. (1998). Using an activities-based approach to teach science to students with disabilities. *Intervention in School and Clinic*, *34*(2), 67–72. doi:10.1177/105345129803400201

Semali, L. M., Hristova, A., & Owiny, S. A. (2015). Integrating Ubunifu, informal science, and community innovations in science classrooms in East Africa. *Cultural Studies of Science Education*, *10*(4), 865–889. doi:10.100711422-014-9640-x

Shea, M. V., & Sandoval, J. (2020). Using historical and political understanding to design for equity in science education. *Science Education*, *104*(1), 27–49. doi:10.1002ce.21555

Stapleton, S. R. (2021). Teacher participatory action research (TPAR): A methodological framework for political teacher research. *Action Research*, *19*(2), 161–178. doi:10.1177/1476750317751033

Suárez, E. (2020). "Estoy Explorando Science": Emergent bilingual students problematizing electrical phenomena through translanguaging. *Science Education*, *104*(5), 791–826. doi:10.1002ce.21588

Sullivan, A. L., & Thorius, K. A. K. (2010). Considering intersections of difference among students identified as disabled and expanding conceptualizations of multicultural education. *Race, Gender, & Class*, •••, 93–109.

Svihla, V., & Reeve, R. (Eds.). (2016). *Design as scholarship: Case studies from the learning sciences*. Routledge. doi:10.4324/9781315709550

Thomas, G. (2013). A review of thinking and research about inclusive education policy, with suggestions for a new kind of inclusive thinking. *British Educational Research Journal*, *39*(3), 473–490. doi:10.1080/01411926.2011.652070

Tosun, C. (2022). Bibliometric and Content Analyses of Articles Related to Science Education for Special Education Students. *International Journal of Disability Development and Education*, *69*(1), 1–18. doi:10.1080/1034912X.2021.2016659

Vakil, S. (2020). "I've Always Been Scared That Someday I'm Going to Sell Out": Exploring the relationship between political identity and learning in computer science education. *Cognition and Instruction*, *38*(2), 87–115. doi:10.1080/07370008.2020.1730374

Vakil, S., & McKinney de Royston, M. (2022). Youth as philosophers of technology. *Mind, Culture, and Activity*, 1–20. doi:10.1080/10749039.2022.2066134

Vernon, A. (1999). The dialectics of multiple identities and the disabled people's movement. *Disability & Society*, *14*(3), 385–398. doi:10.1080/09687599926217

Villa, R., & Thousand, J. (2016). The Inclusive Education Checklist: A Self-Assessment of Best Practices. National Professional Resources, Inc./Dude Publishing.

Villanueva, M. G., Taylor, J., Therrien, W., & Hand, B. (2012). Science education for students with special needs. *Studies in Science Education*, *48*(2), 187–215. doi:10.1080/14703297.2012.737117

Ware, L. (2001). Writing, identity, and the other: Dare we do disability studies? *Journal of Teacher Education*, *52*(2), 107–123. doi:10.1177/0022487101052002003

Warren, C. A. (2021). *About centering possibility in Black education*. Teachers College Press.

Williams-Diehm, K. L., Miller, C. R., Sinclair, T. E., & Wronowski, M. L. (2018). Technology-based employability curriculum and culturally diverse learners with disabilities. *Journal of Special Education Technology*, *33*(3), 159–170. doi:10.1177/0162643417749933

Chapter 2
Science Instruction for Students Identified as Gifted and Talented:
The Efficacy of Makerspace in This Digital Age

Natalie Ann-Marie Johnson-Leslie
https://orcid.org/0000-0003-1654-0266
Arkansas State University, USA

Allen Hays
https://orcid.org/0000-0001-5672-6722
Arkansas State University, USA

Rebekah S. Marsh
Keiser University Online Division, USA

ABSTRACT

The past two decades have ushered changes regarding the national science curriculum for K-12 students. Focusing on gifted and talented (GT) students, the literature reveals they clamor for unique experiences in constructing their knowledge in science education. This chapter explores science instruction for GT in this digital age using a Makerspace learning environment. Guided by the tenets of the Next Generation Science Standards, this chapter is supported by outlined standards and objectives connecting science content to GT learners' unique needs. The theory of socialization, specifically cognitive development coupled with the theory of constructivism, was used to ground this chapter. Results show Makerspace learning environments where GT students are exposed to inductive strategies and yield a greater return on student learning investments compared to the traditional deductive science classroom environment.

DOI: 10.4018/978-1-6684-5585-2.ch002

INTRODUCTION

Historically, giftedness and gifted education have been seen through the lens of high intellectual abilities. Lewis Terman (1922) is recognized as the father of genetic studies of genius, credited for his early work in explaining giftedness. According to Subotnik, et al., (2011) Terman's seminal work, "yielded many valuable insights about cognitive ability and its relationship to academic, vocational, and psychosocial outcomes," (p. 5). Over the years, this perspective on giftedness has been challenged, leading to the development of multiple categories of giftedness as revealed in the works of psychologists Davidson and Sternberg (1986). Therefore, in the 21st Century, giftedness is no longer seen as just being genetic or innate in nature. Conversely, possessing giftedness, as well as achieving outstanding performance, require more than innate intellectual ability (Freeman, 2010; Olszewski-Kubilius, 2000; Olszewski-Kubilius et al., 1988; Renzulli, 1978 Subotnik, et al., 2011; Subotnik & Jarvin, 2005; Terman, 1954a; Winner, 1996; Worrell, 2010a). However, this research is resisted, and instead, the entrenched view of giftedness as mainly intelligence or general mental ability persists among public, educators, policies and their reluctance are reflected procedures such as intelligence tests used to measure giftedness across the United States (Council of State Directors of Programs for the Gifted and the National Association for Gifted Children, CSDPG/NAGC, 2009).

Because of their unique academic abilities, students identified as being Gifted and Talented (GT) see the world and the study of science differently compared to non-GT students (Gifted and Talented Students and the Next Generation Science Standards, 2015). GT students generally need greater access to interactive and advanced techniques and technologies to effectively problem-solve in classrooms of the digital age. Such access will provide them with appropriate challenges and increase their depth of learning. Research has shown that many students love science at an early age, based on its inductive nature and practical applications (Calik & Birgili, 2013). However, as students get older, science tends to get more deductive, requiring reading textbooks with little to no practical applications. According to Tay, et al., (2017), parents have goals for their children, especially regarding STEM education, which is generally a primary factor in the development of students described as gifted. They found that "Young children enjoyed conducting experiments and participating in activities as this allowed them to use all their senses in the learning process," (p. 15). GT students also learn best from highly interactive activities that can be in a physical experimental form or digitally (Siemer, 2009; Sullivan et al., 2013).

In today's digital world, having students learn science from a static textbook can be demotivating, especially for GT students. An interactive learning environment that is inclusive of the Makerspace can deliver an inductive approach in which all students can see, touch, smell, and hear science in action (Prince & Felder, 2006). The Makerspace concept is multi-dimensional, imaginative, engaging, and intellectually stimulating. Makerspace is grounded in the tenets of constructivism, purported by Vygotsky (1962). For example, in the Makerspace environment, GT students can learn how to make models, code programs, and apply additional scientific educational concepts. These learning activities can be done using digital or non-digital formats (Brejcha, 2018). Teachers generally allow GT students to select the format that best fits the student's mode of inquiry as a motivating strategy (Meador, 2019; Subotnik, et al., 2011; Winner, 1996). Effective strategies benefit GT Students who can create new knowledge, apply old knowledge to new situations, and create new products that reflect their enhanced learner experience. In this technologically driven age, Makerspace has many benefits for today's 21st Century students.

Incorporating Makerspace environments in schools allows students to have an extra academic outlet to develop skills that will be crucial for success in the future. Makerspace enables students to develop strengths in areas traditional education does not allow. With Makerspaces, the teacher becomes the guide, mentor, coach, or facilitator (Croft, 2003). At the same time, the students become active learners or educators in this space. Students can then develop areas of interest and use their academic strengths to overcome their academic weaknesses based on the Next Generation Science Standards (NGSS) and academic standards that guide their learning.

NGSS is structured with the learner at the center and makes provision for all students to experience academic rigor. As teachers guide this process, they have the flexibility to utilize the strategy that best ensures "gifted and talented students receive instruction that meets their unique needs as science learners" (National Science Teaching Association [NSTA], 2022, pg. 1). Science teachers imagine the spaces that best facilitate student learning, and in collaboration with GT students, the optimal space could be created that is aligned with standards, assessments, and activities that students will find to be relevant and meaningful to their lives (NSTA, 2022). Each task selected must make sense, relevantly engage the students, and be authentic. Furthermore, GT students must be given the space and time to experiment, revise, revisit, problem solve, evaluate ideas, and create solutions (Meador, 2019; Sullivan, 2013; Tay et al., 2017) Tasks must be meaningful and authentic to challenge GT students and keep them actively engaged in the learning process (Subotnik et al., 2011).

This chapter is arranged into four main sections. *Section 1*, explores the theory of socialization and constructivism used to ground this paper; *section 2* provides an overview of the NGSS science standards; *section 3* focuses on gifted and talented students in general, and *section 4* picks up on the power of Makerspace learning environment. Table 3 contains a list of digital resources for teachers and students to utilize in the GT science classroom. In the spirit of equity and inclusion, we focused on serving the needs of GT students, but there are opportunities for all students to benefit from the information within this chapter.

Chapter Objectives

- Explore the intersection of the theory of socialization and constructivism
- Relate the features of the NGSS
- Analyze the role of GT students in the Makerspace learning environment
- Evaluate the power of Makerspace in the GT science classroom
- Outline the digital resources that are available to enhance learning in the GT science classroom

BACKGROUND AND LITERATURE REVIEW

Section 1: How Students Learn (Theoretical Underpinning)

In today's digital age, the Makerspace environment utilizes technology to enhance learner experience, not only allowing GT students to engage with content but also to create their own content. For example, GT students are encouraged and expected to: create videos, build models, explore the environment around them, collect and analyze data, debate the merits of actions and activities, conduct experiments, develop coding for all aspects of life, design and develop application software, build computers, and solve real-life problems. All of these activities are nested in the wider context in which the students operate. *First*, the theory of socialization will provide context for how GT students were socialized, and *second*, the

Figure 1. Gifted and Talented Students at the center of the Vortex of Influences

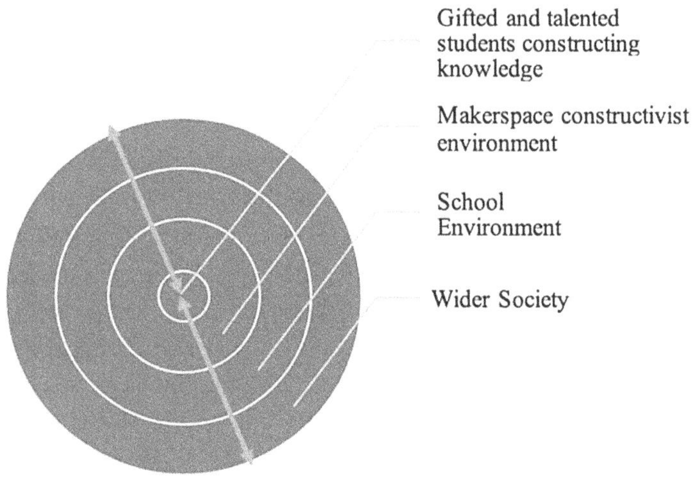

theory of constructivism will show its applicability in motivating and facilitating active hands-on work for GT students. Figure 1 shows GT students at the center of the influences in society.

As shown in Figure 1, Gifted and talented students are at the center of the vortex of influences based on the Makerspace constructivist environment in which they exist as they build knowledge. Student learning *impacts* and is *impacted* by the school and wider society as is justified by the theory of socialization. The double-pointed arrows indicate how environmental and societal changes impact and are impacted when GT students acquire and synthesize knowledge as well as how this takes place.

Constructivism is a blanket term, including a variety of perspectives that are divided into two streams (UKEssays, 2018): (1) cognitive constructivism postulated by Piaget (1953) and (2) social constructivism purported by Vygotsky (1962). Both streams are "watered" by the notion that learners are engaged in constructing knowledge actively based on prior experiences (Birenbaum 2007). In this way, each learner constructs knowledge using their genetic, internal, or intellectual acumen (cognitive constructivism) and by social interaction (social constructivism) supported by the work of (Alexander et al., 1998):

1. **Theory of socialization (cognitive development):** The theory of socialization grounds the work of this chapter. The ways that GT students are socialized greatly impact how they approach learning science in this digital age. Drawing on Piaget's (1953) theory of cognitive development, children acquire a sense of self and their unique personality through nature and nurture as they learn how to think and reason for themselves. Their cognitive development is charted into four distinct stages that are directly related to brain development and socialization. GT students pass through these four stages of development like all other students with the additional ability to understand their environment, as seen in Table 1.

As shown in Table 1, there are four distinct stages identified by Piaget which outline the ways individuals develop cognitively beginning in their early years. Many students are identified as GT by the time they reach STAGE 3 (concrete operational; age 7-11), in the developmental process. These students generally have strong cognitive development in academics. Kool et al., (2018) explored the ways that young children learn and what that reveals about how they learn when provided with developmentally

Table 1. The 4 stages of cognitive development postulated by Piaget (1954)

STAGE 1	*Sensorimotor*	Infants using their senses to figure out the world around them because they unable to formally think or reason
STAGE 2	*Preoperational*	Ages 2 - 7: Children begin to make sense of and use signs and symbols, especially words, to understand their environments
STAGE 3	*Concrete operational*	From age 7 - 11 or 12. Children start thinking in terms of cause and effect. However, they do not fully comprehend the tentacles of fairness, justice, and related concepts.
STAGE 4	*Formal operational*	Begins around 12 years old. Children are able to think critically and abstractly to resolve various challenges they may encounter.

appropriate hands-on activities. These learners enjoy stimulating, challenging, hands-on tasks, that require problem-solving (George Lucas Educational Foundation, 2016; Riley et al., 2009). In intellectually engaging STEM classes, GT students continue to thrive intellectually (Freeman, 1985; Moomaw & Davis, 2010). In fact, "brain activity develops rapidly in the earliest years of life. Stimulation from a child's social environment enhances this development, while a lack of stimulation impairs it [development]" (Sharapan, 2012, p. 2). Early neurological and cognitive stimulations for children enhance their cognitive development and the development of schema. With the emphasis placed on both interaction and socialization, Piaget further emphasized that children need to be cared for, played with, read to, talked with, and sung to for optimal development as they advance through each stage. These activities lead to the development of mental schema. By definition, a schema is the mental structure individuals develop to help them understand and make sense of how things work. That is, the way GT students organize knowledge as they process new information and connect it to other things that they know, believe, or have experienced, which together forms patterns and meaningful connections in the brain (Meta & Fine, 2019). This process is similar for all learners. However, GT students generally go through these processes at a faster pace than other students. Children need to be reared in an enriching environment, and providing GT students with a stimulating learning environment, such as Makerspace, grants them the opportunity to thrive as they think concretely and abstractly, thereby actively constructing their knowledge about various concepts—which is in keeping with the tenets of constructivism.

2. **Theory of social constructivism**: In this chapter, the theory of constructivism, which allows GT students to build on what they know using real-world experiences in the Makerspace environment, complements the theory of socialization. The fact that there are many students with outstanding academic or intellectual abilities who have not experienced an enriching background due to the realities of being marginalized is noteworthy. Additionally, many students who are high achievers may not be identified or selected as GT (Cross & Coleman, 2005; Dai, 2010) because they do not meet the strict qualifications. Thus, teachers in the constructivist classroom must utilize effective strategies to reach all students with varying abilities, from those with special needs to those who are identified as GT.

According to Vygotsky (1962), constructivism focuses on how learners actively construct/build knowledge rather than just passively taking in information. In this light, as GT students experience the

Figure 2. Seven Universal Tenets or Key Ideas about Constrictivist Learning
Adapted from Taber (2006)

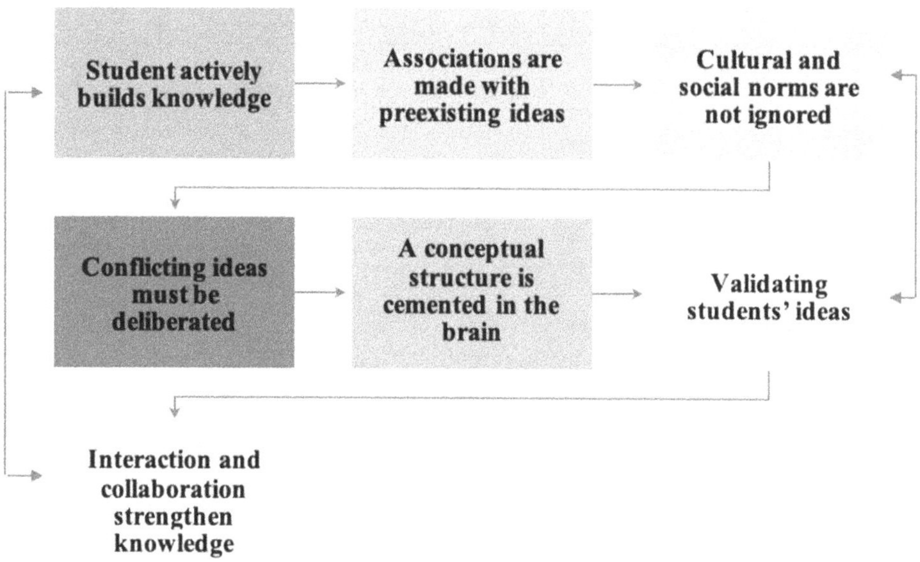

world of Makerspace and reflect upon those experiences, they build their depiction, representation, and interpretation of knowledge as they actively incorporate new information into their pre-existing knowledge.

Furthermore, based on the tenets of constructivism and the Zone of Proximal development, GT students can solve problems that are more challenging and complex than their peers of the same age. Due to their highly developed cognitive skills, these students are generally able to solve problems collaboratively in an active learning environment (Srivastava & Dangwal, 2017; Winner, 2009). This is because knowledge is socially constructed, and GT students build their schemas through interaction, which is amplified in the Makerspace environment. In essence, constructivism allows the students to be seen as active builders and creators of meaning and knowledge (Maksimović, 2020).

In this digital age, the constructivist theory is applicable as it inspires, motivates, and facilitates learning through active hands-on work, collaboration with others, experimenting, and formulating conclusions that are research-driven. In this theory, the student is truly seen as an active learner, whereas the teacher's function is that of being a motivator or guide on the sideline. This is the environment in which GT students thrive, develop, deliberate, question, and shine. In essence, the student is placed at the center of the learning vortex (Pereira & Sithole, 2019); see Figure 1. In Taber's work (2006), seven universal or key ideas that characterize constructivist learning are postulated in Figure 2.

As shown in Figure 2, there are seven tenets of constructivist learning. These tenets are not linear in their structure, but rather they overlap and can be accomplished in any order. This is because constructivist learning is generated from basic concepts learned in the early stages of development and gradually extends into more complex and abstract skills and processes (Morado, Melo & Jarman, 2021). The ultimate learning destination is for students to develop skills, scaffold these skills, and then generate new skills in the learning process. This learning process is dynamic, active, and progressive, with its students becoming knowledge generators rather than simply knowledge consumers. Constructivism, when coupled with effective socialization skills, will facilitate students' development of their view of the world as they seek to create the knowledge, skills, and dispositions that are authentic and meaningful to themselves and others around them (Vygotsky,1962).

Section 2: Role of the NGSS Standards and Student-Centered Learning

Teachers are bound by educational standards that guide what students are expected to know and be able to do in each content area. What do the following, No Child Left Behind (NCLB), Common Core Standards CCS), and Next Generation Science Standards (NGSS), all have in common? They are all national standards introduced over the last 20 years with the intent to reform education in the USA. Though educational standards are the learning goals for what students should know and be able to do at each grade level, they are *not* the same as the curriculum. Local communities and educators choose their curriculum, or the detailed plans for day-to-day teaching. In other words, NCLB, Common Core, and NGSS outline *what* students need to know and be able to do, and the curriculum is *how* students will learn the content (Klein, 2015; Common Core Standards, 2022; Next Generation State Standards, 2022).

The NCLB Act of 2001 was a U.S. Act of Congress that reauthorized the Elementary and Secondary Education Act (ESEA) of 1965 to revise, reauthorize, and consolidate various programs. It included Title I provisions applying to disadvantaged students. "Title I: Improving the Academic Achievement of the Disadvantaged" aimed at improving the Academic Achievement of the Disadvantaged (currently Helping Disadvantaged Children Meet High Standards) and in part A, Improving Basic Programs Operated by Local Educational Agencies. (H.R.1 — 107th Congress, 2001-2002). This act was replaced by the Common Core State Standards in 2009 (CCSS, 2022).

The CCSS was a response to the perceived stagnation of the nation's children as compared to their international peers. One of the chief causes of this stagnation was the lack of consistent expectations for what students should know and be able to do at each grade level. Hence, there was a need for consistent learning goals and outcomes across states. In 2009, the best and brightest coordinated with teachers, school chiefs, administrators, and other experts in a state-led effort to develop the CCSS.As a result, they developed standards that provide a clear and consistent framework for educators (Common Core Standards, 2022). As the name suggests, the CCSS address the core subject areas in schools: English language arts/literacy, and mathematics. At that point, the focus was not placed on science education. In a technologically driven society, The National Science Foundation (NSF) established the Scientific Education Standards that were then used to help inform the CCSS as well as the Next Generation Science Standards (NGSS) because these are research-based, tangible, and measurable educational standards under which students learn the principles of science, along with their core subject areas (NSF, 2012).

The NGSS were developed collaboratively with multiple stakeholders in the broad fields of science, science education, higher education, and industry. In addition, input was sought from science advisory boards, scientific organizations, as well as stakeholders from the business industry. Released in 2013, these research-based K–12 science content standards aimed to improve science education for all students, including GT students. Multiple key drivers were considered when developing the NGSS as shown in Table 2.

As seen in Table 2, there are 18 key drivers for the development of NGSS. These grade-specific standards are designed to help teachers interpret in a way that meets the variety of abilities, needs, learning rates, and achievement levels of students in any given classroom. More importantly, the standards provide clear signposts along the way to the goal of impacting the knowledge base of K-12 students.

Table 2. The 18 key drivers for the development of the NGSS standards. Adapted from Lanoue (2021)

	Activity	Description
S	Supporting critical thinking	Supporting scientific inquiries with abstract reasoning, problem-solving, and critical thinking — essential skills that students can apply to various challenges they face In STEM and in life.
C	Conducting meaningful inquiries	The progression of the practices across K-12, for conducting meaningful investigations includes specific elements of each practice are targeted for students at each grade band.
I	Improving classroom experience	Educators have the flexibility to design and improve hands-on learning experiences that engage students, increase participation, stimulate their interest in STEM, and prepare them for college.
E	Eliminating "teaching to the test"	Eliminate the focus from memorizing scientific facts to the practical application of scientific concepts so students can translate knowledge into applied situations.
N	Noting the need for common standards	Provide a common set of standards guiding the teaching of STEM subjects. These standards ensure students have an internationally benchmarked science education, with clear expectations.
T	Tailoring interest in STEM subjects	STEM curriculum combats a lack of knowledge in scientific topics and engages students, helping them develop a greater interest in STEM and encouraging students to study STEM subjects in college.
I	Investigating scientific notions with empirical evidence	Facilitates all students conducting scientific investigations, solving problems, and engaging in discussions. Emphasis is on science not on isolated facts enabling students to view science as an interrelated not static world of inquiry of scientific disciplines.
F	Fostering depth of knowledge	The NGSS fosters depth of knowledge and allows students to connect scientific knowledge with social and personal experiences. Knowledge grows as students become actively engaged with the scientific content.
I	Instilling a positive attitude toward STEM	Instilling in young students a positive attitude toward STEM subjects by engaging them with exciting material and experiences. This motivates them to pursue the sciences from an early age and helps them develop the evidence-based, lifelong skill to generate ideas, and understand data.
C	Continuity into the educational experience is built	The focus on the disciplinary core ideas that students have continuity built-in as the student progress in school by revisiting important concepts and expanding their understanding and skill development over time. Thus, students connect concepts across and between scientific domains.
E	Encouraging application of scientific principles	The NGSS encouragement of combining knowledge with practice by anchoring learning in real-world phenomena. This inquiry process helps students develop an appreciation of how science applies to everyday life.
D	Developing important life skills	The curriculum helps students develop an in-depth understanding of content and critical skills outside of science education, including communication, collaboration, and inquiry.
U	Understanding the U.S. as a global leader position	The curriculum helps the USA students understand and cultivate future innovators in businesses and academic institutions. This solid science education enables the country to develop new standards that solve the world's most pressing issues.
C	Combating both mis- and disinformation	Students become analytical consumers of scientific information by applying evidence-based reasoning. This is essential for combating conflicting information from multiple sources
A	Accentuating core scientific concepts	The base for students to understand core scientific concepts is accentuated, cultivating a lifelong interest in STEM
T	Tying different domains of science	The crosscutting dimension of the NGSS help students deepen their scientific understandings bridging the engineering disciplines through "crosscutting ideas" or "themes."
I	Instilling equal opportunities	NGSS allows students of all socioeconomic backgrounds and life experiences have the same opportunity to develop their potential.
O	Offering STEM careers	NGSS STEM prepares students for the global economy, to pursue diverse careers critical to society and the planet.
N	Navigating key life decisions	The larger goal is to help students navigate key life decisions preparing them for post-secondary life.

The fundamental reason that K-12 students attend school is to learn at all levels. NGSS promotes active learning in the sciences, and this provides the best return on educational investments. As previously stated, GT students are generally active learners who want to learn in a multi-modal manner (Brainbridge, 2021; Meador, 2019). Research suggests that K-12 students learn and can use scientific concepts more effectively if they are active in developing, creating, modeling, and then are able to explain phenomena in the natural world and argue their ideas with their peers (Klein, 2015; Lanoue, 2021; Mesmer, 2015). Other practices, such as asking questions, analyzing, and interpreting data, and constructing explanations, can be incorporated into the practices of modeling and argumentation. Students are performing science, not just learning the result of scientific research. NGSS is not simplistic, but it is dynamic. The most difficult aspect of the NGSS for teachers in the earlier grades was incorporating the crosscutting concepts.

According to NSTA (2022),

... crosscutting concepts have applications across all the scientific domains, including patterns; cause and effect; scale, proportion, and quantity; systems and system models; energy and matter; structure and function; as well as stability and change (p. 1).

All these activities facilitate learning by doing. Explicit efforts to help GT students in such learning include finding patterns in science data, analyzing structures related to function, and conducting inquiry in many areas they investigated (Mesmer, 2015). The NGSS emphasizes that these science concepts need to be made explicit for students because they provide an organizational schema for interrelating knowledge from various science disciplines, (including paleontology, genetics, chemistry, population ecology, physical science, and many other scientific disciplines) into a coherent and scientifically based view of the world. The NGSS are divided into two main categories: Grades K-8 and high school as seen in Figure 3.

As seen in Figure 3, the NGSS addresses all grade levels. To build a solid understanding of this science content, the K-8 standards are divided into K - 4 (elementary) and 5-8 (middle grades). The high school NGSS are subject-specific so that students can dig deep into their specific areas of study. For all grade levels K-12, the NGSS are written and read similarly. Figure 4 illustrates how to read the NGSS. Understanding how to read the NGSS is particularly important so that teachers can guide student learning in meaningful ways.

As shown in Figure 4, the map for reading the standards is similar for each grade level. Each map is divided into three main sections. Part 1 is the **accessible component** which includes the topic/title and the abbreviation indicating the grade level, the abbreviation of the science topic, the standard number, and the performance expectations for students. Performance expectations are the assessable statements indicating what students should know and be able to do. Some states consider these performance expectations alone to be "the standards."

Below the performance expectations are three **foundation boxes**, which list (from left to right) the specific science and engineering practices, disciplinary core ideas (DCIs), and crosscutting concepts that were combined to produce the performance expectations (PEs) above. These performance expectations provide teachers with a framework or a set of guidelines for teaching science concepts.

The third and final component is the **connection box.** The connection box shows the link between the lower- and upper-grade bands. In addition, specific linkages are made directly between the common core standards for English language arts and mathematics. The adoption of the NGSS requires many teachers, especially those with GT students, to drastically change their classroom instruction by focusing on ac-

Figure 3. The NGSS by Grade Levels

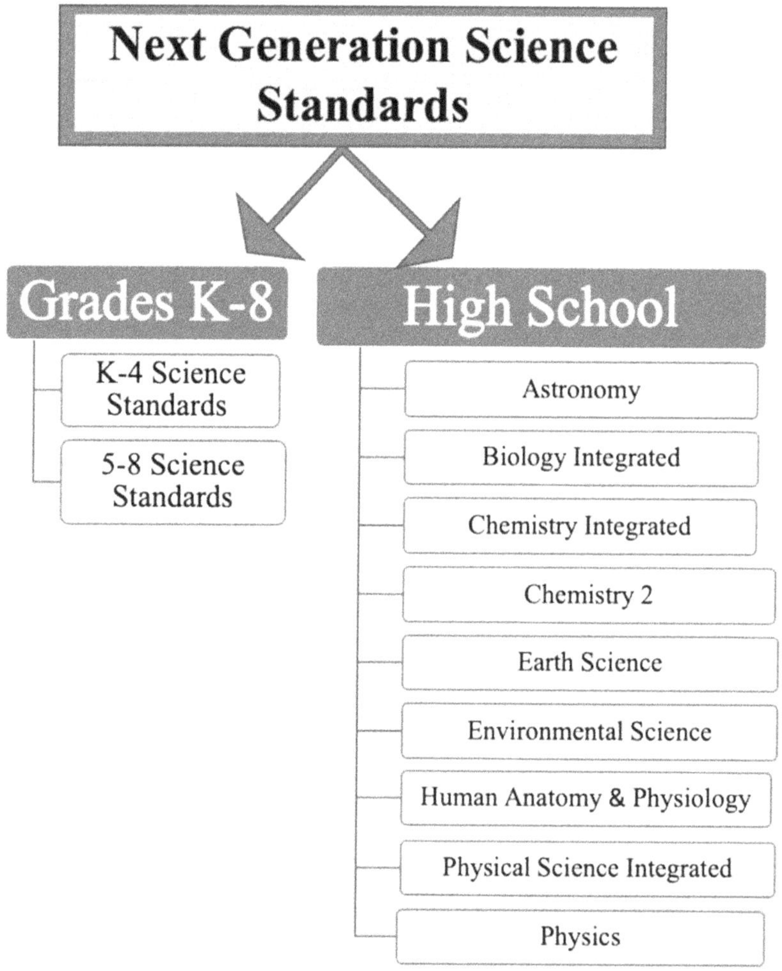

tive learning. NGSS provides guidance, and using those guidelines, teachers need to make instructional resource choices and modifications that best suit their students' learning needs. Appropriate curricular materials must be provided to support and enhance teaching and learning in the science classroom. Having the correct tools and equipment in the science classroom truly reflects the requirements of the NGSS (Martin-Hansen, 2002; McNeill et.al, 2017).

Section 3: Gifted and Talented Students GT Students in the Science Classroom

There is no national rubric used to identify K-12 GT students, and according to the NGSS (2013) release, "a precise figure for the number of gifted and talented students in the United States is not available due to the variation in identification processes from state to state" (p. 1). In other words, selecting GT students is not a standardized process throughout all 50 states and territories of the USA (www.nagc.org). However, the National Association for Gifted Children has a set of standards that some, but not all, states utilize when developing their state GT programs. For example, in the state of Arkansas, students

Figure 4. Map Showing How to Read the Common Core State Standards
Adapted from www.CCSS.org (2022)

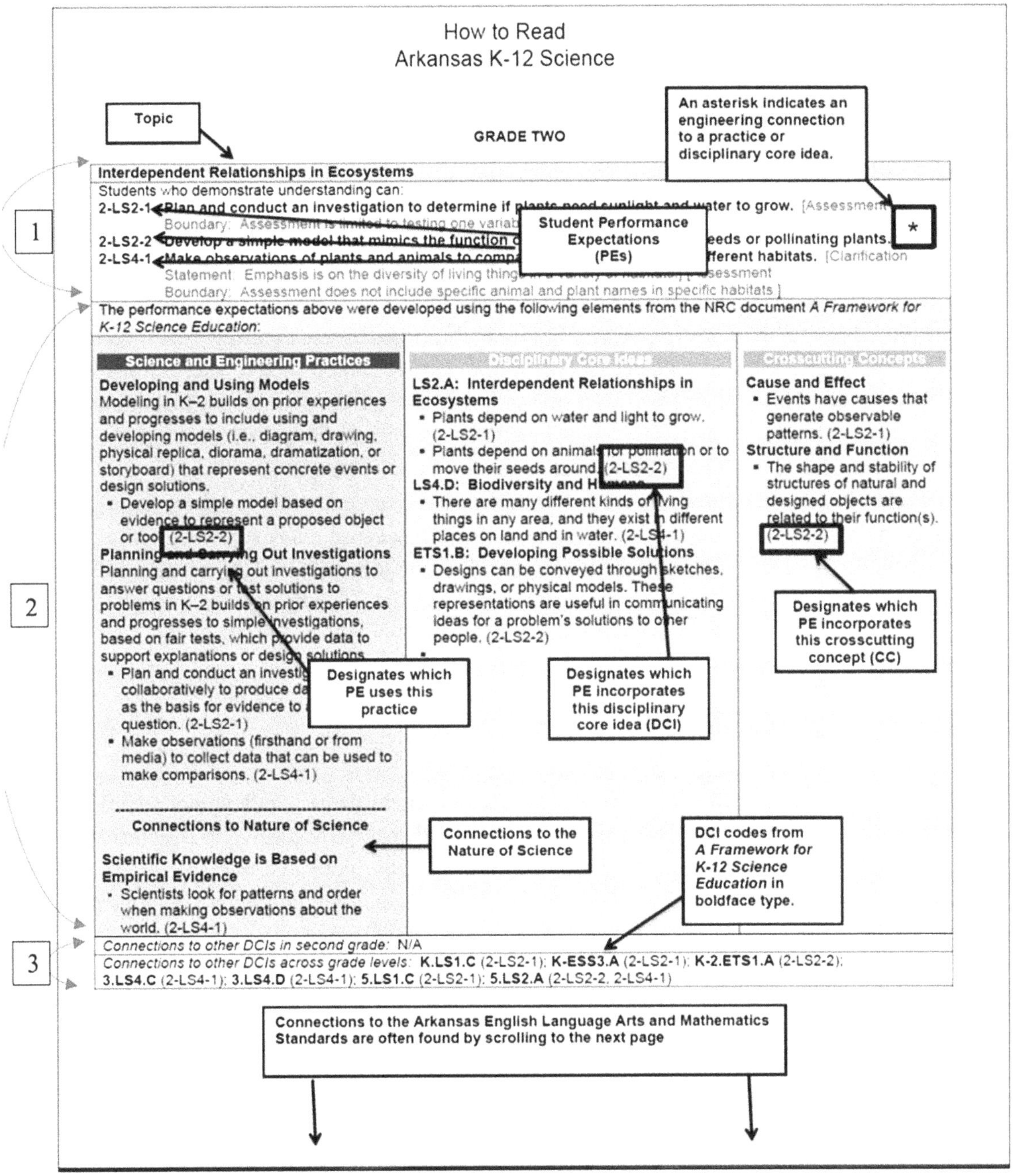

must complete at least two standardized tests, one for critical thinking/logic and the second for creativity to place students in the GT program (www.nagc.org). However, some states and districts use universal screening so that many students are not overlooked. Some states typically rely on their unique academic and nonverbal measures as well as observations and rating scales (www.nagc.org). Furthermore, this process can result in fewer students receiving GT educational services (Gifted and Talented Students and the Next Generation Science Standards, 2015; Croft, 2003). This requires the effort of educators and parents. In many cases, parents are generally the first to notice that their child is performing ahead of their peers, especially in academics. Some of these observable behaviors include the child learning to read or write faster than other children and having an exceptional ability to learn new tasks, especially in the arts and sciences. These are clear indications to many parents who will start questioning whether the child is gifted (Makel, 2022).

Furthermore, the parent-teacher conference is a forum, whereby parents can speak with teachers, and vice versa, about the assumed giftedness of a child in the early grades. Educators may notice many signs of giftedness, and they will share their observations with parents. From their initial observations, teachers will make recommendations to the school counselor or the gifted education facilitator who may employ testing designed to identify potentially gifted children. Parents themselves can be introduced to the Makerspace learning environment, and then Makerspace can be used to successfully enrich the learning experience of GT students by increasing their interest in STEM and focus on critical thinking, creative problem solving, and consistent engagement. By extension, the use of Makerspace in the curriculum links seamlessly to STEM talent development (Subotnik et al., 2011).

On the other hand, some districts also use a universal GT screening process because waiting on a parent or even an educator to notice gifted characteristics may cause many students to be overlooked. Thus, identifying giftedness is not a uniform process. It is the joint effort of many stake-holding constituencies coming together to evaluate distinct types of evidence which indicate a child's giftedness (Croft, 2003; Peters et al., 2019). In essence, it is looking at the expressed needs of these students who enjoy being challenged.

Needs of GT Students to Be Challenged

GT students are generally motivated to be engaged with content that is challenging, meaningful, and relevant to them, especially in (STEM) related subjects. Furthermore, according to Brainbridge (2021), some gifted children may be bored at school or not intrinsically motivated unless they are interested in the task or the material to be learned. This may result in them doing poorly on assignments or even failing to complete them. High achievers may need an educational environment beyond what is offered in the average classroom, but that is not necessarily the same environment that gifted children require to be successful (p. 1). The STEM field helps provide an enriching learning environment requiring higher-order thinking skills, as postulated by Bloom (1956) and reflected in Figure 5.

As shown in Figure 5, GT students thrive on higher-order thinking (HOT) activities, and this is because they like to make connections, analyze, evaluate, and create content (Cohen & Sandberg 1977; Croft 2003; Freeman 1985; Siegle et al., 2014). On the other hand, GT students abhor repetition, slow-paced, lower-order thinking (LOT) skills activities because of their exceptional memory, curiosity, and aptitude to learn. In addition, GT students prefer content that is active and not passive, which is why in the STEM areas, the emphasis is on making connections between subject ideas and subject matter. With this focus in mind, the NGSS are designed to foster active and authentic learning for all students, especially GT students who want to see the bigger picture as they create meaning when learning in a constructivist manner (Davidson, 1984; Guilford et al. 1981; Harrison, 2004; Merrick & Targett 2005; Parkinson,1990).

Figure 5. Bloom's Taxonomy of Thinking Skills Verbs
Adapted from Anderson and Krathwohl (2000)

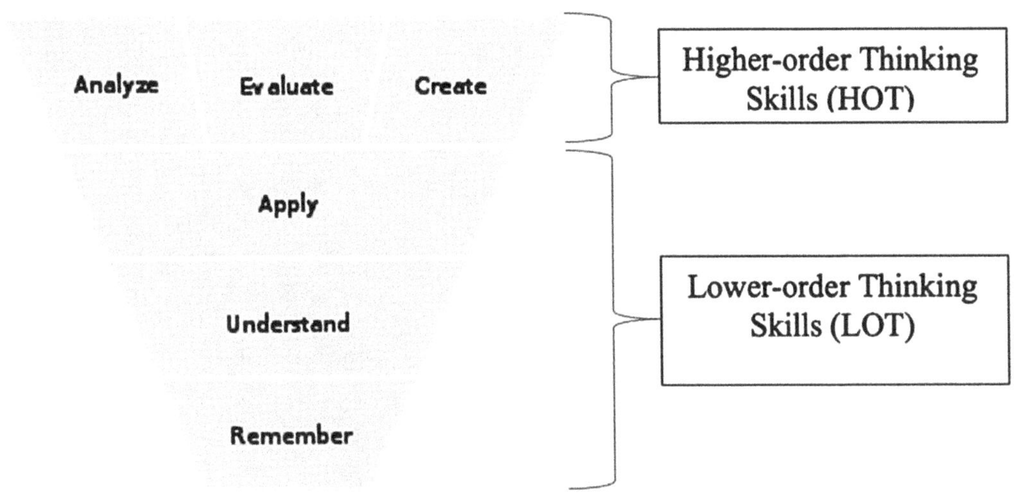

GT students are generally autonomous and independent learners. With guidance, they are highly motivated to dig deep into the subject matter that is of interest to them. They value the freedom to explore the content at their own pace and autonomy over their learning (Mullet et al., 2018). As previously discussed, the NGSS are designed to provide academic rigors for all students, but they are especially beneficial to GT students because these standards outline the content that GT students need to learn while they are making appropriate linkages to STEM. Effective strategies that promote lifelong learning for GT students in STEM can lead to these students becoming highly motivated learners who may not be afraid to take on the challenges they face in the science classroom. However, according to Estes et al. (2011), there also are GT students who may not have been challenged in the past, therefore they may resist challenges by not participating or by choosing to be underachievers if they are not motivated to rise to the challenges before them. Though some students may be unmotivated, most GT students are invested in their learning because they are aware of the vital role they play in constructing their knowledge.

Role of GT Students in the Science Classroom

GT students play a vital role in their learning. Together with teachers, GT students help create the optimal learning environment needed for them to thrive and excel, especially in the modern era. In such a classroom, GT students must show the aptitude and a positive attitude so that as they work through their more challenging content, it reveals what the students are ready to learn as well as the types of enrichments or advanced programming most suitable for them (Northwestern University Center for Talent Development [NUCTD], 2021). Because their presence creates the optimal learning space, students identified gifted are an important resource in today's society (Colesante & Biggs, 2006). As GT students seek to fulfill their roles, they do so with the backdrop of expectations from the various stakeholders who have a vested interest in their success (Brainbridge, 2021).

GT students are at the center of a vortex of activities and support that provide them with the motivation to continue along the path., but they do not make this journey on their own. Constituents have a vested

Figure 6. Stakeholders Vested in the Success of GT students

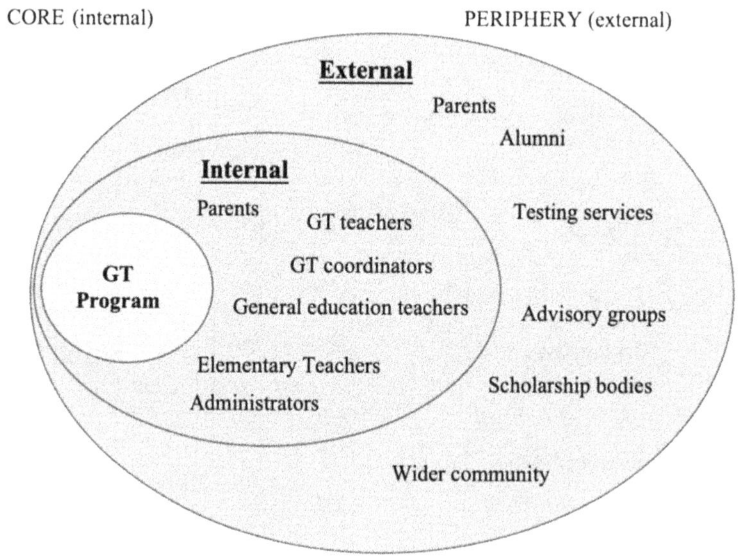

interest or concern in the success and welfare of GT students and thus provide them with much-needed support. These constituents, as identified by Carpenter (2019), are shown in Figure 6.

As shown in Figure 6, both internal and external stakeholders, or constituents, have a vested interest in GT students. Following parents' recommendations, GT coordinators generally proceed with a formal application process that may include testing. During this process, GT teachers and coordinators focus on district and state criteria for gifted identification, which stress assessment test scores as well as each student's potential. As a result, internal stakeholders have high expectations for the success of GT students both inside and outside the science classroom.

Furthermore, external constituencies also have high expectations for GT students' success. Parents are both internal and external constituents, emphasizing the need for their gifted kids to be consistently challenged above and beyond the typical age-based learning environment and curriculum. Testing services review the assessments to evaluate the aptitude of students. Scholarship bodies track the test scores to plan for awarding these students with the appropriate scholarships when they matriculate into higher grades. Advisory groups, alumni, and the wider community all have a vested interest in the success of GT students. These stockholding groups view GT students as having "(a) advanced capacity to learn and reason, (b) high-level performance, (c) the need for challenge, and (d) unique personality characteristics and behaviors" (Carpenter, 2019, p. 1). Due to GT students' high aptitude for learning, the Makerspace science learning environment empowers them to experiment, problem-solve, think critically, construct meaning, as well as create meaningful learning memories with their peers (Colesante & Biggs, 2006; Siemer, 2009).

Empowering GT Students in the Science Classroom

Student empowerment is a vital aspect of attracting, preparing, developing, supporting, and retaining GT students in the STEM fields as shown in Figure 7.

Figure 7. Empowering GT Students
Adapted from Center of Great Teachers and Leaders (2022)

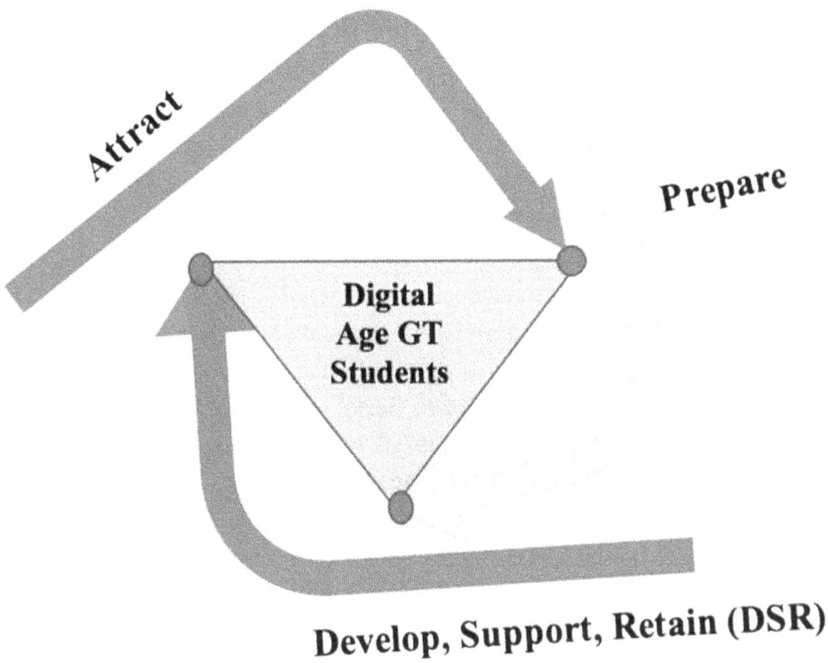

As shown in Figure 7, GT students must be identified to participate in the program. From an early age, students with the aptitude for moving through the curriculum at a faster pace need to be recognized and introduced to GT enrichment programs that meet their needs. GT programs are designed to prepare students for rigorous work that keeps them interested, engaged, and motivated to complete their educational undertaking (Northwestern University Center for Talent Development, 2021). It is through the enriching GT science environment that these students will develop and flourish with strong support from stakeholders. In the long run, with the appropriate amount of support and direction, these students will be encouraged to remain and be retained as part of the GT program in which they are enrolled (Center on Great Teachers and Leaders at the American Institute for Research, 2022).

Section 4: Practical Teaching Strategies for GT Students in the Science Classroom

Makerspace

The development of Makerspaces is a direct result of the Maker Movement, which is a grassroots movement dedicated to hands-on making and technological innovation (Peppler et al., 2016). As a result, Makerspaces have become one of the newest trends in the educational field. What is a Makerspace? For many educators, it is an abstract term that does not have a clear meaning. A Makerspace is a fluid space that is designed specifically to meet students' needs and a school's vision. Consequently, every Makerspace is different. A Makerspace has some aspects of shop classes, art studios, home economics classes, science labs, robotics labs, and digital labs (Brejcha, 2018). A Makerspace allows people to gather in one place for collaboration, tinkering, and critical thinking to complete projects. As such, Makerspaces can be an integral part of a school's science learning environment, and they can help engaged students develop strong interests in their education through exploring the many forms of Makerspace.

The Many Forms of Makerspaces

Makerspaces can have many names and just as many forms. A few examples of common names used for Makerspaces are STEAM Lab, FabLab, Innovation Station, or Techshop. The name can dictate the use of Makerspace, or it can engage the creativity and curiosity of the user. There are several basic forms of a Makerspace: low-tech Makerspaces including little to no technology, high-tech Makerspaces that include a vast array of technology, and Makerspaces that include components of both low and high-tech spaces. Makerspaces can be developed as mobile or stationary spaces.

Mobile Makerspaces usually are smaller than a stationary space and often include portable technologies. Mobile Makerspaces can be as small as a backpack or so large that they must be hauled in a van, bus, or trailer. Mini mobile Makerspaces could include some consumables and basic technologies that are themed toward certain activities. These spaces could be as small as a rolling cart or suitcase, and they are usually themed. Other mini mobile Makerspace examples could include art, coding, basic robotics, sewing, etc. Small stationary Makerspaces are usually located in parts of a school's library or in empty classrooms. Larger Makerspaces may require plumbing and OSHA approval, depending on the equipment used in the space.

Utilization of a Makerspace

A Makerspace is simply a tool to help students deepen their learning. Makerspaces allow the introduction of 21st-century skills and permit students to investigate real-world issues through hands-on inquiry, product development, experimentation, and modeling. Makerspaces can be utilized for any content project. This interdisciplinary inquiry-based approach works well with most subjects and allows projects to become cross-curricular. The versatility of the Makerspace makes it an excellent fit when used with STEM, STEAM, Project-Based Learning, or 5e lessons. The framework of a Makerspace is versatile due to the incorporation of constructivism into its design. This approach allows the learner to create, think critically, experiment, and explore ideas through the construction of tangible objects that model and test students' ideas and deepen learning (Ackermann et al., 2001; Bybee, 2015; ConnectedLib - 1.3 Papert's Constructionism Theory, n.d.; Gravel et al., 2018).

Makerspaces do not require all technology to be set up for students. Incorporating consumables allow the students to engage in engineering, and mathematics and develop the skill necessary for critical thinking (Kurti et al., 2014). Additionally, low-tech items could include basic tool kits, scissors, recyclables, and other general consumables. Conversely, high technology items could include virtual reality systems, computers, 3D printers, laser cutters, greenscreens, vinyl cutters, and multimedia equipment. Large-scale vocational spaces may also include welders, saws, screen-print stations, and Computer Numerical Control (CNC) machines.

Planning a Makerspace

When planning a Makerspace, all stakeholders should be involved. Knowing the requirements of the school, teachers, and students should dictate the design of a Makerspace so that it is a usable, positive learning environment. Next, conducting research should lead to a clear plan and budget. Once a budget is established, finding funding will be the priority. Always start with a small budget for a small space and expand as needed. Once the plan and funding have been established, a timeline for implementation should be developed. It should be noted that the materials and equipment within the space will require periodic maintenance, updating, and timely inventory.

A Makerspace can be extremely rewarding for teachers and students when implemented correctly. Proper training on using the space is integral to its efficacy within the school. If the teacher does not know how to use the space, how to implement inquiry-based learning (Brejcha, 2018), or has poor classroom management skills, the space will not work as intended. Without the structure and creativity of the teacher's 5e planning and collaboration (between the students and other stakeholders), the Makerspace is no more than a room full of equipment.

Using STEM Project-Based Learning (PBL) and STEM Makerspace to Engage Learners

Applying a STEM project-based learning strategy can empower unengaged students to participate in interdisciplinary student-centered collaborative activities (Han et al., 2015a). Project-Based Learning (PBL) is a teaching method in which students learn by actively engaging in real-world projects that are personally meaningful to them (PBLWorks, 2022). A PBL strategy is structured to allow students to research and solve real-world problems. The PBL strategy allows the students an extended period of time to develop their ideas and learn through experience. Using the PBL strategy, students demonstrate the knowledge and skills they have gained through making a public presentation. By incorporating PBL, STEM, and Makerspace, instructors can include many strategies for inquiry-based student-centered learning. These strategies allow the students develop interest in the content in a positive environment that inspires them to learn. Renzulli's (1978) three-ring conception of the giftedness model illustrates task commitment and creativity as characteristics of gifted individuals. The gifted learner characteristics give the gifted learner the ability to enhance learning when being immersed in these strategies. The interaction in PBL, Makerspace, or STEM allows all students to experience real-world learning and problem solving, while also encouraging gifted learners to enrich and enhance their learning (Han et al., 2015a; Renzulli, 1978).

PBL implementation requires teachers to completely comprehend its pedagogical orientation for successful practice (Han et al., 2015b). Issues tend to arise when teachers are not adequately trained in PBL, STEM, or Makerspace. Therefore, it is imperative that appropriate training be provided for teachers in all areas, if they are going to attempt these strategies together. So that K-12 teachers can be effective in the critical analysis of real-world issues, we must provide preservice training for them in the areas of STEM, PBL, and Makerspace (Samarapungavan, et al., 2011). Well-trained teachers plan and implement inquiry-based lessons that are guided by standards, engage students with opportunities for exploration, have clear explanations based on written objectives, allow for elaboration, and have an evaluative component in the science classroom.

5e Lesson Plan for the Science Classroom

The 5e model is an inquiry-based teaching strategy that is grounded in cognitive psychology, constructivist learning theory, and best practices in science teaching (Duran & Duran, 2004). This strategy includes

Figure 8. The Steps in the 5e Lesson Plan

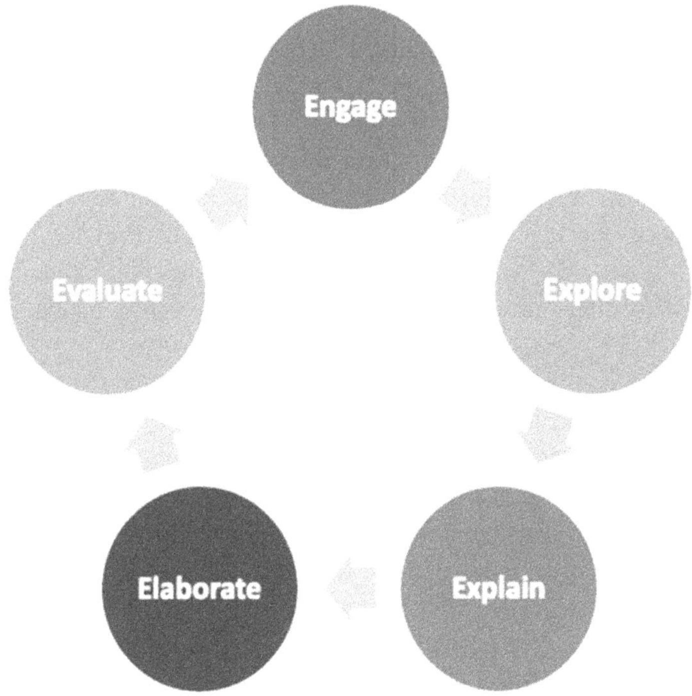

a linear progression of five student-centered scaffolded steps. Figure 8 shows the steps in the 5e lesson: engage, explore, explain, elaborate, and evaluate, in a dynamic process.

As shown in Figure 8, the 5e process is dynamic. The dynamic nature of 5e can benefit GT students in multiple ways. As the GT students make progress, they build on the content that is being taught and eventually reach the final "E" (evaluation), where the GT students are assessed and can show what they have learned. In other words, GT students can move from lower-order thinking skills to reach the higher-order thinking skills outlined by Bloom in Figure 5 and reiterated in Figure 8. So, how can the 5e model be used effectively? First, a working knowledge of the intricacies of the 5e model must be established.

Engage

In the Engage phase, teachers focus on assessing the students' prior knowledge, identifying misconceptions, and motivating the them to learn the content. When introducing new content, teachers usually utilize the hook or attention grabber to build student interest and engagement in the material. Teachers are guided by the NGSS as their hook to launch any activity and/or investigation (National Research Council [NRC], 2012). Relating the content to a phenomenon shows the students both the importance and relevancy of the content, thereby giving the student an opportunity for a deeper understanding of the content.

It should be noted that the learning outcomes must be student-centered and used specifically to assess prior knowledge as well as to gain a high level of student engagement. As a result, this stage of the model should be a point of curiosity and wonder for the students. "Engage" suggests this is not time for lectures, defining terms, or giving explanations for the content of phenomena (Duran & Duran, 2004). Instead, it is the time for student activity, leading them to explore a phenomenon even more.

Explore

Once engaged, students need to have the opportunity to explore the ideas that were created in response to the engagement activities (Bybee et al., 2006; Bybee, 2009; Duran & Duran, 2004). Exploration is a key component in students owning their learning and exploration. Activities in this stage need to be developed based on student-centered inquiry experiences that allow students to investigate and develop new ideas from the information attained during the exploration part of the model. The ideas and experiences gained during exploration will be utilized later within the unit to introduce the actual scientific skills, concepts, and theories that are the roots of the engagement and exploration processes for the unit (Bybee et al., 2006).

The exploration process includes tasks and activities that build on students' critical thinking skills, help them test predictions, develop hypotheses, and observe, question, and investigate to develop a working knowledge based on their own experiences. During the exploration process, students should have common experiences that can be discussed amongst their peers. For example, students can collaborate and perform *Think, Pair, and Share* and develop ideas collaboratively. In any face-to-face or virtual classroom, virtual labs, WebQuests, videos, online field trips, and applicable apps can be utilized for students' exploration. Collaboration and discussion can be performed through discussion boards, social media, virtual meeting spaces, or video conferencing, whereby students can explain their investigations.

Explain

In the first two phases of the 5e model, the students gained a sense of wonder and curiosity through the implementation of a phenomenon and were then encouraged to explore their ideas and hypotheses about the topic. The "Explain" component of the 5e model begins to focus the students' attention. During this phase, students can demonstrate their understanding of the concepts and content. This phase also allows teachers to introduce the required components of the content. Teachers are encouraged to give explanations of the content to deepen the students' learning and conceptual understanding (Bybee, 2015). In the first part of the Explain phase, students should articulate their ideas and explanations. The teachers should act as a facilitator and, through proper questioning techniques, assess the qualitative learning of their students. The second part of the Explaining process is the time for the teacher to inject scientific and technical information directly into the curriculum (Duran & Duran, 2004). Through these experiences, students will gain in the Explain and Explore components, and the teacher should be able to relate the technical information to the experiences the student had (Bybee 2006; Duran & Duran 2004). This explanation allows the students to make genuine connections with the content and develop a deeper level of learning. The Explain phase also allows the teacher to clear up any misconceptions developed during the first two phases of the model.

The Explain phase of the 5e model can be implemented in many formats. In a face-to-face class, teachers could use low-tech approaches such as writing prompts, poster presentations, or speaker presentations. In virtual and face-to-face classes, VoiceThread, FlipShare, DoInk's Greenscreen, or YouTube could all be means of presenting student work. Students can discuss and respond through the comment section of the presentation or classroom platforms (Kool et al., 2018). Through monitoring discussions and directly answering posed questions, teachers can guide students to elaborate on their understanding of the phenomenon at hand.

Elaborate

The Elaboration component of the 5e model is used to enrich and reinforce the students' current knowledge and deepen their understanding. At this point, students are encouraged to share knowledge and check for understanding with their peers or develop new experiments or models based on their new knowledge (Duran & Duran, 2004). The students should be using their new skills to develop products, perform additional inquiries, share information, and investigate cross-curricular issues. The Elaborate component of the model allows students to branch out and develop new areas of interest and learning based on their ideas, interests, and gained skills within the concept and content taught. This process allows students to own their knowledge.

During the Elaboration phase, the students could develop open inquiries, or the teacher could structure project-based learning problems which engage the students in real-world situations that will require 21st Century skills. Through these two types of inquiry, students will be able to learn while engaging in realistic scenarios that are meaningful to each learner. Other enrichment methods could include research projects, projects that link other classes and subjects through teacher-teacher collaboration, and community projects that bring community members and stakeholders into the learning process.

Evaluate

The final "E" of the 5e model is evaluation. There can be many levels to the Evaluation phase within a 5e unit. Formal and informal assessments should be included in this model. Though this "E" is the final component of the model, this does not mean that evaluations should not be performed throughout the learning unit. When using the 5e model, teachers should be aware that many forms of evaluation must take place throughout the process.

Informal assessments are one form of evaluation that should occur throughout the 5e process. Through questioning and proper facilitation, teachers can assess and correct misinterpreted content and misconceptions about what is being learned. Proper questioning tests the students' prior knowledge in order to identify and correct misconceptions as well as to guide students in the right direction. Entrance and exit slips, Kahoot game-based activities, student reflections, journaling, Plickers, and other teacher-made quiz games are a few ways that teachers could informally assess student knowledge throughout the 5e process.

A formal assessment, another form of evaluation, can be performed throughout the process as well. Journaling, portfolios, KWL charts, reflections, presentations, physical models, performance-based assessments, and any assessment that allows students to show growth or change in thought or content can be utilized as formal assessment and implemented within most of the "Es." For example, a pretest implemented at the engagement phase of the process and a summative or post-test implemented at the evaluation phase will allow the teacher to use statistical measures to show student growth (Rovai, 2002). The utilization of the pre- and post-testing, along with the qualitative assessments, can deliver a mixed-methods approach to the learning gained, which allows the qualitative data to support the quantitative assessment data. This process enables the teacher to view an overarching measure of learning.

Hands-on Teaching Strategies

Makerspace, PBL, and STEM all promote a hands-on, inquiry-based approach though Makerspace is a tool that can be used to facilitate STEM and PBL activities. The Makerspace can be developed to fit any group of students or particular focus areas. The fluidity of a Makerspace allows each learner to have a unique, differentiated experience. The 5e structure provides these tools and methods a scaffolded place, allowing the teacher to develop a well-structured curriculum that can be easily differentiated for all learners. Applying these strategies together allows all students to become the owners of their learning. However, the biggest issues with implementing any of these strategies are classroom management and teacher training because all the techniques discussed in this section allow the students to take control of their own learning processes. In order for students to learn and create with these strategies, as the facilitator, the teacher must have a well-developed plan, sound pedagogy, and excellent classroom management. If the teacher is missing one of these requirements, student learning will be lost or will not occur at all. Yet, when implemented properly, these strategies promote critical thinking, complex problem-solving, collaboration, and 21st-century skills. The traits they promote lead to the motivation of the GT student (Chrysovalantis & Drigas, 2020; Han et al., 2015a).

CONCLUSION

Conclusively, GT students thrive on challenges in the classroom. The traditional adage of teacher as sage on the stage has given way to more student-centered learning, which is driven by the technology available in today's digital age. The ubiquity of technology has made it more accessible, affordable, and easily integrated into student learning (Purdue University Online, 2022), and GT teachers and coordinators have responded to the change by encouraging their students to learn multi-modally, anywhere and anyhow, with no borders so as to best meet the needs of today's learners (Morris et al, 2021).

Like never before, pedagogy is shifting towards training students to utilize technology that facilitates literacy, learning, and life skills, thereby preparing them for the continually increasing numbers of technology-driven jobs that are becoming available. The educational landscape has seen and continues to see a multiplicity of changes over the last 25 years in the areas of educational standards and digital technology integration across all levels. 21st Century teachers continue to monitor and adjust to meet the changes, challenges, and successes by delivering dynamic content in science education for all students, including those identified as GT-Engaging students. Using a multi-modal approach is key to creating a GT curriculum that is student-centric. The use of Makerspace is also a key component in the GT teachers' toolkit and one that facilitates the constructivist and socialization theories.

The Makerspace environment is an excellent tool for implementing multiple inquiry approaches such as STEM/STEAM, problem-based learning, and project-based learning. For inquiry-based learning to occur, students must be attracted to the space and be inspired to use it (Kurti et al., 2014). This space should engage the students to arouse their interest and promote their creativity. This is in keeping with both the theory of socialization and the constructivist theory. Within Makerspace, students research, develop questions, design and build prototypes, and develop models to solve a problem. Makerspaces encourage project collaboration, allowing students to build strong social and communication skills and enabling them to share their ideas in a positive, creative setting.

In today's digital landscape, with proper teacher training and implementation of inquiry-based learning activities, GT students can take control of their education and develop high-order thinking skills and levels of engagement. Implementing technology into the curriculum only amplifies the level of student understanding and engagement, and GT students are hindered only by the level of technology they can access and their teachers' ability implement those technologies. In this way, highly trained teachers are an integral part of the GT success formula. Technology is only one tool within a well-designed and implemented GT curriculum (Gravel et al, 2018). Without highly-trained science teachers to students learning, a technology infusion has but a limited impact on student engagement and learning. Instead, the level of student engagement and learning is grounded in the tenets of socialization and constructivist theories. Thus, this work has strong implications for science policy and practices going forward.

Implications for Science Policy and Practice

Preparing K-12 students, including GT students, for STEM careers, means exposing them early to rigorous STEM content that is guided by standards. Teachers can implement this content through scaffolding and the introduction of STEM vocabulary and resources in the early grades. The more and earlier students experience success with STEM concepts, the less challenging STEM courses will be for them in the future. This should translate to middle and high schools offering more STEM-related electives as well as honors and AP-level STEM courses. As part of the schools' policies, teaching K-12 science content using the NGSS's three-dimensional standards leads to students developing critical and divergent thinking skills. In this model, teachers are no longer teaching just science content, skills, and concepts, but broader STEM applications as well. This has implications for practice, in that technology in a Makerspace environment can be used to enhance all students' learning of science. Additionally, digital technology can aid in assisting GT students to become industrious individuals who develop good work habits which allow them to be successful. Through the implementation of STEM content, students can relate real-world situations to pure science, developing much-needed 21st Century skills. The use of Problem-Based Learning (PBL), Makerspace, 5e planning, and 21st Century skills enable GT students to reach their full potential in the science classroom. Furthermore, the ways that all students (including GT students) are socialized from childhood helps to determine how they embrace science (Lynch, 2017).

The theory of constructivism is applicable when motivating and facilitating active hands-on work for GT students. Based on the exploratory nature of constructivism, classroom practices must be supported to foster and maintain a learning environment wherein students feel safe when asking questions and their teacher meets their needs through differentiation. GT students need modification within their classes to engage them in higher-order thinking skills. Because of this, teachers should consider providing supplementary/enrichment work (not just extra work or worksheets) and peer tutoring for GT students. Extra work can be viewed as negative reinforcement and the feeling that GT students are being punished for finishing their work quicker and at a higher quality compared to peers. However, enrichment opportunities can be added to GT students' assignments allowing them to take ownership of what they learn. GT students are at the center of multiple influences in society, and technology, including Makerspace, can be used to enhance all students' learning of science. Digital technology can aid in assisting GT students to become industrious and develop good work habits, and it allows them to be successful. A Makerspace environment rich with digital technologies like WebQuests, virtual field trips, educational social media platforms, and multimedia allows GT students to research and construct their knowledge and ideas in a creative space, all without leaving the classroom (Wagner & Ruch, 2015).

Using robotics and engineering products such as 3D printers, Strawbees, Vex, Micro: bits, Spheros, Circuit Playgrounds, Lego Mindstorms, Lego WeDo, Arduino, Raspberry Pi, etc. in the science classroom allows students to bring their virtual creations into the real world. As such, teachers of GT students need to utilize inquiry-based learning (inductive strategies) that will differentiate student learning through inquiry-based activities. Inductive strategies encourage participation, and collaboration, build critical thinking skills and develop a sense of curiosity within the students (Prince & Felder, 2006). These enrichment opportunities allow students to take ownership of what they learn. The enrichment can be implemented through inquiry-based learning (inductive strategies) that will differentiate student learning through inquiry-based activities. The ways that all students, including GT students, are socialized from childhood determine how well they embrace science. Science educators must recognize the vital roles they play in the way students view science, and they should try to promote perspectives of science for their students.

As a means of policy, school districts need to fund the teaching of science so that a lack of resources or well-trained teachers are not barriers to learning for any students. Proper funding of science within school districts accompanied by quality training in PBL, STEM, Makerspace, and implementation of three-dimensional learning will give teachers the tools to develop rigorous, engaging, differentiated learning within their classroom environments. Funding and proper tools allow teachers to modify GT students' curriculum within their classes using methods that engage the GT learner (Winner, 2009) and lead to the use and refining of higher-order thinking skills.

REFERENCES

Ackermann, E. (2001). Piaget's constructivism, Papert's constructionism: What's the difference. *Future of Learning Group Publication, 5*(3), 438.

Alexander, P. A., Graham, S., & Harris, K. R. (1998). A Perspective on Strategy Research: Progress and Prospects. *Educational Psychology Review, 10*(2), 129–154. doi:10.1023/A:1022185502996

American Institute for Research. (2022). Talent development framework. Center on Great Teachers and Leaders at the American Institute for Research. https://gtlcenter.org/td

Anderson, L., & Krathwohl, D. (Eds.). (2000). *A taxonomy for learning, teaching, and assessing—A revision of Bloom's taxonomy of educational objectives*. Pearson.

Association of Gifted Children (NAGC). (2022). www.nagc.org

Birenbaum, M. (2007). Assessment and instruction preferences and their relationship with test anxiety and learning strategies. *Higher Education, 53*(6), 749–768. doi:10.100710734-005-4843-4

Bloom, B. (1956). *Taxonomy of educational objectives: The classification of educational goals*. Longman.

Brainbridge, C. (2021). *Differences Between Academic High Achievers and Gifted Students*. Available at https://www.verywellfamily.com/high-achiever-1449168

Brejcha, L. (2018). *Makerspaces in Schools: A Month-by-Month Schoolwide Model for Building Meaningful Makerspaces*. Prufrock Press Inc.

Bybee, R. (2009). The BSCS 5E instructional model and 21st-century skills. *BSCS, 24*. HTTPS:// citeseerx.ist.psu.edu/viewdoc/download?doi=10.1.1.674.6559&rep=rep1&type=pdf

Bybee, R., Taylor, J., Gardner, A., Van Scotter, P., Powell, J., Westbrook, A., & Landes, N. (2006). The BSCS 5E instructional model: Origins and effectiveness. *BSCS, 5*, 88-98. https://fremonths.org/ourpages/auto/2006/9/7/1157653040572/bscs5efullreport2006.pdf

Bybee, R. W. (2015). *The BSCS 5E instructional model: Creating teachable moments*. NSTA Press, National Science Teaching Association.

Calik, B., & Birgili, B. (2013). Multiple intelligence theory for gifted education: Criticisms and implications. *Journal for the Education of the Young Scientist and Giftedness*, *1*(2), 1–12. doi:10.17478/JEYSG.201329002

Carpenter, A. (2019*). Stakeholders' conceptions of giftedness* [Doctoral Dissertation, University of Connecticut]. Opencommons@UConn. https://opencommons.uconn.edu/dissertations/2066

Center on Great Teachers and Leaders at the American Institute for Research. (2022). *Talent development framework*. https://gtlcenter.org

Chrysovalantis, K., & Drigas, A. (2020). S.T.E.M.: Inquiry-Based Learning and Gifted Education. *Neurology and Neurobiology, 2020*(4), 1–5. doi:10.31487/j.NNB.2020.04.01

Cohen, R., & Sandberg, T. (1977). Relation between intelligence and short-term memory. *Cognitive Psychology*, *9*(4), 534–554. doi:10.1016/0010-0285(77)90019-6

Colesante, R., & Biggs, D. (2006). Discovering gifted youth in inner-city schools. *Forum on Public Policy*, *1*, 455–465. https://www.researchgate.net/publication/252799923

Common Core State Standards Initiative. (2022). *About the Standards*. Core Standards. http://www.corestandards.org/about-the-standards/

ConnectedLib—1.3 Papert's constructionism theory. (n.d.). Retrieved April 9, 2022, from https://connectedlib.github.io/modules/youth-development/section-1-3.html

Council of State Directors of Programs for the Gifted and the National Association for Gifted Children. (2009). *State of the States in Gifted Education: National Policy and Practice data 2008-2009*. National Association for Gifted Children.

Croft, L. (2003). Teachers of the gifted: Gifted teachers. In N. Colangelo & G. A. Davis (Eds.), *Handbook of gifted education* (pp. 558–571). Allyn & Bacon.

Davidson, J. E. (1984). The role of insight in giftedness. *Gifted Child Quarterly*, *2*, 58–64. https://doi.org/10.1177/001698628402800203

Davidson, J. E., & Sternberg, R. J. (1986). What is insight? *Educational Horizons*, *64*(4), 177–179.

Duran, L., & Duran, E. (2004) The 5E Instructional Model: A Learning Cycle Approach for Inquiry-Based Science Teaching. *The Science Education Review, 3*(2), 2004. https://files.eric.ed.gov/fulltext/EJ1058007.pdf

Estes, A., Rivera, V., Bryan, M., Cali, P., & Dawson, G. (2011). Discrepancies between academic achievement and intellectual ability in higher-functioning school-aged children with an autism spectrum disorder. *Journal of Autism and Developmental Disorders*, *41*(8), 1044–1052. doi:10.100710803-010- 1127-3

Freeman, J. (Ed.). (1985). *The psychology of gifted children*. John Wiley.

Freeman, J. (2005). Permission to be gifted. In R. J. Sternberg & J. E. Davidson (Eds.), *Conceptions of giftedness* (2nd ed., pp. 80–97). Cambridge University Press.

Freeman, J. (2010). *Gifted lives: What happens when gifted children grow up*. Routledge.

George Lucas Educational Foundation. (2016, July 20). *Multiple intelligence: What does the research say?* Edutopia. https://www.edutopia.org/multiple-intelligences-research

Gifted and Talented Students and the Next Generation Science Standards. (2015). Retrieved from https://my.nsta.org/resource/121566

Gravel, B., Tucker-Raymond, E., Kohberger, K., & Browne, K. (2018). Navigating worlds of information: STEM literacy practices of experienced makers. *International Journal of Technology and Design Education*, *28*(4), 921–938. https://doi.org/10.1007/s10798-017-9422-3

Guilford, A., Scheuerle, J., & Schonburn, S. (1981). Aspects of language development in the gifted. *Gifted Child Quarterly*, *25*, 159–163.

Han, S., Capraro, R., & Capraro, M. (2015a). How Science, Technology, Engineering, and Mathematics Project-Based Learning Affects High, Middle, and Low Achievers Differently: The Impact of Student Factors on Achievement. *International Journal of Science and Mathematics Education*, *13*(5), 1089–1113.

Han, S., Yalvac, B., Capraro, M. M., & Capraro, R. M. (2015b). In-service Teachers' Implementation and Understanding of STEM Project Based Learning. *Eurasia Journal of Mathematics, Science and Technology Education*, *11*(1).

Harrison, C. (2004). Giftedness in early childhood: The search for complexity and connection. *Roeper Review*, *26*(2), 78–84.

H.R. 1 - 107 Congress (2001-2002). *No child left behind act of 2001*. https://www.congress .gov/bill/107th-congress/house-bill/1

Klein, A. (2015, April 10). *No child left behind: An overview*. Education Week. https://www.edweek.org/policy-politics/no-child-left-behind-an-overview/2015/04

Kool, A., Mainhard, M., Jaarsma, A., van Beukelen, P., & Brekelmans, M. (2018). Do students with varying academic ability benefit equally from personal qualities? Applying a trait and state perspective. *Research in Higher Education*, *59*(8), 1021–1034. doi:10.100711162-018-9498-y

Kurti, R., Kurti, D., & Fleming, L. (2014). The philosophy of educational maker spaces: Part 1 of making an educational Makerspace. *Teacher Librarian*. https://www.informalscience .org /philosophy-educational-makerspaces-part-1-making-educational-makerspace

Lanoue, N. (2021, October 4). *Why was the NGSS science framework developed?* Know Atom. https://www.knowatom.com/blog/what-is-ngss-a-brief-history-

Lynch, M. (2017). Preparing K-12 students for future STEM careers. *The Edvocate*. https://www.theedadvocate.org/preparing-k-12-students-for-future-stem-careers/

Makel, M. C. (2022). Additional challenges to achieving equity in gifted and talented education. *Gifted Child Quarterly*, *66*(2), 101–102. https://doi.org/10.1177/00169862211039735

Maksimović, J. (2020). The constructivist approach in the research of gifted characteristics. *Significance and Benefits, 3*(25), 265-278. https://www.researchgate.net/publication /342696993

Martin-Hansen, L. (2002). Defining inquiry: Exploring the many types of inquiry in the science classroom. *Science Teacher (Normal, Ill.)*, *69*(2), 34–37.

McNeill, K., Berland, L., & Pelletier, P. (2017). Constructing explanations. In Helping students make sense of the world using Next Generation Science and Engineering practices (pp. 205–228). NSTA Press.

Meador, D. (2019). *Using Effective Instructional Strategies*. ThoughtCo. Available at https://www.thoughtco.com/building-an-arsenal-of-effective-instructional-strategies-3194257

Merrick, C., & Targett, R. (2005). *Gifted and Talented Education: Professional development package for teachers, early childhood: Module 2. Gifted Education Research, Resource and Information Centre (GERRIC)*. University of New South Wales.

Mesmer, K. (2015). My journey to understand and implement the NGSS. *Science Scope*, *39*(4), 18–20.

Meta, J., & Fine, S. (2019). *In search of deeper learning*. Harvard University Press.

Moomaw, S., & Davis, J. A. (2010). STEM comes to preschool. *YC Young Children*, *65*(5), 12.

Morado, M., Melo, A., & Jarman, A. (2021). Learning by making: A framework to revisit practices in a constructionist learning environment. *British Journal of Educational Technology*, *52*(3), 1093–1115. https://doi.org/10.1111/bjet.13083

Morris, J., Slater, E., Fitzgerald, M. T., Lummis, G. W., & van Etten, E. (2021). Using Local Rural Knowledge to Enhance STEM Learning for Gifted and Talented Students in Australia. *Research in Science Education*, *51*(1), 61–79. https://doi.org/10.1007/s11165-019-9823-2

Mullet, D., Kettler, T., & Sabatini, A. (2018). Gifted students' conceptions of their high school STEM education. *Journal for the Education of the Gifted*, *41*(1), 60–92. https://doi.org/ 10.1177/0162353217745156National

National Association for Gifted Children. (n.d.). *Frequently asked questions about gifted education: Is there a definition of "gifted?"* https://www.nagc.org/resources-publications/resources/frequently-asked-questions-about-gifted-education

National Association of Gifted Children. (n.d.). *Gifted children and STEM*. https://www.nagc.org/resources-publications/resources/timely-topics/stem-meeting-critical-demand-excellence/gifted

National Research Council. (2012). *A Framework for K-12 Science Education: Practices, crosscutting concepts, and core ideas.* The National Academies Press. https://doi.org/10.17226/13165

National Science Teaching Association. (2022). *Gifted and Talented Students and the Next Generation Science Standards.* NSTA Press. https://my.nsta.org/resource/121566

Nava, I., & Park, J. (2021). Pre-Service STEM Teachers and Their Enactment of Community-STEM Project Based Learning (C-STEM-PBL). *Journal of Higher Education Theory and Practice*, *21*(9), 217–237. https://www.proquest.com/docview/2579423687/abstract/EA3CCCA80CE145A5PQ/1

Next Generation State Standards. (2013). *Case Study 7: Gifted and talented students and the Next Generation Science Standards.* NGSS Release. https://www.nextgenscience.org/sites/ default/ files/%287%29%20Case%20Study%20TAG%206-14-13.pdf

Next Generation State Standards. (2022). *Improving Through Three-Dimensional Learning Science Education*. NGSS Release. https://www.nextgenscience.org/

Northwestern University Center for Talent Development. (2021). https://www.ctd.northwestern. edu/assessment-resources-education

Olszewski-Kubilius, P. (2000). The transition from childhood giftedness to adult creative productiveness: Psychological characteristics and social supports. *Roeper Review*, *23*, 65–71. doi:10.1080/02783190009554068

Olszewski-Kubilius, P., Kulieke, M., & Krasney, N. (1988). Personality dimensions of gifted adolescents: A review of the empirical research. *Gifted Child Quarterly*, *32*, 347–352. doi:10.1177/001698628803200403

Parkinson, M. L. (1990). Finding and serving gifted preschoolers. *Understanding Our Gifted*, *2*(5), 10–13.

Peppler, K., Halverson, E., & Kafai, Y. B. (2016). *Makeology*. Routledge.

Pereira, L., & Sithole, B. (2019). Towards constructivist learning and teaching in accounting in accounting education. *Journal of Emerging Trends in Educational Research and Policy Studies, 10*(1). http://jeteraps.Scholarlinkresearch.com/articlesTowards%20Constructivist%20Learning%20NEW.pdf

Peters, S., Gentry, M., Whiting, G., & McBee, M. (2019). Who Gets Served in Gifted Education? Demographic Representation and a Call for Action. *Gifted Child Quarterly*, *63*(4), 273–287. https://doi.org/10.1177/0016986219833738

Piaget, J. (1953). *The Origins of Intelligence in Children*. Basic Books.

Prince, J., & Felder, R. M. (2006). Inductive teaching and learning methods: Definitions, comparisons, and research bases. *Journal of Engineering Education*, *95*(2), 123–138. https://doi.org/10.1002/j.2168-9830.2006.tb00884.x

Purdue University Online. (2022). *Gifted Education Resource Institute*. http://geri.education.purdue. edu/academic_programs/Phd_in_gt.html

Renzulli, J. S. (1978). What makes giftedness? Reexamining a definition. *Phi Delta Kappan*, *60*(3), 180–261. https://www.jstor.org/stable/20299281

Riley, D., Klinkner, J. & Ramminger, A. (2009). *Introduction to sociology: Understanding and changing the social world*. https://pressbooks.howardcc.edu/soci101/ chapter/4-2-explaining-socialization/

Rovai, A. P. (2002). Building a sense of community at a distance. *The International Review of Research in Open and Distributed Learning*, *3*(1). https://doi.org/10.19173/irrodl.v3i1.79

Samarapungavan, A., Patrick, H., & Mantzicopoulos, P. (2011). What kindergarten students learn in inquiry-based science classrooms. *Cognition and Instruction*, *29*, 416–470. doi:10.1080/07370008.2011.608027

Sharapan, H. (2012). STEM to STEAM: How early childhood educators can apply Fred Rogers' approach. *YC Young Children*, *67*(1), 36–40.

Siegle, D., Rubenstein, L., & Mitchell, M. (2014). Honors students' perceptions of their high school experiences: The influence of teachers on student motivation. *Gifted Child Quarterly*, *58*, 35–50. https://doi.org/10.1177/0016986213513496

Siemer, E. A. (2009). Bored out of their minds: The detrimental effects of No Child Left Behind on gifted children. *Washington University Journal of Law and Policy*, *30*, 539–560.

Srivastava, S., & Dangwal, K. L. (2017). Constructivism: A paradigm to revitalize teacher education. *International Journal of Applied Research*, *3*, 753–756.

Subotnik, R. F., & Jarvin, L. (2005). Beyond Expertise: Conceptions of Giftedness as Great Performance. In R. J. Sternberg & J. E. Davidson (Eds.), *Conceptions of giftedness* (2nd ed., pp. 343–357). Cambridge University Press.

Subotnik, R. F., Olszewski-Kubilius, P., & Worrell, F. C. (2011). Rethinking giftedness and gifted education: A proposed direction forward based on psychological science. *Psychological Science in the Public Interest*, *12*, 3–54. doi:10.1177/1529100611418056

Sullivan, A., Kazakoff, E. R., & Bers, M. U. (2013). The wheels on the Bot go Round and Round: Robotics curriculum in pre-kindergarten. *Journal of Information Technology Education Innovations in Practice*, *12*, 203–219.

Taber, K. (2006). Beyond Constructivism: The Progressive Research Program into Learning Science. *Studies in Science Education*, *42*, 125–184. https://doi.org/10.1080/03057260608560222

Tay, J., Salazar, A., & Lee, H. (2017). *Parental Perceptions of STEM Enrichment for Young Children.* Department of Education Studies Graduate Student Publications. https://docs.lib.purdue.edu/edstgspubs/1

Terman, L. M. (1922). A New Approach to the Study of Genius. *Psychological Review*, *29*, 310–318. doi:10.1037/h0071072

Terman, L. M. (1954). The discovery and encouragement of exceptional talent. *The American Psychologist*, *9*, 221–230. doi:10.1037/h0060516

UKEssays. (2018). *Implications For Classroom Teachers Education Essay*. Retrieved from https://www.ukessays.com/essays/education/implications-for-classroom-teachers-education-essay.php?vref=1

Vygotsky, L. (1962). *Thought and Language*. MIT Press.

Wagner, L., & Ruch, W. (2015). Good Character at School: Positive Classroom Behavior Mediates the Link Between Character Strengths and School Achievement. *Frontiers in Psychology*, *6*(610). Advance online publication. doi:10.3389/fps.2015.00610

Winner, E. (1996). *Gifted children: Myths and Realities*. Basic Books.

Winner, E. (2009). Toward Broadening our Understanding of Giftedness: The Spatial Domain. In F. D. Horowitz, R. F. Subotnik, & D. J. Matthews (Eds.), *The development of giftedness and talent across the lifespan* (pp. 75–85). American Psychological Association. doi:10.1037/11867-005

Works, P. B. L. (2022). *What is Project-Based Learning?* Available at https://www.pblworks.org/what-is-pbl

Worrell, F. C. (2010a). *Giftedness: Endowment, context, timing, development, or performance? Does it matter?* American Psychological Foundation's Esther Katz Rosen Lecture on Gifted Children and Adolescents was presented at the annual convention of the American Psychological Association, San Diego, CA.

APPENDIX

Table 3. Digital resources for makerspace

Title	Web Application/Address
Chatterpix	App
Daisy the Dinosaur	App
Scratch	App
Scratch Jr.	App
Tynker	App
Code Land	App
codeSpark Academy	App
Gravitrax	App
Green Screen by DoInk	App
Flipshare	App
Voicethread	App
Book Creator	bookcreator.com
Chatter Pix	chatterpix.com
Swift Playgrounds	https://www.apple.com/swift/playgrounds/
Tynker	https://www.tynker.com/
Dogone News	https://www.dogonews.com/
Makecode	https://www.microsoft.com/en-us/makecode
University Corporation of Atmospheric Research: Center for Science Education	https://scied.ucar.edu/
Physics Education Technology Interactive Simulations for Science and Math	https://phet.colorado.edu/
Phenomena for NGSS	https://www.ngssphenomena.com/

Chapter 3
Empowering Students Through Motivational Engagement in the Online Classroom

Effat Zeidan Hatoum
Independent Researcher, USA

ABSTRACT

Student engagement in the online classroom is more important than ever for both traditional and non-traditional students. In this chapter, many strategies will be presented to enhance the student learning experience by empowerment, motivational communication, inclusive learning environments, interdisciplinary teaching, and assessment. The recommendations presented in this chapter are best practices that have positively impacted the learning journey of science and mathematics online at California Baptist University. Instructors are encouraged throughout the chapter to creatively find ways to teach and assess the diverse group of students to better meet their unique needs and learning abilities.

INTRODUCTION

Teaching science in an online learning environment can be a daunting and complex task. Developing rapport and community is a complex task, and the experiential nature of learning science creates a complex set of challenges for online science teachers. Science educators must think outside the box, creatively using strategies to engage learners and build an online community through an inclusive pedagogy and curriculum. Virtual tools for instruction can enhance the learning experience and provide value to online communities in K–12 and higher education (Evans-Amalu & Claravall, 2021) and simulate lab activities in a virtual science classroom. To address the challenges of online classrooms, science teachers should focus on developing inclusive classrooms with motivated students who engage in learning science in authentic and effective ways.

This chapter focuses on inclusive student-centered learning that attends to diverse students' differing and unique needs in online science classrooms. I define the inclusive online learning environment as one committed to meeting the diverse learning needs of a pool of students with various social, cultural,

DOI: 10.4018/978-1-6684-5585-2.ch003

and religious backgrounds. I introduce practices that center on the student's best interests, promoting a sense of belonging and positive interaction with both instructor and peers (MacKay Report on Inclusion in Schools in New Brunswick, 2013).

Students have multiple motives to seek science education in an online setting ranging from the flexibility and convenience of the online classroom to personal and professional enrichment goals. Motivation is the engine of learning as it can influence any part of the learning process (Paris & Turner, 1994). Motivation plays a key role in students enrolling in an online science course and guiding the learning process toward improved retention of learned content. To capitalize on students' initial motivation, instructors must combine effective delivery of subject matter content with consistent motivational support toward improving the student experience. Creating inclusive conditions for students to inquire and learn effectively improves the learning experience. The instructor's role shifts from being a traditional instructor with a primary focus on delivering content to also developing communication strategies that alleviate anxiety and student fears early in the learning process.

This chapter provides strategies that can assist online K–12 science educators, college/university science instructors, and future science teachers in enabling a sense of belonging for their students in the virtual classroom. A list of best practices is provided in this chapter with a focus on creating an inclusive learning environment derived from an interdisciplinary approach to teaching and assessing science content in the online classroom. The tools highlighted under the interdisciplinary approach are central to the concept of integrating knowledge to achieve a more comprehensive understanding of learned concepts (Daniel et al., 2022). Students develop integration skills and learn how to combine their experiences and insights from different disciplines to critically analyze complex concepts beyond the scope of the classroom when presented with examples and exemplars (Stasiewicz, 2022). Additionally, I present supportive communication tools that can be implemented in online science courses. These tools help instructors foster motivational learning and improve academic belongingness (Wenbo et al., 2022), establishing an online community linked to improved student performance fostered by increased social presence and collaboration between instructor and student (Chen & Bogachenko, 2022).

Building Inclusivity: Supporting Students in Online Environments

A fully developed social presence in the online science classroom builds the students' sense of inclusivity. Faculty-to-student communication that occurs early in a course creates the foundation for student agency, which leads to their feeling of inclusivity (Lowenthal & Dunlap, 2020). Furthermore, personalized messaging in asynchronous online classes support the development of students' effective social presence (Elmore, 2022). Sung and Mayer (2012) define social presence in the online classroom as the degree of connectedness with the learning community, including other classmates and the instructor, that the student experiences. Social presence is linked to greater student satisfaction (Richardson et al., 2017), student motivation to learn and participate (Mazzolini & Maddison, 2007), and enhanced student learning and performance (Hughes et al., 2008; Yamada, 2009). Above all, students with a strong sense of connectedness and community are more interactive and engaged in online learning because they feel valued and respected by their instructors and classmates (Al-Dheleai & Tasir, 2019).

In the context of learning science online, some of my students enrolled in an online science course have expressed low motivation toward the nature of the asynchronous learning environment. Furthermore, non-traditional students who have been out of the educational setting for some years have expressed concerns and fear of learning science in a distant setting. Their concerns revolve around the asynchronous nature of the online classroom, leading to fear and anxiety when they feel disconnected from the instructor and their classmates.

ANXIETY IN ONLINE LEARNING

Eliminating the anxiety that accompanies some courses is essential in an online environment. For example, math anxiety is a major pressure on students in introductory mathematics courses that results from negative emotions toward math and the ability to perform math problems effectively (Ashcraft, 2002). Student anxiety over an online format can compound any preexisting anxiety over the subject matter (Ritzhaupt et al., 2022). Online learners have varying learning needs that online, blended learning models can support (Bates, 2015). A well-designed blended online course can be utilized to alleviate the anxiety that online science students experience. Blended models include synchronous and asynchronous elements purposefully blended to support students through an online instructional design model (Power, 2008). As instructors design their online courses, they can utilize technology tools to facilitate digital face-to-face interactions through applications like Zoom or Google Meet and asynchronous learning experiences using structured communication through a learning management system. Two key components of online, blended learning design that alleviate learner anxiety are course organization and instructor communication. Organizing the course in logical, linear patterns and providing multimodal communication so that there is abundant communication between the instructor and students throughout the course can increase student success (Jaggars & Xu, 2016). To reduce anxiety caused by the subject matter, fostering a learning community in the online classroom where students collaborate with the instructor and peers can decrease the negative emotions associated with the subject matter in an online context. In effect, humanizing the virtual classroom to provide an inclusive learning environment can empower students to believe in their learning abilities and boost self-confidence, leading to better performance.

Implementing Best Practices in the Online Science Classroom

Considering previous research on increasing belonging in the online classroom, I implemented multiple strategies into my online science teaching practice. I describe both the context of the courses and how I applied strategies that build students' sense of belonging in the online science classroom. The science courses I teach are eight weeks long and delivered asynchronously. The typical modes of communication between the students and instructor are through email, message boards in our learning management system, and occasional phone calls. All lectures, assignments, and course content are preloaded for students to access as they navigate the course.

Starting the Semester Strong

Students often express signs of discouragement and anxiety within the first two weeks of the session. Before that anxiety creeps in, instructors have the first opportunity to lay the foundation for the online community; to make the student feel welcomed, supported, and heard. Instructors can welcome the students by communicating their support, sense of belonging to the virtual learning experience, and collaboration in their learning journey. The communication can be as simple as a welcome video introducing them to the online classroom, the class structure, and expectations. The welcome videos allow students to see their professor as another human; the presence of the instructor's face promotes interaction with students and feelings of connectedness to the online community for a more facilitated learning experience.

Educators can deliberately use this first opportunity to extend empathy, support, and care to students while encouraging them to pursue and enjoy the learning journey with success. A friendly tone in these videos goes a long way in delivering a welcoming note to the class that, in turn, reduces stress and remedies the students' anxiety. Additionally, instructors can relate to students by sharing real-life experiences and common interests, emphasizing their sense of belonging to the online community (Chen & Bogachenko, 2022).

Recording personalized videos for students who have expressed concerns and hesitations about their online learning journey supports student engagement and belonging. Instructors can record videos or chat with students by offering words of encouragement, by examples of engagement, and by opportunities for interactions with other peers as well. These focused interactions through video lay the foundation for collaborative learning in the online classroom (Oyarzun et al., 2018). My success in building online communication resulted in positive course feedback from students. Students reported that my encouragement and continuous support contributed to their overall success. My efforts to ease students' initial fear and challenges of learning science in the virtual environment have been successful. Ongoing messages of encouragement also contributed to students' success in learning science online.

Weekly Messages of Encouragement

The practice of informal instructor-generated messages of encouragement supports the development of student efficacy and belonging in online science classes. I craft an email before each academic week starts. I purposefully begin the email with a personal anecdote from the previous week's interactions with students, a personal experience, or an interesting informal science fact. Then I focus on academic expectations and specific support for the science content students will experience.

The practice of crafting and sending weekly messages supports student engagement and motivation in the online science classroom. Since beginning the academic week with crafted interpersonal emails, the frequency of student-to-instructor email communication has increased. Since the onset of the pandemic, students will often reply to my weekly emails with special requests for support. This online interaction with students can positively impact both student and instructor, creating a collaborative learning journey in the online learning environment. An example of an email received by one student is shown in Figure 1.

One successful strategy is to add a weekly theme incorporated into the students' encouragement or other supportive messages. These communication tools can incorporate authentic empathy and extend grace as central themes, both of which can positively impact the student's emotional well-being in a virtual learning environment (Cartee, 2021). Extending grace refers to flexibility in assigned task completion and minor alterations to assignment instructions, meeting the needs and students' circumstances.

Social and Emotional Wellbeing Through Check-in Opportunities

Effective online science instructors seek opportunities to develop rapport with the students to support their social and emotional well-being. Traversing new scientific content can be stressful. In addition, students in an online course must navigate and communicate in a virtual digital environment, which can add frustration and stress to the student learning experience. In order to alleviate students' stressors, it is important to offer multiple check-ins that support their social, emotional, and academic well-being.

Figure 1. Sample Student Response to Instructor Support

Sample Student Response

Hello Professor Zeidan,

I am sorry that I did not get everything completed for this last week. I just lost my dad to COVID and have been making arrangements last week for him. Plus taking care of the rest of my family and myself who have been ill with this horrible disease. When I read your email below it really gave me hope today that I can do this and I do have God to lean on. The prayer you provided hit home and has motivated me to get working as I know my father would be proud. I will get everything complete and caught up this week. Thank you for your kind words. This is really really what I needed today. Thank you.

Best,

Student name

Encouraging students to check in helps them develop a working relationship with the instructor and supports their understanding that we are all human and have challenges, and support is always available. During these check-ins, it is helpful for instructors to relate to the student's circumstances by sharing their own experiences and showing empathy as they support navigating stressful and challenging times.

Check-in opportunities can take the form of a virtual meeting, video or voice recording, phone conversation, text messaging, or emails. Variety makes the connection between the instructor and student more personable and supportive. I adopted these strategies to support students more extensively during the COVID-19 pandemic. Instructors can establish boundaries to the offered support presented above by ensuring easy scheduling for one-on-one meetings. During the recent pandemic years, I focused on student well-being as a prerequisite for the effective learning journey in the online classroom. The challenges found in student lack of motivation to complete work increased during the stressful COVID-19 months. Students were forced to develop online learning skills for their personal, professional, and academic life (Hamilton & Gross, 2021). Therefore, it became crucial to ensure a student-centered learning experience that alleviates these challenges; and offers a supportive learning environment. I learned that we could do that effectively in the virtual classroom by honoring our students, recognizing their efforts, and checking on them to ensure they have the emotional support they need. As a result, students were more open about sharing their daily struggles with me, and this feedback encouraged me to continue to be present in their lives through my virtual connection with them. One student sent me an email to express her gratitude; "Professor, it has been a trying eight weeks, but I made it. Thank you again for all your continued support. I was so stricken with illness and anxiety with all that is happening in the world today I almost gave up. I enjoyed learning in this class, and I especially enjoyed the different Lab Assignments."

MONITORING STUDENT PERFORMANCE

Monitoring student performance is crucial in online learning to track at-risk students who need added support from the instructor (Raza et al., 2021). K-20 science educators in online asynchronous learning environments should monitor student progress while considering all other factors influencing the learning journey (Raza et al., 2021). Artificial intelligence applications track students and assist them (Daud et al., 2017; Oladokun et al., 2008). In this study, the authors were interested in collecting learning analytics data from various universities in Pakistan to determine the students' graduation rate. For example, a self-monitoring tool was implemented in an Early Warning System (EWS) Moodle application at the University of the South Pacific, allowing students to self-monitor progress in real-time in the online course (Jokhan et al., 2018). Identifying underperforming learners using technology tools can benefit instructors' timely intervention, enhancing student success (Raza et al., 2021).

Continuous Academic Guidance

Students in online science classrooms must have ongoing academic support and guidance. The complexity of science instruction in the online space requires purposeful communication about each student's academic progress. One way to support this communication strategy is the use of progress reports as a tool to monitor student progress. In my practice, I prepare and distribute individual reports two to three times during the 8-week online course. These reports aim to involve students in their progress by engaging them in the feedback I provide. It is a strategy that encourages students to stay focused during the accelerated online science class. Additionally, it can humanize the virtual classroom by demonstrating connection and academic and emotional support to the student. Students benefit from their progress reports and take them as an opportunity to continue to grow in their learning process. I achieve human connection by including academic and motivational assessments. I highlight students' dedication as a positive drive for continuous development toward success. An encouraging note can be added to the progress report to help students overcome their frustration with the learning process. There is always a critical "next step" in these reports, where the instructor's feedback is provided to direct the student to the specific areas where more work or practice is needed. The example progress report in figure 2 is a simple report highlighting areas of strength that need more attention, recommendations for moving forward, and support lines available to the student.

Engaging students through these progress reports allow them to be involved in their learning progress, which can be a chance to reflect on their past learning experience, increasing their self-efficacy. Therefore, it is important to keep in mind three things in preparation for a student's progress report: reflection on the task itself, feedback on the student's performance and progress, and finally, recommendations for moving forward. Boor et al. (2021) mentioned that one of the levels that feedback can influence is self-regulation, and it does so by developing internal cognitive feedback mechanisms that allow the student to acknowledge their mistakes and know why they made them and how they can avoid making the same mistakes in the future. This self-regulation skill is more important than the feedback on the process or the task itself (Hattie & Timperley, 2007) because it provides practice for students to develop this that they will continue to use throughout their professional lives.

Video Responses to Student Inquiries

Video responses to student questions and concerns are an opportunity to connect with students, demonstrate instructor presence, and promote a growth mindset by providing meaningful answers and feedback to the questions asked in the classroom. I have found that creating a brief video to answer the questions with a step-by-step process has successfully clarified the inquiries while adding a human touch to the virtual assistance provided. To facilitate the productive collaboration between student and instructor, the input from the instructor in a visual modality can provide a safe and positive atmosphere for the student in the online space. Additionally, this provides a personalized and interactive communication experience for the student (Wackenhut & Gillette, 2022; Lowenthal & Dunlap, 2011). For complex responses to

Figure 2. Sample Progress Report Mid-Session

Progress Report: Mid-Session

Student Name:

Areas of Strengths:	Content: Dedication: Engagement:
List of Areas that need more attention:	Item A: Item B:
Recommendations moving forward:	Item A: Item B:
Questions? Would you like to discuss a plan?:	Email: Virtual Meeting: Phone:

students, instructors can break down a problem into its subcategories or provide a step-by-step response using a screencast or virtual whiteboard (Example includes Educreations applications (Educreations, Inc, 2022). In one online Fundamentals of Chemistry course, I had numerous students that struggled with balancing chemical equations. To support student learning and respond to their inquiry, I created videos through Educreations that included a step-by-step tutorial with many examples that clarified the students' questions. One example of a video I created can be found at https://www.educreations.com/lesson/view/chemistry-q1/46924047/?s=4b5CUT.

These are easy ways to engage students with short videos that are interactive and dynamic, and that can be easily shared with the individual student or the online student community. These recorded videos can be replayed anytime. If students encounter difficulty with the same subject or topic, they can replay the recorded video and follow the tips and instructions. Other recordings can include laboratory demonstrations that are difficult to explain in writing or by email. I found this helpful in virtual laboratory courses that utilize lab kits. When students reach out with specific experimental questions from the lab procedure, it is most helpful when a visual demonstration is provided for clarification.

Modalities: Teaching and Assessment

Students in the online classroom can have diverse backgrounds, skills, and educational experiences. Online students expect well-structured online courses that are easy to navigate, and the role of the instructor is to ensure that learner diversity is welcomed throughout the instructional process. Employing multiple assessment modalities is central to creating an inclusive online learning environment in which students find the space to develop and reflect on their unique learning journey. Successful online science instruction and assessment require creativity and innovation, where the instructor tailors the course design to meet the needs of the learners. The learning journey will be perceived as enjoyable when students can relate to the tasks, find meaning in the learning, and connect their experiences to their own educational and professional trajectories. In the next section, I share various strategies that can be applied in the online classroom to ensure the success of the learning experience, both teaching and assessment.

Online Science Assessment Strategies

Instructors tend to standardize weekly course assignments to meet the needs of the assumed student population; however, inclusivity calls for flexibility in assessment and for diverse modes of assessment to meet students' individual needs (Noman & Amrita, 2014). Innovation and flexibility in assessment design are encouraged for the diverse group of students to effectively meet the learning objectives of each task assigned in the course. Instead of creating uniform tasks for each week with specific criteria that should be met, instructors can offer multiple modalities in assessing a certain concept from which students can choose.

A positive implication of this assessment strategy is the reinforcement of a safe online learning environment for accurately assessing student learning outcomes. Not all students prefer writing reports or research papers when presenting a scientific phenomenon or concept. Some students might prefer to present the results of a project using a video or audio recording of themselves or an animated video they creatively designed. Therefore, students can meet the requirements of a specific student learning outcome by completing the task using various modalities; written, oral, or even a combination of audio-visual modalities. Expanding the options for deliverables enhances inclusivity in the online classroom and provides a unique path for students to explore their strengths and areas that need improvement. Additionally, it indirectly involves students in the assessment process, which is crucial in knowledge retention and effective learning.

As mentioned previously, student involvement in the assessment process empowers the student to be more active in the learning process and in receiving feedback and implementing it. One of the unique tasks I have assigned to my students as a critical assignment in the course is the opportunity to re-teach a concept from the course to their online virtual community that includes the instructor and their classmates. I encourage them to get creative in their teaching modes and use tools they would like to see in an online course. This final project emphasizes allowing the student to choose a topic and to relate to the topic based on his or her own professional and personal interests. This task promotes self-assessment, self-reflection, and self-awareness skills that are discussed in the upcoming section. This inclusive assessment allows the student to express their comprehension and learning of a certain concept in their unique ways.

Another strategy that can bring inclusivity to the students' learning experience includes pairing students into groups and assigning the task of designing a meaningful project for their team members based on the content learned in the course while keeping in mind the required student learning outcomes. This opportunity showcases how direct involvement in assessment can promote self-regulated learning towards attaining their learning goals. Students are also required to collaboratively work with their classmates to ensure clear instruction is provided on the project expectations and rubric grading criteria. This exercise not only promotes more effective learning; but also sheds light on the needs of online learners in the virtual community and what works and what does not in effective instruction and assessment (Boor et al., 2021). I encourage instructors to keep in mind the goal of providing a variety of opportunities to demonstrate student learning that are interdisciplinary and engaging in nature.

Building Reflection Skills

Reflection assignments are critical to the inclusive virtual classroom. These tasks can be individual or group reflections involving interactions through the discussion board with classmates. Students are encouraged to participate actively in these reflection assignments because it allows them to evaluate their work, actions, and learning progress. These interdisciplinary tasks create a space for inclusivity in the online classroom and can lead to the development of self-regulated and self-directed learners (Edelbroek et al., 2018). The interdisciplinarity aspect of the reflection assignments is through giving the student the space to talk about what was interesting to them and why it was interesting from the weekly learning content to reflect and provide opinions on the content learned freely. Students can also be allowed to reflect on their own interdisciplinary learning experience and how that can be applied to their future classroom or perhaps professional careers. In all reflection assignments assigned in online science and mathematics courses, students are asked to address the pros and cons of the topics or disciplines covered in the course and to what extent has the interdisciplinary learning approach added value to their learning experience.

Additionally, instructors can adjust the reflection assignments to meet the specific needs of students and get creative in the design of the reflection questions and discussion elements. Furthermore, I recommend preparing students for the reflection task by providing guidance on what the questions entail and by adding examples to encourage them to answer freely, think outside the box, and share their differing perspectives in a safe online environment. Moreover, keep in mind, as you design reflection assignments, the purpose of a reflection is to develop the student's interests in an interdisciplinary way of thinking. These tasks give the student a sense of self-authorship in the learning journey with an interdisciplinary mindset (Boor et al., 2021).

Interdisciplinary Teaching Strategies

To attend to the needs of a diverse pool of students, instructors can get creative in their differentiated instruction or delivery of learned content in the sciences and mathematics. This interdisciplinary approach to teaching can be highly beneficial because it allows neurodiverse students to utilize their unique learning needs in the online classroom (Patrick, 2019). The educational resources provided can take various modalities in the course shell. PowerPoint presentations are recommended in the weekly course learning activities to highlight the important concepts in the weekly readings and guide the weekly recorded lectures, which are another important resource for the online learner. The instructor's recorded lectures are available throughout the course and can be accessed anytime. These lectures are central to enhancing learner engagement because they humanize learning. During these lectures, the instructor delivers the learning content and fosters the learner's empowerment through encouraging messages, relating to real-life issues and applications, and forming connections with the students. The recorded lectures characterized by the active connection with the diverse students lead to a mutual learning experience between instructor and student (Zyngier, 2011). With the increasing surge in open educational resources, instructors are encouraged to add supplemental animations, multimedia posters, case studies, and links to other educational videos. My key recommendation here is for instructors to ask themselves how they can add value to the learning activities because most of the content we are teaching them is already readily available information. Therefore, let this question guide you as you integrate the interdisciplinary approach into creating learning content for your online classroom. Next is an example of how instructors can offer students a blended online learning experience utilizing a hands-on lab experience in a virtual laboratory setting.

APPLYING SCIENCE TO THE REAL WORLD

Despite assumptions that science sometimes exists in a vacuum, science instructors must develop an approach that supports students' connections between science, other disciplines and their lived experiences. Integrating knowledge and perspectives from various disciplines is essential to comprehend a complex phenomenon or concept (Daniel et al., 2022). This integration can be introduced to the virtual classroom in different forms and modalities. The instructor teaching from an integrative standpoint tends to discuss the links between various concept elements, provide examples from the real world, and combine insights from different disciplines to ensure the effective delivery of the knowledge and learning content. This strategy in teaching can attend to the diverse nature of the student body by creating opportunities for students to think outside the box. Students develop their critical thinking and analysis skills through these integration examples. Developing personal and professional socio-emotional skills is necessary for the future workforce. These so-called "soft skills" require an authentic and creative approach to learning that starts in the classroom (McKinsey Global Institute, 2017). The interdisciplinary approach in providing real-life examples and issues to students better prepares them for reality in the professional space after graduation. Furthermore, this integration of knowledge can be included in the authentic assessment design, as discussed in the section below.

My recommendation for integrating knowledge in teachings stems from personal educational and professional experience as a nanoscientist. As a nanoworld researcher, my knowledge background was interdisciplinary in the sciences, which had to include an overview of disciplinary perspectives and concepts from nanochemistry, nanobiology, nanophysics, and nanomathematics necessary for conducting research at the nanoscale. My work focused on developing ultra-sensitive diagnostic tools to detect disease biomarkers in human serum early in the disease stages. When beginning my journey in education, I appreciated the interdisciplinary approach from my experience and used it as a source of motivation for delivering science content creatively and inclusively to virtual students. I was always looking for examples and strategies to connect the sciences in explaining a biological, chemical, or physical concept. Moreover, a good example of when adopting this interdisciplinary and inclusive approach is useful is when illustrating the coherent functioning units of the human body. To provide a comprehensive understanding of one of the most complex systems in the world, the human body, one must pull concepts from all the sciences and mathematics in the learning process. These are examples that I encourage you to reflect on as you think of designing and implementing authentic learning resources for your online classroom.

Laboratory Experiences

Science educators and future science teachers should consider exploring the hands-on lab experience through an online laboratory. Traditional laboratory experiences offer a unique way to engage the student with the scientific process and reinforce the troubleshooting skills needed in real-world laboratory settings (Massey et al., 2021). Traditional "wet-laboratory" exercises can include a structured set of experiments that are not necessarily attending to a diverse group of learners. Combining virtual guidance, online assistance, and lab kit experiences leads to the effective exploration of scientific phenomena in the virtual classroom. These labs occur in a risk-free environment (one's home), are widely accessible, and promote a hands-on approach (Faulconer & Gruss, 2018). Guided virtual labs resulted in students achieving higher levels of general inquiry skills and increased ability to design investigations across various contexts (Wen et al., 2020). Efstathiou et al. (2018) investigated technology-based feedback loops and content control in a digital lab activity finding that younger learners increased their inquiry skills and experiment design abilities when using a digital tool over paper and pencil in science lab simulations. In addition, Chang et al. (2020) investigated middle schoolers' development of scientific knowledge, scientific literacy, and motivation in a digitally mediated inquiry lab condition. The students who engaged digitally with simulations showed a greater increase in knowledge, motivation, and science literacy than their counterparts. Conversely, another study conducted by Wen et al. (2018) found that virtual labs do not guarantee that the student develops scientific inquiry and literacy, indicating the need for a hands-on lab experience when learning science in both online and in-person settings. Although virtual laboratories can help students visualize the scientific process, students' mental models of the process might not change after simply viewing the experience through a screen (McElhaney et al., 2015).

Online Hands-On Lab Experiences

It is common for students to expect that enrolling in a virtual science lab entails only digital or online lab experiences. However, the hands-on experience in conducting labs and experiments can positively enhance learning retention and provide a community of learning even in the fully asynchronous online setting, as argued by the learning science by doing science model. I began to provide hands-on labs for undergraduate science courses by adopting hands-on lab kits for science labs, including a lab kit that can be purchased before the start of the course (Science Interactive, 2022). It is worth mentioning that this model for utilizing the hands-on lab kits was adopted for introductory biology, chemistry, and environmental courses and not higher-level major labs, which are not offered virtually due to the need for a traditional wet-lab setting. These lab kits are safe to use at home, and the experience can be shared with other family members and, often, with small children who may be in the home, which can be an exciting experience.

Instructors should offer a blended learning experience in virtual laboratory science courses, which can be accomplished by providing experimental protocol videos, recorded learning guides for each exercise, and other readily available resources on the subject matter. A good resource provided each week can be a recorded illustration or demonstration of the assigned lab uploaded to their weekly learning activities before starting their week. This can prepare the student for the lab and provide a visual illustration that mirrors the face-to-face interaction in a traditional laboratory setting. Additionally, if students are enrolled in both the course and corresponding lab, their lab experience runs parallel to the recorded lectures on the concept of each lab, and demonstrations, which enhances the practical learning experience. Once students complete their weekly assigned lab, they complete a lab report that includes data tables, short answer questions, images and demonstrations, and specimen slide images retrieved from a virtual microscope and then submit it for grading. Figure 3 is an example of an uploaded image to the lab report titled "Senses" in human anatomy and physiology labs representing the dissection of a cow's eye the student performed on her kitchen table. Figure 4 is retrieved from an introductory undergraduate biology lab conducted virtually using a similar kit. The figure represents the use of colored beads to assemble chromosomes in their various mitotic and meiotic phases.

Figure 3. Dissection of Cow Eye

Despite the presumption that a virtual learning experience can be lonely, instructors can bridge that gap and be more involved in the learning process through these engagement tools. A leading visual component in the learning process can motivate students to perform the labs despite certain hesitations that can arise in some instances and lack of motivation due to previous experiences in the lab setting.

CONCLUSION

Students come to an online science course with varied life experiences and goals for engaging in science learning. Learning about the students' backgrounds and future goals benefits the instructor and the learners. Instructors can employ communication tools, integrative modes of learning and assessment, and reflective exercises mentioned in this chapter to impact the learners. Integrating these strategies can support online K-20 science instructors in meeting each student's individual learning needs. These strategies can be adopted into the practice of online teaching to empower the student, provide a safe environment to learn and develop, and ultimately enhance student engagement in an asynchronous learning modality. I recommended that online science teachers focus on involving students in their learning process as much as possible and promoting peer interaction through peer assessment strategies, group work on complex problems, and online discussion boards. Utilizing these strategies has positively impacted my online science students' learning and application of science to the world around them.

Figure 4. Mitosis: Metaphase

REFERENCES

Al-Dheleai, Y., & Tasir, Z. (2019). Web 2.0 for fostering students' social presence in online learning-based interaction. *Journal of Technology and Science Education*, 13-19.

Ashcraft, M. H. (2002). Math anxiety: Personal, educational, and cognitive consequences. *Current Directions in Psychological Science*, *11*(5), 181–185. doi:10.1111/1467-8721.00196

Bates, A. T. (2015). *Teaching in a digital age.* https://opentextbc.ca/teachinginadigitalage/

Boor, I., Gerritsen, D., de Greef, L., & Rodermans, J. (2021). *Meaningful assessment in interdisciplinary education*. Amsterdam University Press. doi:10.5117/9789463729048

Cartee, J. (2021). Strategic empathy in virtual learning and instruction: A contemplative essay about teacher-student rapport during times of crisis. *Journal of Instructional Research*, 12-19.

Chang, C. J., Liu, C. C., Wen, C. T., Tseng, L. W., Chang, H. Y., Chang, M. H., Fan Chiang, S.-H., Hwang, F.-K., & Yang, C.-W. (2020). The impact of light-weight inquiry with computer simulations on science learning in classrooms. *Computers & Education*, *146*, 103770. doi:10.1016/j.compedu.2019.103770

Chen, J., & Bogachenko, T. (2022). Online community building in distance education: The case of social presence in the blackboard discussion board versus multimodal voicethread interaction. *Journal of Educational Technology & Society*, 62–75.

Daniel, K., McConnell, M., Schuchardt, A., & Peffer, M. E. (2022). Challenges facing interdisciplinary researchers: Findings from a professional development workshop. *PLoS One*, *17*(4), e0267234. doi:10.1371/journal.pone.0267234 PMID:35439277

Daud, A., Daud, A., Aljohani, N., Abbas, R., Lytras, M., Abbas, F., & Alowibdi, J. (2017). Predicting student performance using advanced learning analytics. *26th International Conference on World Wide Web Companion*, 415-421. 10.1145/3041021.3054164

Edelbroek, H., Mijnders, M., & Post, G. (2018). *Interdisciplinary learning activities.* Amsterdam University Press.

Educreations, Inc. (2022). *Teach anything to anyone from anywhere.* https://www.educreations.com

Efstathiou, C., Hovardas, T., Xenofontos, N., Zacharia, Z. C., deJong, T., Anjewierden, A., & van Riesen, S. A. N. (2018). Providing guidance in virtual lab experimentation: The case of an experiment design tool. *Educational Technology Research and Development*, *66*(3), 767–791. doi:10.100711423-018-9576-z

Elmore, R. A. (2022). *Teaching presence in a fully online asynchronous undergraduate mathematics course and its impact on social and cognitive presence* [Doctoral dissertation, Michigan State University]. ProQuest Information & Learning.

Evans-Amalu, K., & Claravall, E. B. (2021). Inclusive online teaching and digital learning: Lessons learned in the time of pandemic and beyond. *Journal of Curriculum Studies Research*, *3*(1), i–iii. doi:10.46303/jcsr.2021.4

Faulconer, E., & Gruss, A. (2018). A review to weigh the pros and cons of online, remote and distance science laboratory experiences. *International Review of Research in Open and Distributed Learning*, 155-168.

Hamilton, L., & Gross, B. (2021). *How has the pandemic affected students' social-emotional well-being? A review of the evidence to date.* University of Washington, Bothell. Center on Reinventing Public Education (CRPE).

Hattie, J., & Timperley, H. (2007). The power of feedback. *Review of Educational Research*, *77*(1), 81–112. doi:10.3102/003465430298487

Hughes, J., Luo, W., Kwok, O., & Loyd, L. (2008). Teacher-student support, effortful engagement, and achievement: A 3-year longitudinal study. *Journal of Educational Psychology*, *100*(1), 1–14. doi:10.1037/0022-0663.100.1.1 PMID:19578558

Jaggars, S. S., & Xu, D. (2016). How do online course design features influence student performance? *Computers & Education*, *95*, 270–284. doi:10.1016/j.compedu.2016.01.014

Jokhan, A., Sharma, B., & Singh, S. (2018). Early warning system as a predictor for student performance in higher education blended courses. *Studies in Higher Education*.

Lowenthal, P., & Dunlap, J. (2011). Investigating students' perceptions of various instructional strategies to establish social presence. In *Annual Meeting of the American Educational Research Association.* Google Scholar.

Lowenthal, P., & Dunlap, J. (2020). Social presence and online discussions: A mixed method investigation. *Distance Education*, 490-514.

Massey, A., Zhang, W., & Amar, A. (2021). A comparison of non-traditional online and traditional wet-lab experiences in human anatomy and physiology: An innovative approach for pre-licensure nursing education. *Nurse Education Today*, *107*, 105149. doi:10.1016/j.nedt.2021.105149 PMID:34592576

Mazzolini, M., & Maddison, S. (2007). When to jump in: The role of the instructor in online discussion forums. *Computers & Education*, *49*(2), 193–213. doi:10.1016/j.compedu.2005.06.011

McElhaney, K. W., Chang, H.-Y., Chiu, K. L., & Linn, M. (2015). Evidence for effective uses of dynamic visualizations in science curriculum materials. *Studies in Science Education*, *51*(1), 49–85. doi:10.1080/03057267.2014.984506

McKinsey Global Institute. (2017). *Jobs Lost, jobs gained: Workforce transitions in a time of automation*. McKinsey & Company.

Noman, M., & Amrita, K. (2014). Differentiated assessment: a new paradigm in assessment practices for diverse learners. *International Journal of Education and Applied Sciences*, 167-174.

Oladokun, V., Adebanjo, A., & Charles-Owaba, O. (2008). Predicting students' academic performance using artificial neural network: A case study of an engineering course. *Pac J Sci Tech*, 72-9.

Oyarzun, B., Barreto, D., & Conklin, S. (2018). Instructor Social Presence Effects on Learner Social Presence, Achievement, and Satisfaction. *TechTrends*, *62*(6), 625–634. doi:10.100711528-018-0299-0

Paris, S., & Turner, J. (1994). *Student motivation, cognition, and learning: Essays in honor of Wilbert J. McKeachie*. Academic Press.

Patrick, F. (2019). Second life and virtual learning: An educational alternative for neurodiverse students in college. *College Student Journal*, 33–41.

Power, M. (2008). The emergence of a blended online learning environment. *Journal of Online Learning and Teaching*, *4*(4), 503–514.

Raza, S. H., Sharma, B. N., & Chaudhary, K. (2021). A new pair of watchful eyes for students in online courses. *Frontiers in Applied Mathematics and Statistics*, *7*, 620080. doi:10.3389/fams.2021.620080

Richardson, J. C., Maeda, Y., Lv, J., & Caskurlu, S. (2017). Social presence in relation to students' satisfaction and learning in the online environment: A meta-analysis. *Computers in Human Behavior*, *71*, 402–417. doi:10.1016/j.chb.2017.02.001

Ritzhaupt, A. D., Rehman, M., Wilson, M. L., & Ruggles, K. (2022). Exploring the factors associated with undergraduate students' online learning anxiety: Development of the online learner anxiety scale. *Online Learning, 26*(3), 111–131. https://doi-org.libproxy.calbaptist.edu/10.24059/olj.v26i3.2810

Science Interactive. (2022). *Digital Lessons Hands On Experiments*. Retrieved from Science Interactive: https://www.scienceinteractive.com

Stasiewicz, C. G. (2022). Interdisciplinary Education Equips People to Face Unique Challenges. *Journal of National Collegiate Honors Council*, 89-91.

Sung, E., & Mayer, R. (2012). Computers in Human Behavior Five facets of social presence in online distance education. *Computers in Human Behavior*, 1783–1747.

Wackenhut, A. F., & Gillette, M. B. (2022). Covid-19 and (re)learning teaching: Never let a crisis go to waste. *Högre Utbildning*, *12*(1), 52. doi:10.23865/hu.v12.3562

Wen, C. T., Chang, C. J., Chang, M. H., Chiang, S. F., Liu, C. C., Hwang, F. K., & Tsai, C. C. (2018). The learning analytics of model-based learning facilitated by a problem-solving simulation game. *Instructional Science*, *46*(6), 847–867. doi:10.100711251-018-9461-5

Wen, C. T., Liu, C. C., Chang, H. Y., Chang, C. J., Chang, M. H., Chiang, S. F., ... Hwang, F. K. (2020). Students' guided inquiry with simulation and its relation to school science achievement and scientific literacy. *Computers & Education*, *149*, 103830. doi:10.1016/j.compedu.2020.103830

Wenbo, D., Weina, L., Xipei, G., Xinyi, L., Wenshuang, G., & Weiping, H. (2022). Effects of regulatory focus on online learning engagement of high school students: The mediating role of self-efficacy and academic emotions. *Journal of Computer Assisted Learning*, 707–718.

Yamada, M. (2009). The role of social presence in learner-centered communicative language learning using synchronous computer-mediated communication: Experimental study. *Experimental Study. Computers & Education*, *52*(4), 820–833. doi:10.1016/j.compedu.2008.12.007

Zyngier, D. (2011). (Re)conceptualizing risk: Left numb and unengaged and lost in a no-man's lands or what (seems to) work for at-risk students. *International Journal of Inclusive Education*, *15*(2), 211–231. doi:10.1080/13603110902781427

Section 2

Incorporating Engineering and Computational Thinking

Chapter 4
Engineering Design-Based Arduino Activities in STEM Education

Hatice Yildiz Durak
https://orcid.org/0000-0002-5689-1805
Necmettin Erbakan University, Turkey

Sedef Canbazoglu Bilici
Gazi University, Turkey

Evrim Baran
Iowa State University, USA

ABSTRACT

The purpose of this chapter is to present the use of Arduino in engineering design-based STEM activities at the secondary school level in K-12 education. Examples of the use of Arduino and the free Arduino Science Journal mobile application are also presented along with different stages of the design process in an engineering design-based thematic activities (e.g., greenhouse design, noise pollution, etc.). The widespread use of the user-friendly Arduino Science Journal App in STEM education activities will contribute to the development of students' scientific process skills (e.g., observation, measurement, data recording, reading graphics, and interpreting data). This chapter presents examples of using Arduino in engineering design-based activities for science teacher educators.

INTRODUCTION

STEM (Science, Technology, Engineering, and Mathematics) education is the application of disciplinary knowledge and skills together with an interdisciplinary approach to solving real-world problems (Akarsu et al., 2020; Daugherty & Carter, 2017). The main goal is to enable interdisciplinary integration

DOI: 10.4018/978-1-6684-5585-2.ch004

by improving learning and STEM career development opportunities for students (Bybee, 2013; Chang & Chen, 2020; Hansen, 2014).

Despite different definitions towards the nature of STEM integration, there is an intersection between different approaches including engineering pedagogy, engineering design, and inquiry-based science education as teaching strategies (Psycharis & Kalovrektis, 2022). There are cross-cutting concepts in these approaches that can act as integration agents and mediate STEM practices in the classrooms. This integration is the result of the need to improve students' participation in engineering practice, creative, and innovative thinking abilities as well as problem-solving competencies (Perignat & Katz-Buonincontro, 2019). Engineering facilitates the integration of disciplines in different ways in line with the "content integration" or "context integration" models in STEM education. The context integration model, while focusing on science concepts, is generally considered as the context in engineering (Moore et al., 2014a). Engineering design-based STEM activities contribute to the development of students' skills such as creative secondary school science courses (Perdana et al., 2021; Sari et al. 2022).

Arduino is an open-source electronic prototyping platform that enables users to create interactive electronic objects. Arduino has increasingly been used in K-12 STEM learning contexts spanning different disciplines and grade levels. The use of Arduino through mobile technologies offers opportunities such as encouraging students to explore science in real life, recording the necessary real data instantly, and conducting experiments anytime and anywhere. The purpose of this chapter is to examine the use of Arduino in STEM activities based on engineering design at the secondary school level. This chapter presents several examples of Arduino integration from the STEM education literature. Mobile learning integration with Arduino Science Journal is also discussed and illustrated with examples.

BACKGROUND

Engineering Design-based STEM Activities

Engineering design is a process used to solve engineering problems and develop products. Engineering design can act as a binder to learn the content of mathematics and science in a meaningful way (Moore et al., 2013; Moore et al., 2014a; 2014b; Shahali et al., 2016). Siverling et al. (2019) define engineering design-based STEM education as a pedagogical structure that guides the learning of four STEM disciplines: science, mathematics, engineering, and technology. In STEM integration activities, students can learn content related to STEM disciplines through participation in engineering design processes. According to Bryan et al. (2016), one of the touchstones of STEM learning experiences is the integration of engineering practices and engineering design of technologies as a context or component of the target content. Engineering design can also be considered a pedagogical strategy for solving open-ended problems, developing alternative solutions, and creative decision-making. Design-based STEM integration results in positive learning outcomes in science education (e.g., Apedoe et al., 2008). Using the engineering design process as a context for STEM problems can be an effective way for students to learn through STEM integration (Grubbs & Strimel, 2015). Engineering design activities can support the knowledge and attitudes toward science, mathematics and technology, and improve the competencies for problem-solving and creative thinking (Brophy et al., 2008; Moore et al., 2013).

The engineering design process, which is a sequential and/or iterative process, may differ in terms of education level, the steps followed, the order of the steps, and the elements of the engineering design process (Alemdar et al., 2017; National Academy of Engineering [NAE] and NRC [National Research Council], 2009; NRC, 2013). For example, the Massachusetts Department of Education (2006) defined the engineering design process consisting of eight stages as a student-focused model. In this model, the process starts with "identifying the need of the problem" and ends with the redesign (8th step). While this model can be used for students studying at the K-12 level, the Museum of Science-Boston (2009) indicates that a five-step cycle of "ask, imagine, create, test, and improve" is the design process used in elementary level activities. The National Aeronautics and Space Administration (NASA), on the other hand, structured the engineering design process by adding "imagine" to the planning phase, such as "ask, imagine, plan, create, test, improve" (NASA, 2011). Although there are different engineering design processes, the common feature of these models is that the process starts with the definition of the problem (Mentzer, 2011; NRC, 2009).

Using Arduino in Engineering Design-based STEM Activities

One of the technologies frequently preferred by science teachers in STEM activities in recent years is Arduino (Chu et al., 2021; Psycharis & Kalovrektis, 2022). Arduino is a simple, functional, and widely used microcontroller in STEM education developed for everyone (non-engineers). Arduino provides easy-to-use and accessible functions for engineering design-based activities such as sensors, motor control, programming, and wireless communication. With Arduino, students can integrate and apply interdisciplinary skills through hands-on activities (Psycharis, 2020). Arduino kits (https://www.arduino.cc/education/middle-school) offer learning opportunities with Lilypad Arduino and Arduino Science Journal.

For example, Sarıçam (2020) gives students a problem situation in astronomy and space. He tells students that they work at NASA's Jet Propulsion Laboratory (JPL). He also highlights the need for a system that can digitally measure the temperatures of liquids for a new vehicle to be sent to Mars in 2030. Using the SPACECRAFT AR mobile application developed by NASA JPL, students examine and research the spacecraft and design a digital temperature sensor using Arduino. Before designing with real materials, the circuit diagram can be created online using Arduino components in Tinkercad (ttps://www.tinkercad.com/). Then Arduino kits (https://www.arduino.cc) are distributed to the student groups to find the best solutions to the given problem considering the criteria and the limitations. In this activity, it is important that students follow the engineering design process with Arduino, and learn the science concepts (for example, life on Mars, heat and temperature) at the same time. Students can also be encouraged to review the examples on the Arduino website (https://www.arduino.cc).

In another example, within the scope of the Green Innovation Project organized by The Southeastern Anatolia Project (GAP), secondary and high school students carry out projects aiming to find solutions to the environmental problems on the subjects of "renewable energy, sustainability, and energy efficiency". Figure 1 shows examples of greenhouses designed by secondary school students in the GAP Green Innovation 2019 Project (Ministry of Industry and Technology, The Southeastern Anatolia Project Regional Development Administration, 2019).

The Arduino system used in the smart greenhouses shown in Figure 1 measures the humidity of the environment and operates the ventilation fan according to the desired need. The humidity sensor automatically measures the moisture of the soil at any time. When the soil needs water, it irrigates the soil in the form of drip irrigation until the soil reaches its saturation with water. In addition, Arduino can heat the environment according to the needs of the produce to be grown by measuring the temperature of the

Figure 1. The Southeastern Anatolia Project (GAP) - Examples of greenhouses designed by secondary school students within the scope of the Green Innovation 2019 Project

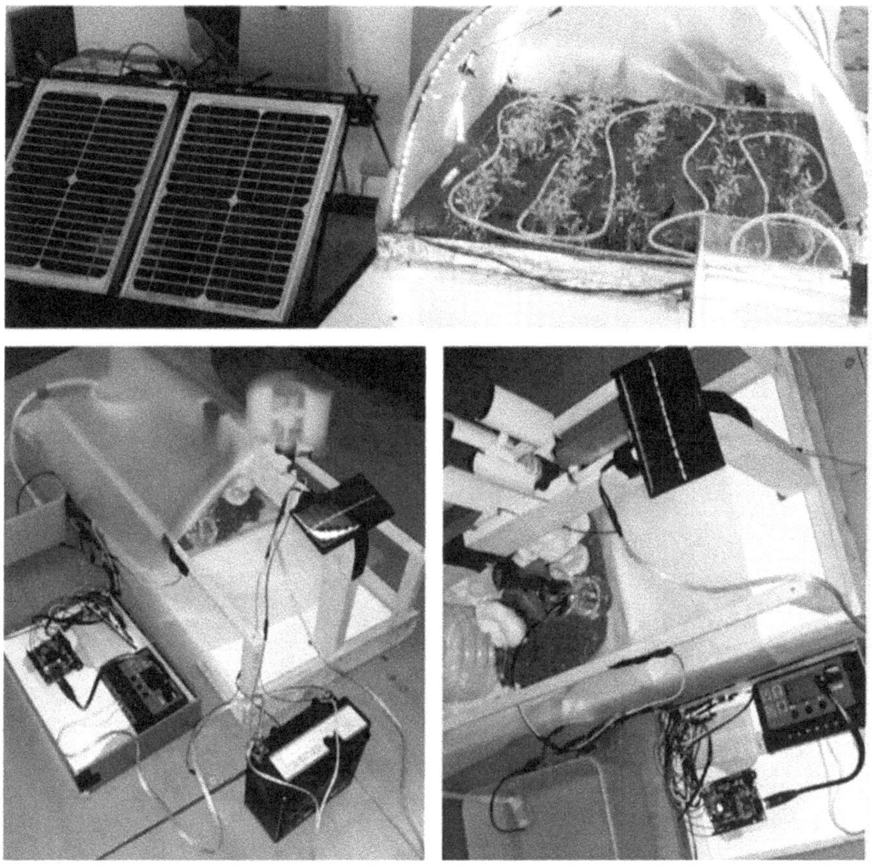

environment. The LDR sensor placed in the smart greenhouse mechanism can automatically measure the ambient light and operate the light energy needed by the environment.

Another type of Arduino is the Arduino LilyPad. Arduino Lilypad is a wearable Arduino model. It works with a logic similar to the other Arduino boards. It was developed for e-textile and wearable projects. It offers 2 different card options with 10 and 20 pins. It has a rounded design for easier integration into wearable devices. By sewing microcontrollers (LilyPad Arduino USB, LilyPad Arduino Simple Board) and sensors (light, temperature, and color sensor) to fabric and textile items, engineering design-based wearable technology projects can be implemented (Buechley et al., 2008). Wearable technology projects provide an opportunity for students to learn about the engineering design process, support students' interdisciplinary learning and STEM attitudes, and increase their interest in electronics and programming (Nugent et al., 2019; Qiu et al., 2013; Sat & Cagiltay, 2022). Wearable technologies provide an important multidisciplinary environment for the integration of engineering, computing, and aesthetics (Nugent et al., 2019). Students also engage in authentic learning experiences about different science concepts such as electric voltage, current, current flow, resistance, and polarity during the wearable e-textile activities.

Arduino-based activities can be integrated into STEM teacher professional development programs to help teachers develop their capacities for engineering design and technology integration in the classrooms.

Figure 2. Thermometer hair band

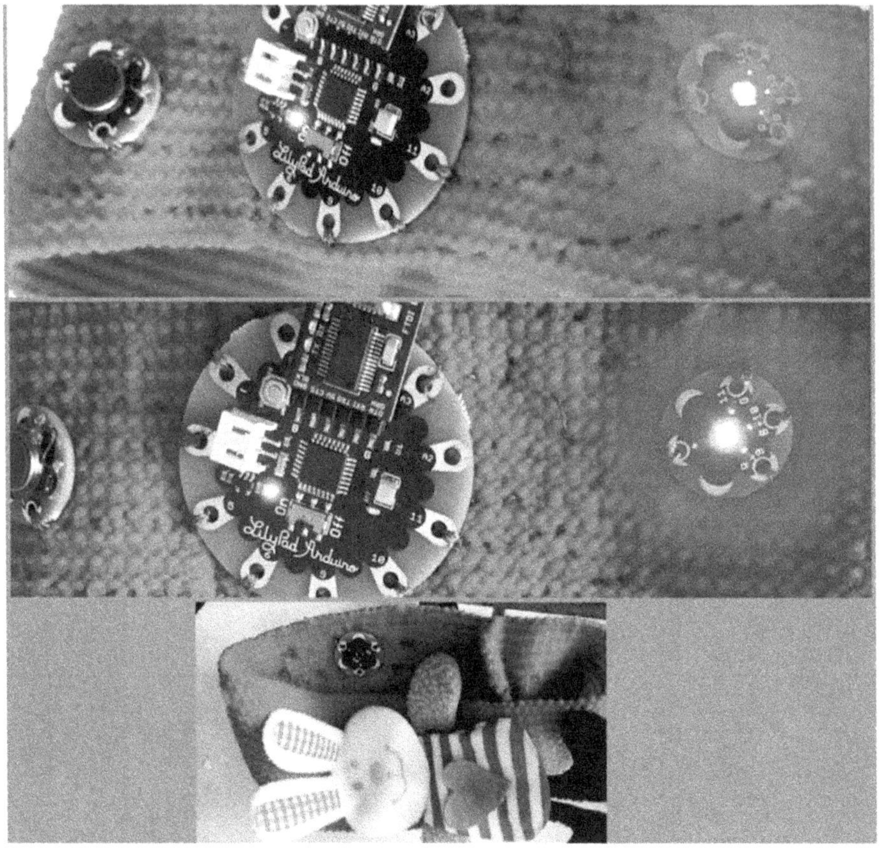

For example, a technological pedagogical content knowledge (TPACK) based PD program funded by the Scientific and Technological Research Council of Turkey under the Science and Society Innovative Educational Practices Program included probeware and Arduino activities for science teachers (Baran et al., 2016). One of the participant teachers continued to develop and implement Arduino activities in his classroom. For example, the headband shown in Figure 2 is designed to measure the temperature of children during the Covid 19 pandemic (Conductive thread, LilyPad battery holder, Lily Tiny, and LilyRGB LED, temperature sensor combined with aesthetic design on a headband).

In this activity, students created prototypes of circuit diagrams on paper before working with real materials following the engineering design process. Once they received feedback from their teachers on their prototypes, students designed their products using Arduino Lilypad. In this design, one LED lights up at the normal room temperature, and the number of LEDs increases as the temperature rises. The temperature sensor beeps when the ambient temperature rises.

Figure 3 presents another example of students designing smart textile products by sewing.

A study by Buechler et al. (2008) presented an example with the LilyPad Arduino, a fabric-based construction kit that allows students to design and build their own soft wearables and other textiles. This kit consists of a microcontroller and a set of sensors and actuators in sewn-in packages. In Buechler et al.'s (2008) study, female students participated in Arduino LilyPad activities with great enthusiasm.

Figure 3. Students design smart textile products by sewing

Female students continued to explore computer science and electrical engineering both during class at school and in their after-school projects. Unlike female students, male students did not worry about the appearance of the e-textile design but focused more on functionality (Buechler et al., 2008).

Engineering design-based Arduino activities can be integrated into the robotics and engineering club activities in schools. These activities can also be organized as workshops where students can attend with their families and friends in out-of-school learning environments (Buechley & Eisenberg, 2008). Peppler and Glosson (2013) carried out e-textile projects in the after-school program (twenty-hour) with secondary school students. Conductive threads and circuits used in e-textile projects enabled students to learn the concepts of electricity. These kinds of STEM-focused after-school programs (e.g., meeting with engineers and scientists, field trips to science-related places, science camps) also help students develop their interests in STEM subjects and careers (Baran et al., 2019; Dailey et al., 2018).

Mobile Learning With Arduino Science Journal

The recent affordances of mobile computing expand learning and teaching opportunities in educational settings (Díez et al., 2017). Crompton (2013) defines mobile learning as learning in multiple contexts through social and content interactions using mobile devices. Mobile learning can facilitate anytime anywhere learning through portable devices that provide context sensitive and immediate feedback (Azizi & Khatony, 2019; Harris, 2001; Traxler, 2005). Within STEM learning contexts, learning activities can be enriched with different applications such as coding and robotics, animation, simulation, augmented reality, and probeware (Bender, 2017). Mobile STEM teaching can create learning opportunities in an integrated learning environment with real and virtual components (Chu et al., 2021). While mobile devices have unique possibilities to transform learning (Lindsay, 2016), good teaching practices of mobile learning need to be evaluated to properly implement mobile devices and achieve positive learning outcomes for students (Aznar-Díaz et al., 2020).

Students may underperform in STEM-related subjects due to several challenges, such as limited learning materials, insufficient science labs, and lack of motivation (Schulze & Bosman 2018; Makgato, 2007; Mutambara & Bayaga, 2021). Mobile Arduino applications can address some of these challenges and offer accessible learning opportunities for diverse learners. Arduino Science Journal (https://www.arduino.cc/education/science-journal) allows students to conduct scientific experiments by measuring environmental stimuli with sensors, documenting and comparing data, developing and verifying hypotheses. For example, by collecting data such as heat, sound, temperature, voltage, magnetic field, and light intensity with Arduino, students can engage in STEM learning experiences in schools that do not have science or STEM laboratories. In addition, the portability of Arduino sensors can enable students to use them in out-of-school learning environments. Arduino can also contribute to the development of students' coding skills in ICT classes. In science lessons and STEM activities, Arduino can be used with an interdisciplinary approach for solving a problem encountered in daily life. For example, students can design a system using Arduino to water their flowers when they are on vacation. Exemplary activities are needed to guide practitioners and students about the effective integration of Arduino activities in STEM learning contexts.

With the Arduino Science Journal mobile application, students develop their ability to autonomously explore and solve problems with technology applications. This is closely related to the student's knowledge about scientific inquiry and engineering design. Several studies revealed positive learning outcomes related to the integration of the Arduino Science Journal in STEM activities. For example, in Vinluan's (2021) study conducted with secondary school students, the Arduino Science Journal mobile application was used to support students' self-learning with Home-Lab Kits. Nawawi and Dafrita's (2022) study revealed that STEM learning materials based on Arduino Science Journal helped develop biology teacher candidates' competencies. Chu et al.'s (2021) study aimed to develop and evaluate the effectiveness of a STEM practice kit by using the Arduino Science Journal application for middle school students. The purpose was to examine students' perceptions regarding the application of the developed practical kit. Participants agreed that running an experiment using mobile devices was a positive experience. Students developed a positive attitude toward the application of the Arduino Science Journal STEM practice kit in secondary school (Chu et al., 2021). However, several weaknesses were also discussed, such as students showing cyberloafing behaviors, the lack of smartphone ownership, experiments producing erroneous results due to the noisy environment or notifications from other applications, and the difficulty for the teacher to control or monitor the students.

Studies emphasized the importance of using the Arduino Science Journal application in promoting engineering applications from various perspectives. For students who cannot reach real experimental conditions, this application offers important learning opportunities. Educators need to establish strategies for facilitating and tracking students' progress as they practice engineering design with the Arduino Science Journal application.

Example of Engineering Design-based Activity With Arduino Science Journal

The STEM activity presented in this chapter is structured in line with the engineering design process followed in the PictureSTEM Project (http://picturestem.org) (Tank et al., 2017).

Table 1. Noise pollution in Zoo

Grade Levels	Secondary School
Approximate Time Needed	Four 50-minute class periods
Objectives	With this activity, the students develop a solution to the noise pollution problem faced by the animals living in the zoo in the city center by using technology. **The students will be able to:** **Science:** ● Explain the relationship between sound intensity and distance. ● Indicate that loud sounds can cause hearing loss. ● Question the causes of sound pollution. ● Explain the negative effects of sound pollution on human health and the environment. ● Produce solutions to reduce sound pollution. ● Give examples of reflection and absorption of sound. ● Make predictions and test predictions to prevent sound propagation. ● Explain the importance of sound insulation. **Technology:** ● Use mobile applications to solve daily life problems. ● Explain the basic features of the Arduino Science Journal application. ● Recognize the interface and features of the spreadsheet program and create a table. ● Visualize data using appropriate chart types. **Engineering:** ● Define a real-world problem. ● Generate possible solutions for the problem and select the appropriate one within the criteria. ● Design and present the product. **Mathematics:** ● Use ratios to compare quantities and display the ratio in different formats. ● Recognize area measurement units. ● Solve area problems.
Key Concepts	Sound intensity, sound pollution, sound insulation
Standards Alignment	**Next Generation Science Standard** ● 5-PS1-3: Make observations and measurements to identify materials based on their properties. ● MS-ETS1-1: Define the criteria and constraints of a design problem with sufficient precision to ensure a successful solution, taking into account relevant scientific principles and potential impacts on people and the natural environment that may limit possible solutions. ● MS-ETS1-2: Evaluate competing design solutions using a systematic process to determine how well they meet the criteria and constraints of the problem. ● MS-ETS1-3: Analyze data from tests to determine similarities and differences among several design solutions to identify the best characteristics of each that can be combined into a new solution to better meet the criteria for success. ● MS-ETS1-4: Develop a model to generate data for iterative testing and modification of a proposed object, tool, or process such that an optimal design can be achieved. ● MS-PS4-2 Develop and use a model to describe that waves are reflected, absorbed, or transmitted through various materials.
Materials	Mobile device, Arduino Science Journal app, carton/box, styrofoam, cotton, sponge, fabric, glass wool, felt, aluminum foil, egg carton, newsprint, utility knife, scissors, liquid glue, tape. * Students can choose any of the above-mentioned materials or use other materials. Care should be taken to use recyclable waste materials at the event.
Define the problem	The activity starts with a real world problem (https://theconversation.com/noise-pollution-is-hurting-animals-and-we-dont-even-know-how-much-186408). Students read the current news and identified a real-world problem faced by the animals in the zoo located in the city center. *"…. The noise pollution in the city center negatively affects the living conditions of the animals living in the zoo. For example, construction projects, fireworks affected the living activities of animals such as feeding, breeding and sleeping…."*
Learn about the problem	Students individually and in groups research the answers to the questions related to the problem such as "How does sound pollution harm the natural life of animals?", "What are the activities that cause noise pollution?", "How much is the sound intensity due to activities in the city center?", "What can be done to reduce noise pollution?", "How many decibels disturb animals?" At this stage, attention is drawn to developments in sound technologies. Students are guided in the use of reliable sources when conducting research.
Plan a solution	In line with their research, students produce solutions that will provide sound insulation and reduce sound pollution in the zoo. Students use their creativity and imagination to find solutions to sound pollution and produce possible solutions in line with criteria and limitations. They choose and plan a solution that is most suitable for them among possible solutions. They can make use of diagrams, budget tables and storyboards while planning.
Try a solution	From this stage, a prototype can be built with a square prism, rectangular prism or a cube-shaped cardboard box, for example for an animal shelter. Small squares can be drawn on the box with the help of a ruler. Perimeter and area calculations are made for each surface of the box, and the materials needed for insulation are calculated.
Test a solution	*The Arduino Science Journal* mobile application opens. As shown in Figure 4, a new activity is started by clicking the plus (+) icon in the lower right corner of the mobile application, and the name of the activity is defined by clicking the pencil sign. *Figure 4. Introduction to Arduino Science Journal Application and Screenshots of Sound Sensor* The sound sensor turns on. The sound intensity of the environment is recorded. The minimum-average and maximum values of the sound intensity are noted. *Figure 5. The sound intensity of the environment* In the second stage, the mobile device is recorded in the designed box by applying the sound intensity to the subjects. *Figure 6. Sound intensity measured inside the box* As shown in Figure 6, the sound intensity of the environment is 41.8 dB on average. When all the surfaces of the box were covered with styrofoam and then with fabric, the average value of the sound intensity first decreased to 20.7 dB. At this stage, the sound intensity can be measured by adding the insulation materials to the box in order. For example, the measurement is made by placing styrofoam foam on the inner surfaces of the box first. In the second application, the second measurement is taken by sticking the fabric on the styrofoam. Students can write their notes about each test in the journal created by the mobile application, take a photo with the "camera" feature or add the photos they recorded before with the "gallery" feature.
Decide if the solution is good enough	The resulting graphic and table regarding the effect of each material in the design to reduce sound pollution on the sound intensity in the environment are evaluated. The visuals related to the design taken during the application can also be used in the evaluation phase of the results. The materials used in terms of providing sound insulation are evaluated based on data, and it is decided which material should be used in the design.
Communication and Teamwork	Students prepare a report or presentation explaining their work at all stages of the activity, and the ideas and products developed are discussed. Graphics and photos saved in the Arduino Science Journal application can be used for reporting and presentation. The presentation can be made in Microsoft PowerPoint as well as in the form of videos to be prepared in the form of brochures, posters and/or public announcements. Students present their solutions to other groups within the scope of the activity, and the strengths and weaknesses of the solutions are discussed. In addition, brochures, posters and/or public service announcements prepared to raise environmental awareness on sound pollution can be shared with different stakeholders to raise awareness.

CONCLUSION

This chapter presents research and examples of the integration of Arduino and the free Arduino Science Journal mobile app within engineering design-based STEM learning contexts. To prepare students for such activities, orientation and try-out sessions can be conducted with students about Arduino (Arduino, Lilypad Arduino, Arduino Science Journal) and its basic components. Students can be given opportunities to experience Arduino with simple activities. During these warm-up activities, engineering and engineering design process steps can be discussed as well as the relationship between the activity and the science concepts.

The use of Arduino in engineering design-based activities can help science teachers enrich the learning and teaching processes with technology. Obtaining Arduino kits and Arduino Lilypad materials can be a barrier for those schools with limited resources. Arduino Science Journal's free mobile application can be used to make learning more accessible. Future work is needed to generate authentic examples of Arduino integration with mobile tools through design-based learning activities at the K-12 level. Teacher professional development programs can be implemented to provide teachers experiences for designing lessons with Arduino mobile application. Observing the critical contextual factors that impact the success of such implementations will be critical in identifying best practices.

REFERENCES

Akarsu, M., Okur Akçay, N., & Elmas, R. (2020). *STEM eğitimi yaklaşımının özellikleri ve değerlendirilmesi* [Characteristics and evaluation of STEM education approach]. Boğaziçi Eğitim Fakültesi Dergisi, Özel Sayı.

Alemdar, M., Lingle, J. A., Moore, R., & Wind, S. A. (2017a). Developing an engineering design process assessment using think-aloud interviews. *International Journal of Engineering Education*, *33*(1(B)), 441–452.

Apedoe, X. S., Reynolds, B., Ellefson, M. R., & Schunn, C. D. (2008). Bringing engineering design into high school science classrooms: The heating/cooling unit. *Journal of Science Education and Technology*, *17*(5), 454–465. doi:10.100710956-008-9114-6

Azizi, S. M., & Khatony, A. (2019). Investigating factors affecting on medical sciences students' intention to adopt mobile learning. *BMC Medical Education*, *19*(1), 1–10. doi:10.118612909-019-1831-4 PMID:31638977

Aznar-Díaz, I., Hinojo-Lucena, F. J., Caceres-Reche, M. P., & Romero-Rodríguez, J. M. (2020). Analysis of the determining factors of good teaching practices of mobile learning at the Spanish University. An explanatory model. *Computers & Education*, *159*, 104007. doi:10.1016/j.compedu.2020.104007

Baran, E., Canbazoglu-Bilici, S., Mesutoglu, C., & Ocak, C. (2019). Investigating the impact of an out-of-school STEM education program on students' attitudes towards STEM and STEM careers. *School Science and Mathematics*, *119*(4), 223–235. doi:10.1111sm.12330

Baran, E., Canbazoglu Bilici, S., & Uygun, E. (2016). Investigating the impact of a Technological Pedagogical Content Knowledge (TPACK)-based professional development program on science teachers' TPACK. In M. Herring, M. J. Koehler, & P. Mishra (Eds.), *Handbook of Technological Pedagogical Content Knowledge* (2nd ed., pp. 271–283). Routledge.

Bender, W. N. (2017). *20 strategies for STEM instruction*. Learning Sciences International.

Brophy, S., Klein, S., Portsmore, M., & Rogers, C. (2008). Advancing engineering education in P-12 classroom. *Journal of Engineering Education*, *97*(3), 369–387. doi:10.1002/j.2168-9830.2008.tb00985.x

Bryan, L. A., Moore, J. T., Johnson, C. C., & Roehrig, G. H. (2016). Integrated STEM Education. In C. C. Johnson, E. E. Peters-Burton, & T. J. Moore (Eds.), *STEM Road Map: A Framework for Integrated STEM Education* (pp. 23–37). Routledge, Taylor & Francis Group.

Buechley, L., & Eisenberg, M. (2008). The LilyPad Arduino: Toward wearable engineering for everyone. *IEEE Pervasive Computing*, *7*(2), 12–15. doi:10.1109/MPRV.2008.38

Buechley, L., Eisenberg, M., Catchen, J., & Crockett, A. (2008, April). The LilyPad Arduino: Using computational textiles to investigate engagement, aesthetics, and diversity in computer science education. In M. Czerwinski & A. Lund (Chairs), *Proceedings of the SIGCHI conference on human factors in computing systems* (pp. 423–432). New York: Association for Computing Machinery.

Bybee, R. W. (2013). *Case for STEM education: Challenges and opportunities*. National Science Teachers Association.

Chang, C. C., & Chen, Y. (2020). Using mastery learning theory to develop task-centered hands-on STEM learning of Arduino-based educational robotics: Psychomotor performance and perception by a convergent parallel mixed method. *Interactive Learning Environments*, 1–16.

Chu, W. W., Ong, E. T., Ayop, S. K., Azmi, M. S. M., Abdullah, A. S., Abd Karim, N. S., & Tho, S. W. (2021). The innovative use of smartphone for sound STEM practical kit: A pilot implementation for secondary classroom. *Research in Science & Technological Education*, 1–23. Advance online publication. doi:10.1080/02635143.2021.1978963

Crompton, H. (2013). A historical overview of mobile learning: Toward learner-centered education Z.L. Berge. In L. Y. Muilenburg (Ed.), *Handbook of mobile learning* (pp. 3–14). Routledge.

Dailey, D., Jackson, N., Cotabish, A., & Trumble, J. (2018). STEMulate engineering academy: Engaging students and teachers in engineering practices. *Roeper Review*, *40*(2), 97–107. doi:10.1080/02783193.2018.1434709

Daugherty, M. K., & Carter, V. (2017). The nature of interdisciplinary STEM education. In M. J. de Vries (Ed.), *Handbook of technology education* (pp. 159–171). Springer.

Díez, L. F., Valencia, A., & Bermúdez, J. (2017). Agent-based model for the analysis of technological acceptance of mobile learning. *IEEE Latin America Transactions*, *15*(6), 1121–1127. doi:10.1109/TLA.2017.7932700

Grubbs, M., & Strimel, G. (2015). Engineering design: The great integrator. *Journal of STEM teacher. Education*, *50*(1), 8.

Hansen, M. (2014). Characteristics of schools successful in STEM: Evidence from two states' longitudinal data. *The Journal of Educational Research*, *107*(5), 374–391. doi:10.1080/00220671.2013.823364

Harris, P. (2001). Going mobile. Learning circuits, *ASTD Online Magazine*. http://www.learningcircuits.org/2001/jul2001/harris.html

Lindsay, L. (2016). Transformation of teacher practice using mobile technology with one-to-one classes: M-learning pedagogical approaches. *British Journal of Educational Technology*, *47*(5), 883–892. doi:10.1111/bjet.12265

Makgato, M. (2007). Factors associated with poor performance of learners in mathematics and physical science in secondary schools in Soshanguve, South Africa. *Africa Education Review, 4*(1), 89-103.

Massachusetts Department of Education. (2006). Massachusetts science and technology/engineering curriculum framework. Author.

Mentzer, N. (2011). High school engineering and technology education integration through design challenges. *Journal of STEM Teacher Education*, *48*(2), 103–136. doi:10.30707/JSTE48.2Mentzer

Ministry of Industry and Technology, The Southeastern Anatolia Project Regional Development Administration. (2019). *GAP green.* http://www.gapgreen.org

Moore, T. J., Glancy, A. W., Tank, K. M., Kersten, J. A., Smith, K. A., & Stohlmann, M. S. (2014b). A framework for quality K-12 engineering education: Research and development. J*ournal of Pre-College Engineering Education Research, 4*(1), Article 2. . doi:10.7771/2157-9288.1069

Moore, T. J., Stohlmann, M. S., Wang, H.-H., Tank, K. M., Glancy, A. W., & Roehrig, G. H. (2014a). Implementation and integration of engineering in K-12 STEM education. In J. Strobel, S. Purzer, & M. Cardella (Eds.), *Engineering in precollege settings: Research into practice* (pp. 35–60). Sense Publishers. doi:10.2307/j.ctt6wq7bh.7

Moore, T. J., Tank, K. M., Glancy, A. W., & Kersten, J. A. (2013, January). A framework for implementing engineering standards in K-12. Paper to be presented at the *International Conference of Association for Science Teacher Education*.

Mutambara, D., & Bayaga, A. (2021). Determinants of mobile learning acceptance for STEM education in rural areas. *Computers & Education*, *160*, 104010. doi:10.1016/j.compedu.2020.104010

National Academy of Engineering & National Research Council. (2009). *Engineering in K-12 education understanding the status and improving the prospects.* Washington, DC: National Academies Press.

National Aeronautics and Space Administration. (2011). *Engineering design process*. Retrieved from https://www.nasa.gov/audience/foreducators/best/edp.html

National Research Council. (2013). Next generation science standards: For states, by states. Washington, DC: The National Academies Press. doi:10.17226/18290

Nawawi, N., & Dafrita, I. E. (2022). STEM Teaching Materials Integrated with Arduino Science Journal For Biology Prospective Teachers. *Jurnal Pendidikan Sains Universitas Muhammadiyah Semarang*, *10*(1), 36–44. doi:10.26714/jps.10.1.2022.36-44

Nugent, G., Barker, B., Lester, H., Grandgenett, N., & Valentine, D. (2019). Wearable textiles to support student STEM learning and attitudes. *Journal of Science Education and Technology*, *28*(5), 470–479. doi:10.100710956-019-09779-7

Peppler, K., & Glosson, D. (2013). Stitching circuits: Learning about circuitry through e-textile materials. *Journal of Science Education and Technology*, *22*(5), 751–763. doi:10.100710956-012-9428-2

Perdana, R., Apriani, A. N., Richardo, R., Rochaendi, E., & Kusuma, C. (2021). Elementary students' attitudes towards STEM and 21st-century skills. *International Journal of Evaluation and Research in Education*, *10*(3), 1080–1088. doi:10.11591/ijere.v10i3.21389

Perignat, E., & Katz-Buonincontro, J. (2019). STEAM in practice and research: An integrative literature review. *Thinking Skills and Creativity*, *31*, 31–43. doi:10.1016/j.tsc.2018.10.002

Psycharis, S. (2020). Editorial: A new era with STEM education? *Hellenic Journal of STEM Education*, *1*(2), 43–44. doi:10.51724/hjstemed.v1i2.14

Psycharis, S., & Kalovrektis, K. (2022, March). Assessment and integrated STEAM in engineering education. In *2022 IEEE Global Engineering Education Conference (EDUCON)* (pp. 695-703). IEEE. 10.1109/EDUCON52537.2022.9766654

Qiu, K., Buechley, L., Baafi, E., & Dubow, W. (2013). A curriculum for teaching computer science through computational textiles. In *Proceedings of the 12th International Conference on Interaction Design and Children* (pp. 20-27). ACM. 10.1145/2485760.2485787

Sarı, U., Çelik, H., Pektaş, H. M., & Yalçın, S. (2022). Effects of STEM-focused Arduino practical activities on problem-solving and entrepreneurship skills. *Australasian Journal of Educational Technology*, *38*(3), 135–149. doi:10.14742/ajet.7293

Sarıçam, U. (2020). 5. sınıf açık kaynak kodlu yazılımlar i̇çeren stem aktivite planları [5th grade- open source software integrated STEM activities]. In M. Yıldırım & G.Y. Uz (Eds.), Açık kaynak kodlu yazılımlar içeren STEM aktiviteleri: Ortaokul Fen Bilimleri Öğretim Programı Kazanımlarına Uygun [Open source software integrated STEM activities for middle school science curriculum objectives] (pp. 15-32). Anı yayıncılık.

Sat, M., & Cagiltay, K. (2022). The design of educational wearable technologies with teachers: Issues, challenges, and suggestions. *2022 International Conference on Advanced Learning Technologies (ICALT)*, 201-205. 10.1109/ICALT55010.2022.00066

Schulze, S., & Bosman, A. (2018). Learning style preferences and mathematics achievement of secondary school learners. *South African Journal of Education*, *38*(1), 1–8. doi:10.15700aje.v37n1a1276

Shahali, E. H. M., Halim, L., Rasul, M. S., Osman, K., & Zulkifeli, M. A. (2016). STEM learning through engineering design: Impact on middle secondary students' interest towards STEM. *Eurasia Journal of Mathematics, Science and Technology Education*, *13*(5), 1189–1211.

Siverling, E. A., Suazo-Flores, E., Mathis, C. A., & Moore, T. J. (2019). Students' use of STEM content in design justifications during engineering design-based STEM integration. *School Science and Mathematics*, *119*(8), 457–474. doi:10.1111sm.12373

Tank, K. M., Moore, T. J., Pettis, C., & Gajdzik, E. (2017). *PictureSTEM designing paper baskets*. PictureSTEM @ 2017 Purdue University Research Foundation. http://picturestem.org/wp-content/uploads/2017/07/PictureSTEM-Designing-Paper-Baskets_July-2017.pdf

Traxler, J. (2005). Mobile learning: It's here, but what is it. *Interaction*, *9*(1), 1–12.

Vinluan, M. V. (2021). *(Holks) Home-Lab Kits Utilizing Arduino Science Journal App: Enhancing scientific attitudes in a new normal grade 7 classroom of biñan secondary school of applied academics*. Retrieved from https://www.depedbinan.com/media/research/SDO-RES_001.433_V786_2021.pdf

Chapter 5
Computational Thinking: The Bridge Between the Engineering Design Process and Project-Based Learning

Lorraine A. Jacques
University of Tampa, USA

Heather Howle
Louisiana State University, USA

ABSTRACT

When integrating computational thinking (CT) skills with science education lessons, we thought that the engineering design process (EDP) would connect CT with the science content. The EDP has been included in science teacher training because it helps structure the engineering practices of the Next Generation Science Standards (NGSS) as well as provide a framework for project-based learning, a highly recommended instructional approach for full realization of NGSS. However, many teachers are still having difficulties with both PBL and the EDP. By mapping CT skills to steps in the EDP, each step better reflects how engineering truly occurs, which in turn better reflects authentic PBL in science and provides an easier-to-manage focus for PBL planning.

INTRODUCTION

When the authors wrote modules integrating computational thinking (CT) skills into the regular curriculum (Jacques & Howle, 2020), they assumed that the engineering design process (EDP) would serve as a bridge between CT and science content. Since then, they have learned that these assumptions were wrong, at least regarding science teachers implementing project-based learning (PBL). Instead, the authors have found that CT is what actually connects EDP with science PBL. By mapping CT skills to steps in the EDP, each step better reflects how engineering truly occurs, which in turn better reflects authentic PBL. Thus, this chapter will discuss: (1) Why PBL and the EDP are needed to fully imple-

DOI: 10.4018/978-1-6684-5585-2.ch005

ment the Next Generation Science Standards (NGSS), (2) What difficulties teachers are having when applying the EDP in science education, and (3) How teachers can use CT to connect the EDP's structure with PBL's openness.

This connection between the EDP and PBL is especially important when one considers the digital tools available today. Students today are more aware of issues facing the world and their communities because of what they see on their social media platforms (Greenhow & Lewin, 2016). Such awareness provides them with some knowledge, possibly erroneous, and interest for projects teachers could introduce in the classroom (Höttecke & Allchin, 2020). They also have a variety of resources such as mobile apps that support science inquiry (Liu et al., 2021), scientific datasets (e.g., Schwortz & Burrows, 2021), citizen science initiatives (Roche et al., 2020), data collection tools (e.g., *Arduino Science Journal*), and easier ways to build their own simulations (e.g., *AccelleratAR*) or programs (e.g., *Thunkable*). What students are missing is the knowledge that connects their interests with these resources; specifically, many students do not know how to solve a complex science problem using available digital tools. The use of the EDP in science education is intended to provide some structure for students (Yu, Wu, & Fan, 2019), but the authors have found that without attention to the CT skills embedded within the EDP, teachers and students still struggle with PBL in science.

The authors' work with secondary science teachers forms the basis of their conclusions and spans several projects from the past five years. These projects include delivering professional development (at the local and national level), reviewing teaching portfolios, instructional coaching, and research studies. The original focus for each project addressed CT, engineering, or both.

WHY PBL AND THE EDP ARE NEEDED IN NGSS

The Framework for K-12 Science Education, from which the Next Generation Science Standards (NGSS) are based, proposed engaging students in engineering and science practices to "deepen their understanding of crosscutting concepts and disciplinary core ideas" (National Research Council, 2012, 217). This engagement needs to include a phenomena or design challenge that places the scientific idea in context, opportunities for students to develop their own lines of inquiry and their own investigations, and public accountability for their findings, both within and outside the classroom (Penuel & Reiser, 2018), which are also important aspects of PBL (Thomas, 2000) and the EDP (Lloyd, 2000).

PBL is considered a necessary pedagogical approach for implementing NGSS with fidelity (Holthuis et al., 2018; Penuel & Reiser, 2018). When applied to science education, PBL increases student achievement (e.g., Craig & Marshall, 2019), engagement (e.g., Carrabba & Farmer, 2018), and equitable participation (e.g., Deutscher et al., 2021). Authentic PBL is different than just "doing projects" in that is meets the following criteria (Thomas, 2000):

1. The students are learning the central concepts and ideas through the project (rather than doing a project for enrichment or to apply prior learning),
2. The project is centered on an ill-defined problem or driving question,
3. The project results in the construction of new knowledge for the students (rather than considering already-learned material in a different way),
4. The project is student-driven, and
5. The project is realistic.

The Framework for K-12 Science Education has three requirements for design challenges (Penuel & Reiser, 2018) that can be met with PBL: (1) Solving the problem requires developing or applying core ideas and crosscutting concepts, (2) Students find the problem interesting, and (3) The problem must address enough performance expectations to be worth investing the time to investigate and to support students recognizing the connections between various scientific ideas. This first requirement is addressed in the first and third PBL criteria, learning and constructing knowledge through the project. The second is addressed when the teacher identifies a realistic, ill-defined problem that the students can connect with (PBL criteria 2 and 5) and when the project is student-driven (PBL criteria 4). The remaining criterion Thomas (2000) does not include as a PBL criteria, but research that has successfully implemented PBL in science shows students making such connections through the integration of several practices, core ideas, and cross-cutting concepts as delineated in NGSS (e.g., Craig & Marshall, 2019; Deutscher et al., 2021).

The EDP is considered a necessary tool for PBL in science because it mediates between critical thinking and the final product (Yu, Wu, & Fan, 2019) for complex, open-ended problems (Mentzer, Huffman, & Thayer, 2014), which further improves student achievement (Fidai et al., 2020). According to Teach Engineering (n.d.), the EDP consists of seven phases:

1. Ask: Identify the needs and constraints of the problem.
2. Research: Gather information about the problem and what solutions already exist.
3. Imagine: Brainstorm possible solutions.
4. Plan: Select a possible solution and develop an implementation plan.
5. Create: Build a prototype of the chosen solution.
6. Test and evaluate: Determine what works, what does not, and what to improve.
7. Improve: Revise the prototype to improve the solution.

Additionally, training in the EDP is necessary for science teachers' self-efficacy with NGSS's engineering standards, since few have engineering experience or education (Christian, Kelly, & Bugallo, 2021), but training in the EDP alone is insufficient for authentically applying PBL to science education.

WHAT GOES WRONG

With the adoption of NGSS, and because of its inclusion of engineering standards, many teachers have received training on the EDP and its use in the classroom (e.g., Christian, et al., 2021; Capobianco, DeLisi, & Radloff, 2018; Kelley et al., 2020). However, teachers are having difficulties applying their training. In searching the literature and in our work with teachers, the authors have identified two broad reasons for these difficulties: Teachers miss parts of the EDP and/or teachers use the EDP trivially.

Missing Parts of the EDP

The literature has found that when teachers use the EDP, they tend to allow more time for one part at the expense of another (e.g., Christian et al., 2021; Capobianco et al., 2018; Kelley et al., 2020). Early grades tend to spend the most time on idea generation, secondary grades on predictive analysis, and technology classes on building the prototype (Lin et al., 2021). the authors have also observed these trends. Skipping or shortening parts of the EDP, especially research and testing, often leads to poorer final projects and less customer satisfaction (Atman et al., 2007). Unfortunately, the literature and the authors' work both suggest that those are the two steps that get the least attention.

Test/Evaluate and Improve were the two steps of the EDP the authors saw have the least amount of consideration in teachers' plans. At best, teachers left very little time for students to test their solution and no time for revisions to be made, instead simply asking students to write an explanation for why their solution did not work as intended. At worst, they skipped testing completely. When testing was planned for, teachers provided few opportunities for peer feedback (<25% of all lesson plans and portfolios) and none for public or professional feedback. Time for improvements or revisions were included even less, with variations of "we don't have time for revisions" and "we needed to move on to the next unit" as the most common reasons given. When pressed further, however, many also admitted to not knowing how to handle classroom management during these steps. These teachers believed that it was too difficult to handle when some groups were finished and some were not, suggesting that they did not see projects as engineers do, who commonly believe that if the project is not broken, then it does not have enough features (e.g., Adams, S., n.d., as cited in Kwan, M., 2012), or do not know how to transfer that belief to their students.

Another portion of the EDP that teachers tend to shorten or eliminate is the research step. Compared to professional engineers, high school students spend very little time researching before generating ideas and tend to ask very general questions of their research instead of specific ones (Atman et al., 2007). In lesson plans and portfolios the authors have reviewed, nearly all teachers allowed at most one class session for research. The impression that guided these decisions was that research consisted only of searching published media. None of the teachers' work that the authors reviewed considered other forms of research common to engineering, such as reverse engineering existing solutions, interviewing invested parties, collecting detailed information about the problem, identifying constraints, observation, and data collection (e.g., Markula & Aksela, 2022; Zarske, Cooper, & Carlson, 2009).

One possible reason why these two steps become reduced or eliminated in teachers' plans could be due to their not anticipating a need for data collection/analysis plans. Expert engineers collect data during research to better understand the problem they are trying to solve and to identify what constraints they need to work within. They also collect data during testing to ensure the solution works under various conditions, to be very specific about points of underperformance, and to prioritize needed improvements (Atman et al., 2007). In the lesson plans that the authors reviewed, nearly half were missing plans for collecting and analyzing data, even though teachers were asked to include such plans.

Trivial Applications of the EDP

Even when teachers try to do the EDP in full, they often have students approach it like a worksheet or checklist, which they then apply to trivial or overly restricted projects. For example, in the authors' work, they observed several teachers using the traditional egg drop project, where students design a case for a raw egg so that the egg remains intact after a long drop. In each case, teachers simply told the students what the outcome should be rather than presenting a driving question or problem, such as designing a structure that will safely carry an astronaut to "splashdown." In some cases, teachers added unnecessary restrictions, such as what materials to use, resulting in student solutions that appeared very similar. Similar problems occurred when teachers had students build a tower from straws, because they did not present it as a real-life problem, and when they had students design a space colony, because the project requirements included unnecessary or unexplained restrictions (i.e., for the shapes of the buildings). These "cookbook" projects reduce the benefits students gain from PBL (De Vivo, 2022).

Each of those projects applied important scientific principles but render the first parts of the EDP trivial. There is nothing for students to ask about the problem when the needs and constraints are given, nor is there much to research beyond how other people have solved the problem. Additionally, there are several solutions available online to these common projects; for example, searching online for "egg drop" resulted in over 400,000,000 results in less than one minute as of this writing. It is difficult to picture students trying to find their own solution when so many options are already available, thus making the Imagine step of the EDP also trivial. Changing the project to something realistic, such as surviving an airplane crash, poses additional constraints that the students would have to discover themselves. For example, the comfort of the passenger would render many of the egg drop solutions online as impractical.

Trivial projects are not unique to the authors' data. Mentzer et al. (2017) found that teachers regularly failed to include driving questions or used unnecessary restrictions with their projects until they have had more than a year's coaching in PBL. Markula and Aksela (2022) reported that many teachers in their study also failed to use driving questions and used several small tasks that they called projects, such as creating a weather map, that were disconnected from each other even when they were in the same instructional unit.

Reasons for the Disconnect

The teachers that the authors worked with have provided many reasons to not use authentic PBL in the classroom, including a lack of materials, pressure to teach for the state exam, and a belief that PBL is too difficult for their students. Teachers new to PBL often share these concerns (Mentzer et al., 2017). The inclusion of the EDP was intended to structure PBL for teachers and alleviate some of these concerns (Fidai et al., 2020). In the authors' work, however, they have identified three major gaps in understanding teachers have that may be causing the disconnect between the EDP and PBL:

- Teachers struggle to pose a realistic student-driven engineering problem that can be solved computationally.
- Teachers are unclear of the various forms data can take or how to apply data to engineering problems.
- Teachers do not recognize the sub-tasks that are implied within each step of the EDP or how to guide students through these.

When computational thinking skills are applied to and considered within each step of the EDP, however, they provide teachers with a more-manageable framework from which they can design authentic PBL.

COMPUTATIONAL THINKING WITH THE EDP

Computational thinking (CT) is an NGSS science and engineering practice (NGSS, 2013), which Ted Willard, editor of The NSTA Quick-Reference Guide to the NGSS (NSTA Press, 2014), claims is "the least understood of the science and engineering practices" (personal communication, Feb. 11, 2022).

If true, this lack of understanding is problematic because CT is applicable to complex problems in any area (Lockwood & Mooney, 2017; Shute, Sun, & Asbell-Clarke, 2017; Wing, 2006).

There are several skills associated with CT. Shute, Sun, and Asbell-Clarke (2017) identify four as the core skills of CT:

- **Decomposition:** Breaking down a complex problem into smaller, more manageable parts.
- **Abstraction:** Identifying and using the essential parts of a complex system.
- **Algorithm design:** Creating a series of ordered and logical steps to achieve a solution.
- **Debugging:** Finding, identifying, and fixing errors to improve the solution.

Other researchers, including Hsu, Chang, and Hung (2018), also include as skills:

- **Data collection/analysis:** Gathering and making sense of information.
- **Pattern recognition:** Identifying regularities in the data.
- **Modeling:** Creating a representation of the solution or part of the solution.
- **Conditional logic:** Identifying patterns between different events.

Hsu et al. (2018) include other skills as well, for a total of nineteen CT skills, but for simplicity the authors will focus on the eight listed above.

As figure 1 shows, these CT skills map to the phases of the EDP, with each phase requiring students to apply multiple CT skills to reach that phase's goals (Jacques, 2020; Yang, Swasnon, Chittoori, & Baek, 2018).

Note. The *Ask* step is where an understanding of the problem develops. This understanding requires that the problem be considered as four elements, each of which need to be fully understood. According to Flemming and Johnston (2020), these elements are:

- **Goals:** Desired outcomes, both in the result and during the intermediary parts of the problem.
- **Variables:** Information directly impacting the problem and information desired with the solution.
- **Constraints:** Restrictions that may affect the variables, materials that can be used in the solution, or other aspects of the problem.
- **Strategies:** Identifying actions that are intended to change the state of the variables to reach the desired outcomes.

Breaking the problem into these four elements uses the CT skill of decomposition, but each element can be decomposed further. For example, to determine the goals for the intermediary parts of the problem requires one to decompose the problem into intermediary parts. Algorithms and pattern recognition then apply as one recognizes parts that are similar to another, parts that are similar to problems solved in the past, and parts that are new to one's experience (Flemming & Johnston, 2020). Digital resources such as videos, online articles, or simulations of the phenomena would help students to decompose the problem into these four elements and find patterns for themselves as they manipulate the media.

The **Research** step is where expert engineers spend significantly more time than students (Atman et al., 2007). In that study, experts requested more information and more types of information about the problem to better define the variables and constraints. In other words, they spent more time with the CT skills of collecting and analyzing data. Researching the problem using interactive tools (e.g.,

Figure 1. Mapping of Computational Thinking Skills to the Engineering Design Process

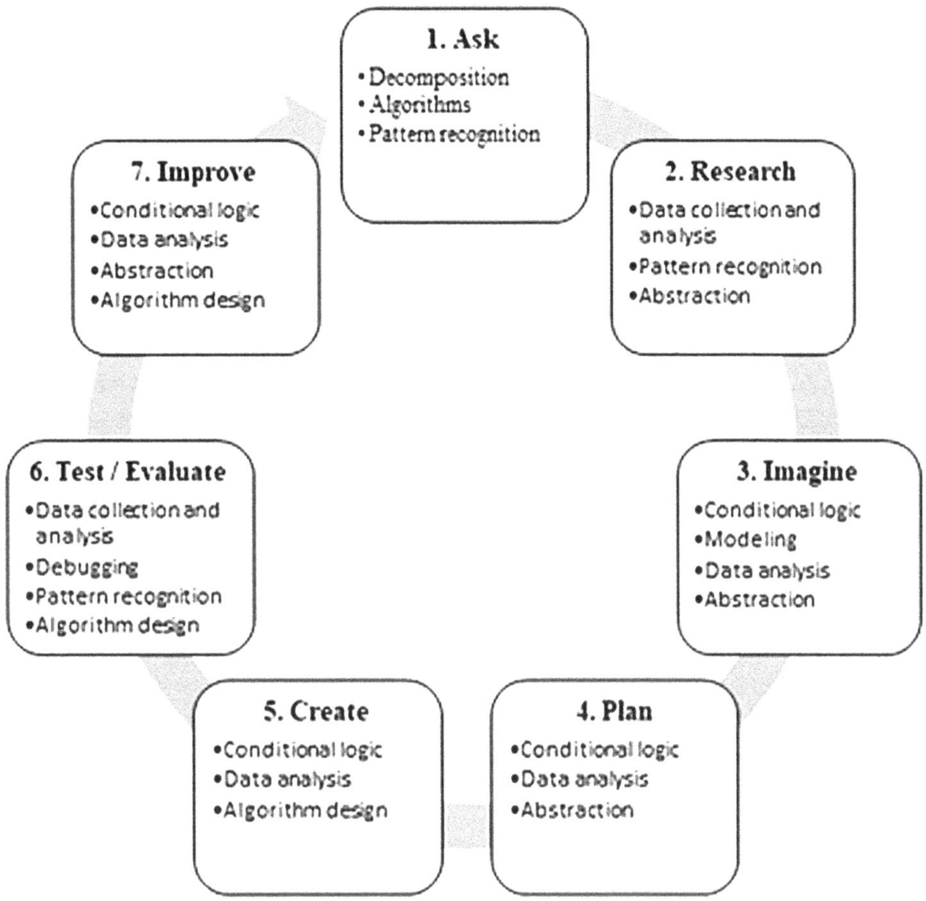

simulations), sensors on their devices, and video interviews with outside interested parties helps students spend more adequate time in this step. Also embedded in the Research step are the CT skills of pattern recognition and abstraction in order to identify what information is relevant to the problem or similar to other collected information.

Modeling, data analysis, and conditional logic are essential CT skills during the **Imagine** step because one needs to sketch or outline possible solutions, conduct feasibility analysis on these possibilities, and decide which would be the most promising path to explore further (Atman et al., 2007; Lin et al., 2021). To model these possible solutions, one also uses the CT skill of abstraction by including only the most relevant parts of the problem and which parts can or should be reused within the model and with other models. Creating rough models with 3D printers, animation programs (e.g., *Pencil2D*), or virtual reality programming (e.g., *Unity*) further support abstraction as students will only use the most essential information to develop possible solution paths. The **Plan** step of the EDP also uses the CT skills of modeling, conditional logic, and abstraction by choosing one of the possible solutions and developing the idea further, using the same digital tools as before, to determine if it still meets the goals and constraints identified during the Ask step.

The **Create** step is where one develops a prototype solution, which requires a development plan. That plan is an application of the CT skill of algorithm design. Even with an excellent plan in place, however, constraints anticipated and unanticipated will affect development, such as materials no longer being available or working as expected. Information from these difficulties requires one to re-examine their approach and make new decisions, which use the CT skills of data analysis and conditional logic.

Testing/Evaluating can be considered as a problem within the larger problem. To test or evaluate, one must *ask* how the product or solution will function in a real setting, *research* how others will use it or what additional conditions it must function in, *imagine* ways to test these conditions, *plan* how to test these conditions, and *create* test cases. Thus, testing or evaluating could apply all of the previously used CT skills, although the emphasis is on algorithm design (how will one test the solution), data collection/analysis (what worked, what did not, and why), and pattern recognition (what weaknesses could be from common causes). Additionally, the test/evaluate step uses the CT skill of debugging in that one identifies and fixes problems found. The **Improve** step can also be considered a problem of its own, since one is essentially posing a new problem to solve: How to make an existing solution better. However, during the Improve step, the emphasis is on the CT skill of conditional logic, since decisions need to be made on which improvements are feasible or necessary.

Mapping the appropriate CT skills to each step in the EDP reveals in more detail how engineering works. These details can then be used to help teachers create more authentic PBL experiences for their students by providing smaller, more concrete actions within the EDP. For example, instead of simply planning time for student research, teachers focus on what data students will need to collect before trying to solve the problem, possible ways for them to collect that data, essential knowledge they should gain from the data, and what forms of analysis could guide students to that knowledge.

COMPUTATIONAL THINKING AS THE BRIDGE

CT skills are not simply items that can be fit into the EDP, they elucidate the many considerations engineers include as they solve real problems. Teachers may have difficulties anticipating the same considerations, especially if they lack an engineering background, but they can structure their PBL unit to include CT within the EDP so that the students will be able to identify these considerations themselves. Explicitly stating the relevant CT skill(s) as students apply them or using CT vocabulary to help scaffold each phase of the EDP for students, will further help them understand how engineering approaches problem-solving. This inclusion of CT supports authentic PBL in the following manner:

PBL Criterion 1: The Students are Learning the Central Concepts and Ideas Through the Project

To gain a clear understanding of the problem, students must decompose it, collect and analyze existing data through research, identify patterns from their research, and with those patterns, abstract the core ideas necessary for their solution (Condliffe et al., 2017). These CT skills, applied in the first two phases of the EDP (Ask and Research), give students the foundational knowledge needed to learn the key concepts and ideas that the project will address. Teachers can support students during these first two phases by asking for evidence or artifacts showing their research, asking students for the key take-aways they learned through their research, and asking students what essential information the research suggests they will need to apply to a solution. When such questions are asked of students in their working groups, the teacher can formatively assess their understanding and suggest areas of further exploration to fill knowledge gaps.

They can further develop students' understanding by asking for *data collection* plans, *data analysis*, and what conclusions students can draw from the data (*pattern recognition* and *abstraction*), which are CT skills included in nearly every step of the EDP. In doing so, teachers will be requiring students to support their reasoning with data, which is a key NGSS standard, instead of using intuition to determine if their solution is adequate, as most will do without such requirements (Yin et al., 2021). An additional benefit of requiring students to work with data is the teacher can then pose scenarios that might affect the data collected to deepen students' understanding of the core concepts addressed in the project. For example, once students have collected data from their egg drop solution, the teacher can ask them to use that data to hypothesize if their solution would still work for larger or smaller eggs, different heights, or different landing surfaces; each of these scenarios would deepen students' understanding of the physics involved as they determine which variables would be affected and how. By using the CT skills of data collection, data analysis, pattern recognition, and abstraction regularly in the EDP, teachers change students' focus on the project from "just an assignment" to a more complete learning activity.

PBL Criterion 2: The Project Is Centered on an Ill-defined Problem or Driving Question

One of the difficulties teachers have with PBL, based on the authors' observations and the work of others (e.g., Christian et al., 2021; Markula & Aksela, 2022; Yin et al., 2021), was that they do not understand how to develop PBL lessons based on ill-defined problems or driving questions that are open-ended enough to be ill-defined. A likely reason for this lack of understanding is that teachers do not recognize the need for a driving question or ill-defined problem as a focus (Markula & Aksela, 2022; Mentzer et al., 2019), thinking instead that hands-on activities are the same as PBL (Mentzer et al., 2019).

To identify appropriate ill-defined problems or driving questions, such that the goals lack clarity or conflict with each other (Flemming & Johnston, 2020), teachers should look at real science issues (e.g., those found at livescience.com or @ngssphenomena on *Twitter*) and consider how students will use the CT skills applicable to the third and fourth phases of the EDP. Specifically, how will students apply the essential information gained during research to *model* potential solutions? How will students *abstract* the essential components a viable solution must have? How can students use *conditional logic* to determine the behavior of each potential solution under the conflicting goals of the problem? What *data analysis* can they use to support their reasoning as they model, abstract, and apply conditional logic? If only a limited number of possibilities present themselves when considering how students will apply these CT skills, then the problem is likely not ill-defined, leaving the teacher with two options: Continue looking or modify the problem at hand to be more ill-defined. Making a problem more ill-defined may be the easier option, especially for teachers who are required to use certain activities as part of their curriculums and can be accomplished by connecting the activity to a natural phenomenon or real-world problem. For example, the egg drop activity supports students' sense-making and problem-solving when connected to surviving an airplane crash. Rather than simply stating the connection, however, invite students to research survivor's accounts so they may gather the information needed to apply the CT skills of abstraction, conditional logic, data analysis, and modeling.

PBL Criterion 3: The Project Results in the Construction of New Knowledge for the Students

Teachers tend to avoid using PBL to teach content knowledge and instead use it to develop students' soft skills (Markula & Aksela, 2022). However, teachers with deeper training and experience in PBL use guiding questions throughout the project to help students discover the knowledge for themselves (Mentzer et al., 2019). All science disciplines allow for deep content learning through student sense-making or problem-solving driven by ill-defined yet relevant problems or guiding questions. For example, students can build their knowledge and understanding of endothermic and exothermic reactions through a challenge to design an egg transportation device to incubate and transport sea turtle eggs from a threatened nesting site to a safe location. The authors have found that the Research and Test/Evaluate steps of the EDP lend themselves well to learning the content through guiding questions when students are asked to apply CT skills to support their reasoning.

As students create and test or evaluate their solutions, phases 5 and 6 of the EDP, they will be tempted to simply state that their solution works or does not work. Teachers can have students explore deeply their solutions by requiring justifications with evidence for why their solution is viable and under what conditions. Such justification would require students to *abstract* the significant conditions that their solution might encounter, develop *algorithms* using *conditional logic* not only for implementing their solutions but for testing it under various conditions, *collect and analyze data* on their solutions under each condition, and apply *pattern recognition* to identify the overall strengths and weaknesses of their solution. Using the chemistry example mentioned in the previous paragraph, after students have researched the optimal temperature to transport sea turtle eggs and safe chemical reactions that meet the design criteria, they will test their design solution to determine the exact ratio of substances needed to transport the eggs. The student artifacts that could result from these justifications would -provide a detailed presentation of new knowledge that the students have gained, more than simply an evaluation of their solution can give.

PBL Criterion 4: The Project Is Student Driven

It is easy to argue that when students design and implement a solution of their own choosing to a problem, then the project is, by definition, student driven. But when students create on their own time, they also seek peer feedback and revise their creations, especially with the use of social media (e.g., Pires, Masanet, & Scolari, 2021). In the authors' experiences, teachers seldom include such opportunities in their PBL units, and when they do, students often fail to give or use meaningful feedback.

Having students use the CT skills of *data collection* and *analysis*, *conditional logic*, and *abstraction* during the final two phases of the EDP (Test / Evaluate and Improve) by using the conditions they identified for their own solutions would help them learn to give meaningful feedback to their peers. This form of peer review is very important in engineering (National Society of Professional Engineers, 2022) and provides a structure for students to give informed critiques to each other. Further, it places authority on the students rather than the teacher for determining success, which increases the student-driven aspect of the project.

PBL Criterion 5: The Project Is Realistic

Much of the research conducted in science PBL and what the authors have seen in their own work shows that teachers seldom have students share their solutions outside of the classroom, thus reducing the realism of the project and the efficacy of the EDP. The EDP is a social activity, where discussions with stakeholders and peers are equally important as the technical requirements for creating the final product (Lloyd, 2000). Thus, to increase the realism of a project, students must communicate not only with their peers but with people outside of the classroom who are knowledgeable about or invested in the problem (Condliffe et al., 2017).

CT skills can help prepare students for these communications when teachers have them identify the essential information that their audience needs to know (*abstraction*), how to present that information (*data representation* and *modeling*), and in what order the information should be presented (*algorithm design*). Additionally, when their audience includes stakeholders of the central problem, students will be receiving information. These stakeholders should be included in both the Ask and Test/Evaluate steps of the EDP to help students understand that the project addresses a real problem rather than simply another school assignment. Meeting with the stakeholders during these two steps means that students will need a data collection plan (e.g., interview questions) to garner the essential information they will need for development and improvement. Teachers can help students develop their plan with the CT skills of *decomposition* (what categories of information is needed from the stakeholders), *abstraction* (what is the essential information needed), and *algorithm design* (how will stakeholders provide this information).

LESSON EXAMPLE: USING CT TO BRIDGE BETWEEN THE EDP AND PBL

To demonstrate how CT enhances the EDP, the authors present a middle school PBL lesson addressing NGSS (2013) standard MS-LS2-5: Evaluate competing design solutions for maintaining biodiversity and ecosystem services. This lesson will also address all four middle school engineering design standards:

- **MS-ETS1-1:** Define the criteria and constraints of a design problem with sufficient precision to ensure a successful solution, taking into account relevant scientific principles and potential impacts on people and the natural environment that may limit possible solutions.
- **MS-ETS1-2:** Evaluate competing design solutions using a systematic process to determine how well they meet the criteria and constraints of the problem.
- **MS-ETS1-3:** Analyze data from tests to determine similarities and differences among several design solutions to identify the best characteristics of each that can be combined into a new solution to better meet the criteria for success.
- **MS-ETS1-4:** Develop a model to generate data for iterative testing and modification of a proposed object, tool, or process such that an optimal design can be achieved.

Problem Statement and Driving Question

For this lesson, students are told about the Puaiohi bird, a critically endangered bird that lives in the forest on Hawaii's Kauai Island that many native island plants depend on for seed dispersal. Computational ornithologists, scientists who study data about birds, create spectrograms of the bird sounds to track the species' health in their environment without needing to approach or disturb them. Similar techniques are being used throughout northern America with the help of citizen scientists. For example, the Cornell Lab of Ornithology has developed a free app to help identify birds and to collect photos and sound recordings from citizen bird watchers. This data provides evidence not only of local bird species' health but of the health of local ecosystems. However, in many places it is not easy to collect these photos or recordings without first attracting the birds. What could people create that would safely and ethically attract birds?

In this PBL unit, students will not only build something to attract birds, they will also share their plans publicly so that other citizen scientists and bird watchers can build it, too. During the unit, students will learn about birds, the ecosystems they inhabit, and how to affect their environment. The process they will experience applies the EDP, where they will learn additional scientific content, although the exact content will depend on the solution path they choose.

Step 1: Ask

Using the CT skill of decomposition helps students realize that this problem has more considerations and constraints than what appear on the surface. As a whole class discussion, students will learn there are two perspectives they need for understanding the problem: The ornithologists' recommendations and the citizen scientists who will implement the students' solutions. Further decomposition of the driving question will help students identify what they need to learn from these stakeholders, such as cost to build, ease of use, reliability under various weather conditions, environmental impact, and birds' emotional and physical safety. Pattern recognition then helps students categorize these smaller learning goals and gain an initial understanding of the constraints of the problem, while algorithm design helps them develop their research plans, including interview questions, observations of the environments their devices will function in, and what to learn from published resources.

Step 2: Research

In applying their research plans, students will use the CT skills of data collection and analysis. Data from interviews, observations, and published sources will need to be organized so students can determine if they have enough information in each of the areas they need to learn about or if they need to collect additional information. Students will also likely discover what is already available for attracting birds; collecting data on these solutions, such as through reverse engineering, will help students identify their strengths and weaknesses. During this process, students will be applying pattern recognition and abstraction because some of the information they gather will likely be contradictory. For example, online resources may suggest installing a bird bath, but in some areas, these can become mosquito breeding grounds or miniature ice rinks. Guiding questions from the teacher and further conversations with stakeholders will help students recognize not only when they have enough information to proceed but also what information or open questions are less important at this point in the project.

Steps 3 and 4: Imagine and Plan

As students brainstorm ideas, teachers can encourage them to model their ideas on paper or with technology. This CT skill can be further enhanced if the teacher suggests students include with each model how the essential data that they gathered during the research step is applied in each design, which is the CT skill of abstraction. Such inclusion then leads to the CT skills of data analysis and conditional logic as students identify the strengths and weaknesses of each model. For the Plan step, data analysis, abstraction, and conditional logic will be emphasized if the teacher has students use a decision matrix or similar structured tool to help them choose which model they will develop into a prototype.

Step 5: Create

While tempting to simply allow students to build their prototypes at this stage, having them write a plan first (algorithm design) helps them identify what materials they need and how to share responsibilities during the build. Reminding them that other citizen scientists, like the people they interviewed, need to follow their plans will encourage them to include details they might otherwise skip or only verbalize among themselves. Data analysis and conditional logic may be used as they write their plans, depending on their designs and materials, but may be further enhanced if the teacher has them peer review each other's plans. All three of these CT skills will be further applied as students build their prototypes following their plans since there will likely still be problems in those plans, requiring students to analyze the information they receive during the build, make decisions, and alter their plans.

Step 6: Test/Evaluate

Testing is more than simply seeing if the prototype functions; in engineering, testing uses a variety of scenarios to determine how well the prototype withstands changes in conditions and what its tolerances are for the more important conditions. For example, students' prototypes should work in some wind, but how much wind it can handle is information others will need to use their idea. Creating test cases requires students to apply algorithm design; running the test cases requires data collection as students identify and record the results of the tests. Data analysis and pattern recognition will help students understand why failures occurred or weaknesses emerged, which they may then address by debugging their prototype and adjusting their design plans.

Once the prototypes work to the students' satisfaction, they should seek additional feedback through peer review and user review. A "gallery walk", where students test each other's solutions, is a simple way to conduct peer review. During the walk, students should be directed to share one thing they liked and one thing they believe would improve each prototype, which again applies the data collection CT skill. For user review, students could ask the citizen scientists they interviewed or other stakeholders to try and build their solution using their plans and recording any problems they encountered.

Step 7: Improve

Students are not only improving their prototypes during this step, they are improving their plans. First, they will need to analyze the data they collected during testing and apply abstraction and conditional logic to determine which suggestions or problems they should prioritize. Addressing these issues will require a plan, but students will also need to adjust their original plan to include these changes, both of which apply the CT skill of algorithm design.

Finalizing the Project

From the start of this project, students have been aware that their plans will need to be made public. How this sharing occurs should be decided by the teacher and students together ideally, although some districts may have regulations to consider as well. Possibilities include a student-created public website, posting on an existing site such as Young Birder Blogs (Cornell, n.d.), or presenting and making their plans available at the local public library.

CONCLUSION

Through digital media, students today are more aware of the bigger issues in science, but that exposure without background knowledge may also make them more susceptible to misinformation. Exploring these same issues in science classes, using digital tools and PBL, allows students to determine for themselves the truth of what they see and the possible solutions they might support. Only after years of practice and training, however, do teachers comfortably implement project-based learning (PBL) in their science classrooms (Mentzer et al., 2017). Without such experience or support, the fidelity of NGSS and their connected academic gains lie in jeopardy (Holthuis et al., 2018; Penuel & Reiser, 2018). The engineering design process (EDP) has been taught to science teachers as a way to structure PBL and to include engineering practices in their instruction, but without an engineering background, teachers are having difficulties using the EDP fully or in meaningful ways (e.g., Christian et al., 2021). In the authors' experiences with teachers, mapping computational thinking (CT) skills to the EDP makes the thought processes of engineers during each step more transparent and enhances the authenticity of the PBL. Teachers then can consider how students will apply the mapped CT skills within each step, scaffolding the lesson planning process without overly restricting what students can do with the project or making the project so open-ended that the science learning cannot be assessed.

REFERENCES

Atman, C. J., Adams, R. S., Cardella, M. E., Turns, J., Mosborg, S., & Saleem, J. (2007). Engineering design processes: A comparison of students and expert practitioners. *Journal of Engineering Education*, *96*(4), 359–379. doi:10.1002/j.2168-9830.2007.tb00945.x

Capobianco, B. M., DeLisi, J., & Radloff, J. (2018). Characterizing elementary teachers' enactment of high-leverage practices through engineering design-based science instruction. *Science Education*, *102*(2), 342–376. doi:10.1002ce.21325

Carrabba, C., & Farmer, A. (2018). The impact of project-based learning and direct instruction on the motivation and engagement of middle school students. *Language Teaching and Educational Research*, *1*(2), 163–174.

Christian, K. B., Kelly, A. M., & Bugallo, M. F. (2021). NGSS-based teacher professional development to implement engineering practices in STEM instruction. *International Journal of STEM Education*, *8*(1), 1–18. doi:10.118640594-021-00284-1

Condliffe, B., Quint, J., Visher, M. G., Bangser, M. R., Drohojowska, S., Saco, L., & Nelson, E. (2017). *Project Based Learning: A Literature Review*. MDRC.

Craig, T. T., & Marshall, J. (2019). Effect of project-based learning on high school students' state-mandated, standardized math and science exam performance. *Journal of Research in Science Teaching*, *56*(10), 1461–1488. doi:10.1002/tea.21582

De Vivo, K. (2022, Jan. 24). A new research base for rigorous project-based learning. *Phi Delta Kappan*. Retrieved from: https://kappanonline.org/research-project-based-learning-de-vivo/

Deutscher, R., Holthuis, N., Maldonado, S., Pecheone, R. L., Schultz, S. E., & Wei, R. C. (2021). *Learning Through Performance Project-Based Learning as a Lever for Engaging the Next-Generation Science Standards*. Retrieved from: https://scienceeducation.stanford.edu/sites/g/files/sbiybj8661/f/ltp_science_report_2.8.21_final.pdf

Fidai, A., Barroso, L. R., Capraro, M. M., & Capraro, R. M. (2020, October). Effects of Engineering Design Process on Science and Mathematics. In 2020 IEEE Frontiers in Education Conference (FIE) (pp. 1-4). IEEE. doi:10.1109/FIE44824.2020.9274167

Flemming, S. A., & Johnston, C. R. (2020). The Nature of a Problem, Problem Diagnosis, and Engineering Design. *Proceedings of the Canadian Engineering Education Association (CEEA)*. 10.24908/pceea.vi0.14156

Greenhow, C., & Lewin, C. (2016). Social media and education: Reconceptualizing the boundaries of formal and informal learning. *Learning, Media and Technology*, *41*(1), 6–30. doi:10.1080/17439884.2015.1064954

Holthuis, N., Deutscher, R., Schultz, S. E., & Jamshidi, A. (2018). The New NGSS Classroom: A Curriculum Framework for Project-Based Science Learning. *American Educator*, *42*(2), 23–27.

Höttecke, D., & Allchin, D. (2020). Reconceptualizing nature-of-science education in the age of social media. *Science Education*, *104*(4), 641–666. doi:10.1002ce.21575

Hsu, T., Chang, S., & Hung, Y. (2018). How to learn and how to teach computational thinking: Suggestions based on a review of the literature. *Computers & Education*, *126*, 296–310. doi:10.1016/j.compedu.2018.07.004

Jacques, L. (2020, April). Using the engineering design process to teach computational thinking. *Proceedings of Society for Information Technology & Teacher Education International Conference*, 33-37.

Jacques, L., & Howle, H. (2020). *Computational thinking high school course.* Cyber Innovation Center. Retrieved from https://cyber.org/

Kelley, T. R., Knowles, J. G., Holland, J. D., & Han, J. (2020). Increasing high school teachers self-efficacy for integrated STEM instruction through a collaborative community of practice. *International Journal of STEM Education*, *7*(1), 1–13. doi:10.118640594-020-00211-w

Kwan, M. (2012, Jan. 15). Sunday snippet: Scott Adams (Dilbert). *Beyond the Rhetoric*. Retrieved from: https://btr.michaelkwan.com/2012/01/15/sunday-snippet-scott-adams-dilbert/

Lin, K., Wu, Y., Hsu, Y., & Williams, P. J. (2021). Effects of infusing the engineering design process into STEM project-based learning to develop preservice technology teachers' engineering design thinking. *International Journal of STEM Education*, *8*(1), 1. doi:10.118640594-020-00258-9

Liu, C., Bano, M., Zowghi, D., & Kearney, M. (2021). Analysing user reviews of inquiry-based learning apps in science education. *Computers & Education*, *164*, 104–119. doi:10.1016/j.compedu.2020.104119

Lloyd, P. (2000). Storytelling and the development of discourse in the engineering design process. *Design Studies*, *21*(4), 357–373. doi:10.1016/S0142-694X(00)00007-7

Lockwood, J., & Mooney, A. (2017). Computational thinking in education: Where does it fit? A systematic literary review. *International Journal of Computer Sciences and Engineering Systems*, *2*(1), 41–60.

Markula, A., & Aksela, M. (2022). The key characteristics of project-based learning: How teachers implement projects in K-12 science education. *Disciplinary and Interdisciplinary Science Education Research, 4*(2).

Mentzer, G. A., Czerniak, C. M., & Lisa, B. (2017). An examination of teacher understanding of project-based science as a result of participating in an extended professional development program: Implications for implementation. *School Science and Mathematics*, *117*(1–2), 76–86. doi:10.1111sm.12208

Mentzer, N., Huffman, T., & Thayer, H. (2014). High school student modeling in the engineering design process. *International Journal of Technology and Design Education*, *24*(3), 293–316. doi:10.100710798-013-9260-x

National Research Council. (2012). *A framework for K-12 science education: Practices, crosscutting concepts, and core ideas*. National Academies Press.

National Society of Professional Engineers. (2022). *Peer Review Legislation*. Retrieved from: https://www.nspe.org/resources/issues-and-advocacy/professional-policies-and-position-statements/peer-review-legislation

NGSS Lead States. (2013). *Next Generation Science Standards: For States, By States*. Washington, DC: The National Academies Press. Retrieved from: https://www.nextgenscience.org/

Penuel, W. R., & Reiser, B. J. (2018). Designing NGSS-aligned curriculum materials. *Committee to Revise America's Lab Report*, 1-51.

Pires, F., Masanet, M. J., & Scolari, C. A. (2021). What are teens doing with YouTube? Practices, uses and metaphors of the most popular audio-visual platform. *Information Communication and Society*, *24*(9), 1175–1191. doi:10.1080/1369118X.2019.1672766

Roche, J., Bell, L., Galvão, C., Golumbic, Y. N., Kloetzer, L., Knoben, N., Laakso, M., Lorke, J., Mannion, G., Massetti, L., Mauchline, A., Pata, K., Ruck, A., Taraba, P., & Winter, S. (2020). Citizen science, education, and learning: Challenges and opportunities. *Frontiers in Sociology*, *5*, 613–814. doi:10.3389/fsoc.2020.613814 PMID:33869532

Schwortz, A. C., & Burrows, A. C. (2021). Authentic science experiences with STEM datasets: Postsecondary results and potential gender influences. *Research in Science & Technological Education*, *39*(3), 347–367. doi:10.1080/02635143.2020.1761783

Shute, V. J., Sun, C., & Asbell-Clarke, J. (2017). Demystifying computational thinking. *Educational Research Review*, *22*, 142–158. doi:10.1016/j.edurev.2017.09.003

Teach Engineering. (n.d.). *Engineering design process*. University of Colorado Boulder. Retrieved from: https://www.teachengineering.org/k12engineering/designprocess

The Cornell Lab of Ornithology. (n.d.). *Young birder blogs*. eBird. Retrieved from https://ebird.org/about/resources/for-young-birders/young-birder-blogs

Thomas, J. W. (2000). *A review of research on project-based learning*. Autodesk Foundation.

Wing, J. M. (2006). Computational thinking. *Communications of the ACM*, *49*(3), 33–35. doi:10.1145/1118178.1118215

Yang, D., Swanson, S. R., Chittoori, B. B., & Baek, Y. (2018, June). Board 70: Work in Progress: Integrating Computational Thinking in STEM Education through a Project-based Learning Approach. *2018 ASEE Annual Conference & Exposition*. 10.18260/1-2--30091

Yu, K., Wu, P., & Fan, S. (2019). Structural relationships among high school students' scientific knowledge, critical thinking, engineering design process, and design product. *International Journal of Science and Mathematics Education*. Advance online publication. doi:10.100710763-019-10007-2

Zarske, M. S., Cooper, L., & Carlson, D. (2009). Design step 2: Research the problem. In *Teach Engineering*. Regents of the University of Colorado. Retrieved from: https://www.teachengineering.org/activities/view/cub_creative_activity2

KEY TERMS AND DEFINITIONS

Complex Problem: A problem that can be addressed through various approaches and that has multiple viable solutions. Also known as a "wicked problem" or "ill-defined problem."

Computational Thinking: A problem-solving strategy that originated in computer science but is applicable to any complex problem.

Engineering Design Process: A series of steps used in engineering to create solutions to complex problems.

Guiding Question: A question posed to students to encourage interest, deeper thinking, and/or an alternate perspective.

Project-Based Learning: An instructional approach where students learn content through in-depth engagement with realistic, complex problems.

Realistic Problem: An issue that affects, or will affect, people's well-being. Also known as a "real-world problem."

Student-Driven: A learning environment that is co-created with the students.

Section 3

Technology Integration in Practice

Chapter 6
Integrating ArcGIS Digital Technologies for Learning:
Three Case Studies From University Design Partnerships With Teachers

Kate Popejoy
https://orcid.org/0000-0002-5184-5105
Popejoy STEM LLC, USA

Thomas Hammond
https://orcid.org/0000-0002-5718-8134
Lehigh University, USA

Danielle Malone
https://orcid.org/0000-0002-0561-7064
Washington State University Tri-Cities, USA

Judith Morrison
Washington State University Tri-Cities, USA

Jonah Firestone
Washington State University Tri-Cities, USA

Alec Bodzin
Lehigh University, USA

Doug Leeson
Lehigh University, USA

Kristen A. Brown
Texas Christian University, USA

Curby Alexander
https://orcid.org/0000-0002-3888-6930
Texas Christian University, USA

Molly Weinburgh
Texas Christian University, USA

ABSTRACT

This chapter presents three cases of high school science teachers collaborating with personnel at three different universities to design and implement innovative, curriculum-aligned geospatial lessons involving hands-on data collection, following the authors' Socio-Environmental Science Investigation (SESI) model. Each case is presented in the context of professional development activities conducted at each university, the curriculum of the collaborating teacher, the students' engagement with the instructional sequence, and the teacher's reflections following implementation. All teacher professional development, collaborative curriculum development, and classroom implementation were conducted using ArcGIS Online and its affiliated tools. Implications and conclusions for other teachers' and teacher educators' development of similar, geospatially-integrated instruction are discussed.

DOI: 10.4018/978-1-6684-5585-2.ch006

INTRODUCTION

Teaching science utilizing digital technologies requires teachers to integrate technologies within their content area and instructional practices that lead to learning experiences for their students. Thus, a first step is to develop an approach for integration of digital technologies and then provide professional development (PD) focused on increasing teachers' use of practical teaching strategies aligned with their content and district requirements. Socio-environmental science investigations (SESI) provide a method to utilize geospatial technologies for curriculum aligned learning experiences. SESI investigations are designed to prepare students to gain important STEM workforce skills. The learning activities provide opportunities for students to collaborate, seek evidence, problem-solve, master technology, develop geospatial thinking and reasoning (GTR) skills, and practice communication skills, all of which are essential for the STEM workplace (National Science Board, 2015).

Geospatial technologies such as GIS (Geographic Information System) are extensively used in agriculture, urban and regional planning, environmental resource management, surveying and cartography, conservation, national resource management, public health, transportation, and wildlife ecology, among others. A GIS consists of map information, a database, and a computer interface that ties them together to create a visualization tool for spatial analysis. A GIS is often regarded as a computer system that can capture, store, query, analyze, and display geographical information of two types, vector or raster. GIS allows users to select different layers of information to construct a map. The map is then displayed on a computer screen or Web browser and the data and information can be examined spatially. Coverages could include political boundaries, demographic information, rivers and streams, soil coverage, topography, water quality, or census data to support multidisciplinary science education.

Most of the information we have about our world contains a location reference, placing that information at some point on the globe. For example, when point source pollutant discharge information is collected, it is important to know where the discharge is located. This is done by using a location reference system, such as longitude and latitude, and perhaps elevation. Comparing the discharge information with other information, such as the location of marshes that may serve as breeding grounds for shellfish, may show that certain marshes receive higher levels of pollutant discharges that may not be beneficial for the survival of newly born shellfish. This inference can help us make the most appropriate decisions about how pollution discharge in the marsh might be regulated. A GIS, therefore, can reveal important new information that leads to better decision-making.

GTR skills are essential for occupations in which there is a heavy reliance on cognitive thinking skills that include knowledge about geospatially-referenced data and their relationships (Goodchild & Janelle, 2010; NRC, 2006). These skills involve important scientific practices highlighted in the Next Generation Science Standards [NGSS] (NGSS Lead States, 2013), and include data manipulation, analysis, data mining, computational thinking, and modeling that provoke and *require* critical thinking and problem solving that are connected to data referenced to Earth's surface or to the Earth's representation through map and globe visualizations (Huynh & Sharpe, 2013). GIS is now the standard for spatially referenced data management, but STEM curricula often contain learning experiences that do not match the analytic practices that are critical for success in STEM-based occupations (Aikenhead, 2005; Chin et al.,2004). Science curricula that engage students to collect and analyze data, and solve problems provide important skills that help prepare students for career opportunities and lifelong learning (National Research Council, 2011; National Science Board, 2015).

Socio-environmental science investigations focus on social issues related to environmental science. SESI involves inquiry-based investigations that are open-ended with students engaging in data collection and geospatial analysis. SESI activities are multi-disciplinary and involve decision-making based on the analysis of georeferenced geospatial data, examination of relevant social science content, and consideration of social equity implications. SESI are based on the pedagogical frameworks of place-based education and socio-scientific issues-based instruction. Place-based education focuses on local or regional investigations, is designed around engaging students in examining local problems (Sobel, 2004), and utilizes fieldwork to gather evidence in that local setting (Semken, 2005; Semken et al., 2017). SESI investigations can provide opportunities to empower students to address important socio-scientific issues in their community. Socio-scientific issues are socially relevant, real-world problems that are informed by science and often include an ethical component (Sadler et al., 2006). They are sometimes controversial in nature but have the added element of requiring a degree of moral reasoning or the evaluation of ethical concerns in the process of arriving at decisions regarding possible issue resolution (Zeidler & Nichols, 2009). These issues require the use of evidence-based reasoning and provide a context for understanding scientific information using an active approach to learning, placing science content within a social context in a way that fosters motivation for, and the ownership of, learning by the student (Sadler et al., 2006; Zeidler & Nichols, 2009).

We have developed a design partnership model (Bodzin & Cirucci, 2009) for geospatial curriculum development that has been implemented for multiple Earth and environmental science projects (Bodzin, 2011; Bodzin et al., 2014; Bodzin et al., 2013; Hammond et al., 2019). In this model, university and school personnel work together from the very earliest stages, identifying topics that both align with existing curricula and involve understanding geospatial patterns and relationships in real-world data. The collaborative design group then identifies opportunities for students' hands-on use of geospatial data and analysis tools to master the curriculum-specified content through inquiry learning. During this stage, we adhere to our geospatial curriculum approach (Figure 1), which is informed by our work as well as the NSF Geotech Center's Geospatial Technology Competency Model (DiBiase et al., 2010). This design approach focuses on GTR and calls for a sequence of student engagement, moving from investigative questions through a cycle of examining data visualizations, performing geospatial analysis, constructing explanations, and supporting claims based on evidence through argumentation. This approach focuses on developing learners' GTR skills and is supported through PD to promote teachers' geospatial pedagogical content knowledge.

Note. An essential feature of this project is our hybrid, curriculum-linked PD model. This process incorporates both face-to-face and online learning and follows a design partnership model (Bodzin & Cirucci, 2009; Hammond et al., 2018). By integrating the teachers' PD into the curriculum design and development activities, we are able to advance teachers' geospatial pedagogical content knowledge (Bodzin et al., 2012) in the authentic context of their curricular practice. Teachers with geospatial science pedagogical content knowledge (PCK) have a more developed understanding of the complex interplay between science PCK and geospatial PCK and can teach science and related disciplines using appropriate pedagogical methods and geospatial technologies (Bodzin et al., 2012). This involves understanding how to model geospatial data exploration and analysis techniques and effectively scaffold students' geospatial thinking and analysis skills. The idea of geospatial PCK transcends content disciplinary boundaries since geospatial technology can interact with other discipline-based pedagogical content (e.g., geography and history) in ways that may produce effective teaching and student learning opportunities (Hammond et

Figure 1. Key components of the geospatial curriculum approach

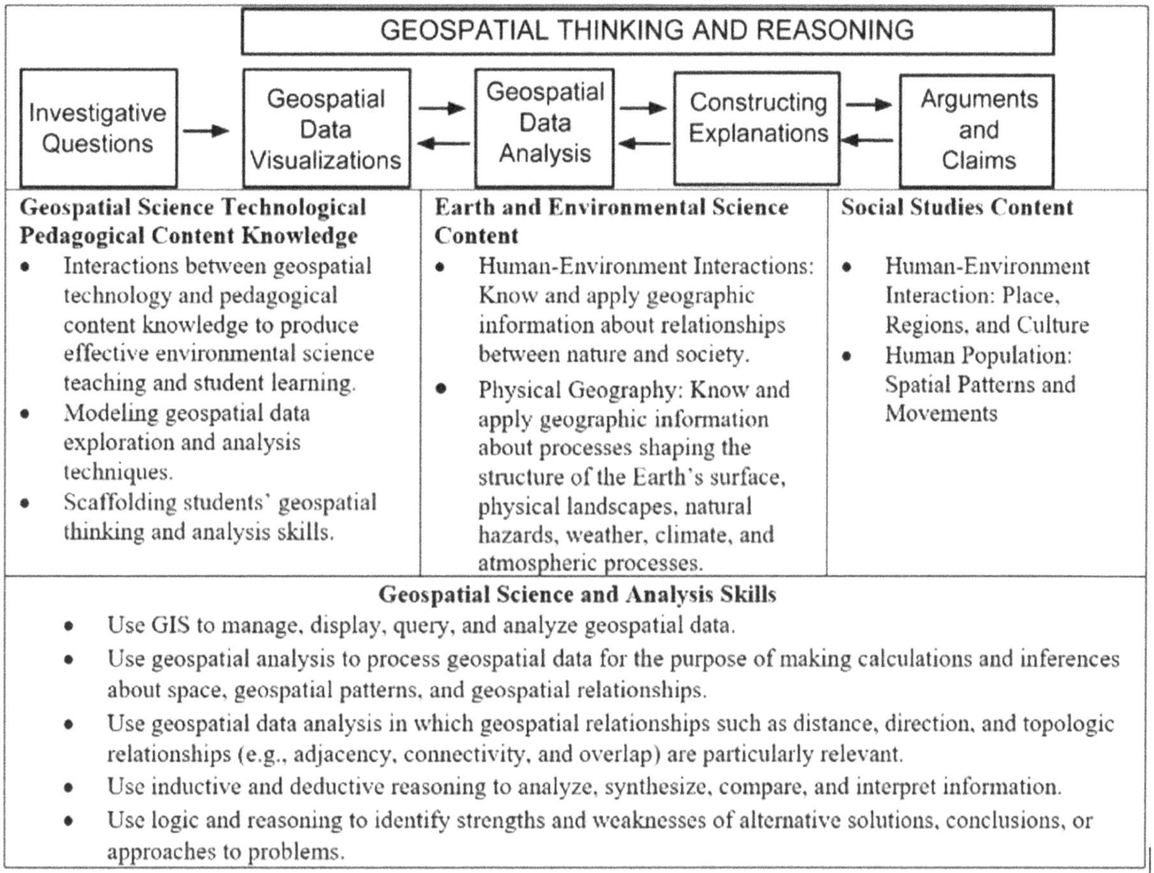

al., 2018). Our PD approach also acknowledges that classroom teachers are pedagogical experts capable of adapting curriculum materials to meet the needs of their students (Penuel et al., 2007).

During the PD, teachers learned to use a suite of tools developed by ESRI. These included ArcGIS Field Maps, ArcGIS Online, and ArcGIS StoryMaps. ArcGIS Field Maps is a mobile technology app that is used for gathering georeferenced data (see https://www.esri.com/en-us/arcgis/products/arcgis-field-maps/overview). ArcGIS Online is a dynamic mapping application that utilizes Web and cloud-based services for viewing, editing, and analyzing collected data along with other available georeferenced data (see https://www.esri.com/en-us/arcgis/products/arcgis-online/overview). An ArcGIS StoryMap is a web-based application that allows a user to share their maps in the context of narrative text and other multimedia content (see https://www.esri.com/en-us/arcgis/products/arcgis-storymaps/overview).

In this chapter, we present three different cases of the SESI curriculum-linked PD approach that was designed to support secondary teachers' development and implementation of SESI investigations and projects in three different high schools located in different geographic locations in the USA. These included Ft. Worth, Texas, Tri-Cities, Washington, and the Philadelphia, Pennsylvania, and Wilmington, Delaware areas. Each SESI investigation and project was collaboratively designed and developed between classroom teachers and science teacher educators, social studies teacher educators, and faculty with expertise in learning technologies. Each case illustrates the effectiveness of the curriculum-linked PD approach in providing teachers with professional growth experiences that enabled them to implement SESI investigations and projects in their classrooms with diverse high school students.

CASE ONE: RIVER WATER QUALITY IN TEXAS

The Texas site includes two high schools in a large urban school district and a project team of two professors and one doctoral student. School A, built in 1957, is a typical high school with a student population of 2,301, while School B, renovated in 2019, is a STEM Academy with a student population of 600. The teaching assignments include psychology, history, computer science, robotics, environmental science, biology, and ethnic studies. This case study focuses on the journey of one science teacher in a STEM-focused school. Tara was a second career educator with a Ph.D. in biochemistry and prior experience working in cancer research. Even with her own experience working in a research lab, she indicated that her campus was limited with their science lab equipment, technology support, and chemistry lab supplies. Because her campus focused on instructional technologies required for blended and virtual learning rather than scientific data collection and analysis, the adoption of geospatial technologies with data collection capabilities was new to her. Even with these campus-level limitations, she was still interested in using ArcGIS Online and mobile data collection to enrich a water quality unit in her chemistry course by making it place-based and related to current environmental issues. Although Tara was new to using geospatial technology, working with students to use iPads to collect field data, and integrating geospatial thinking and reasoning within her science content, she wanted to enhance her students' experience using technology to collect water quality data in the field.

During the first year of PD sessions with the university, Tara learned the basics of ArcGIS Online and the overall capabilities and embedded tools of this dynamic mapping platform. She was introduced to curriculum materials curated by ESRI (e.g., GeoInquiries and StoryMaps) and mobile applications for collecting geo-referenced data (e.g., Survey123 and Field Maps app). The intent of the PD was to provide experiences with ArcGIS Online allowing teachers to interact with geospatial inquiry in a way similar to their students. Tara also reviewed basic cartography elements such as boundaries, points and lines, distance, geometric shapes, density, and patterns. Because in-service teachers have likely not had previous exposure to geospatial reasoning (Bourke & Lidstone, 2015), it was important for the PD to demonstrate the importance of geography and geospatial thinking in order to appreciate GIS as a learning tool (Bednarz & van der Schee, 2006). Tara was able to view existing maps, see how the map layer features worked, and add her own MapNotes to a map. This led to an activity where she collected her own data using ArcGIS Field Maps, and she manipulated her data within ArcGIS Online. During the first year of PD, there was no requirement for her to implement geospatial technologies into her curriculum, only that she would increase her confidence with ArcGIS Online and consider possible applications for geospatial thinking in her chemistry course.

Field Trip

The university team provided technology and curriculum support for Tara as part of the NSF funding for the collaborative research grant project. In addition to monthly PD meetings, the university team was available to meet individually with teachers to assist with instructional planning and technology integration as needed during the school year. Early in the school year, Tara emailed the university team to discuss her plans to revise the existing water quality curriculum for her chemistry course and integrate the geospatial concepts and technologies she had learned in the PD. Although she was confident with her science content and pedagogy, she was apprehensive about incorporating technology into a field trip for water quality field testing. During several planning meetings, Tara expressed her concerns about the field trip and being able to facilitate her students using geospatial technology, collecting geo-referenced data with iPads, and analyzing water samples for chemical elements.

The university team assisted her with the logistics of the field trip, including collecting permission forms, accessing the river site near the school, reviewing safety protocols for students, and having students use the grant-funded iPads for the first time. The technology demands included creating student accounts for ArcGIS Online, writing a survey to collect chemical water quality data, and adding a map for students to access on their iPads while at the river sampling location. Students also needed to log in to the iPads while on the school's WiFi account to download the map that would later be used to view the map at the river site. Tara decided that using water quality test strips purchased by another teacher on her campus would be the easiest and most efficient way for her students to quickly test water samples. Learning this, the university team advised her on the methods and equipment for collecting water samples, limitations of using chemical test strips with the water samples, considerations for safe handling of samples, and alignment of chemistry content with state curriculum standards.

As a result of the collaborative planning with the university team, Tara finalized plans for the field trip and even arranged for an industry expert from the local water district to enrich the field trip experience. On the morning of the field trip, the industry expert met Tara, her students, and the university team on the riverbank. After the expert described his own career pathway in STEM and explained how to properly collect, label, and transport water samples, Tara distributed the water sampling equipment to her students. Students worked in pairs to carefully dip collection jars into the river and marked their GPS location for their sample site using the ArcGIS Field Maps app, which was loaded on the iPads. Students also labeled their jars with their names, date, and time the water was collected.

When the group returned to their classroom and the iPads were back on the school's WiFi network, the students used chemical test strips to measure the properties of their water samples. The test strips Tara purchased were difficult for students to read because they measured multiple water quality parameters, and students were unfamiliar with how to read these strips. Adding to the complexity, the text fields in the ArcGIS Field Maps app did not exactly match the elements on the chemical test strips, so students had to search the app for the correct place to input their data. Students worked in pairs to make estimates of the readings for each chemical parameter before entering the data into the ArcGIS Field Maps app, which then connected their data to the GPS location where they collected the sample. As often happens with a field trip experience, students did not have enough time to enter their information into the iPads and analyze their data in ArcGIS Online before the bell rang to dismiss students to their next class period. Students abruptly left their water samples, test strips, and iPads before Tara was able to debrief the students about geospatial trends and patterns in their data and what this means for water quality at the river site. Due to school schedules and time constraints, Tara was not able to follow up with the students about how the water data and corresponding map locations provided insight into river conditions.

Reflection

After the field trip, Tara initiated a meeting with the university team to discuss the field trip and, specifically, using the iPads and test strips. Although she had been apprehensive about taking the students to the nearby river, she was comfortable with how they behaved and their general learning experience. She found that integrating geospatial technologies into the field trip was easier than she anticipated, and the field trip added rigor to the water quality unit because students participated in authentic fieldwork similar to the industry expert. Although she expressed concerns about the accuracy of the student collected data, she sought ideas for ways her students could view the data in ArcGIS Online and ways for students to analyze the data. The university team realized the post-field trip experience did not provide the students with opportunities for geospatial reasoning, even though students had used technology in the field and collected their own authentic data.

Tara began thinking about how she could use geospatial technologies to expand her unit to include water samples from the students' homes and other city locations. The data collected from the field trip was limited because of student measurement errors with the test strips and the fact that all of the samples were collected from a 15-meter stretch of the riverbank. Expanding the geographic area of samples would provide students opportunities to use geospatial reasoning to compare patterns and trends in water quality data from multiple locations using a dynamic map of the city using ArcGIS Online. The GIS map used by students also had data layers that identified storage towers, pump stations, treatment plants, and reservoirs, which could be used to show the interconnected nature of the water supply system in their city. In future iterations of this activity, students could use these visualizations to increase their spatial understanding of where their water originates and the many steps it takes before it comes out of their tap. Tara also inquired about the possibility of purchasing probes as a way to increase the accuracy of the data collection and to give students the opportunity to use other STEM technology tools. She thinks the revised water quality unit will allow her to integrate geospatial thinking and technology applications that will make the content more relevant to her students' lives.

CASE TWO: CAMPUS TRASH IN WASHINGTON STATE

At the Washington State site, the project team consisted of three faculty members, a doctoral student, a program coordinator, and eight teachers from two local high schools. The high schools are a small project-based alternative school and a large comprehensive school—one of the largest in the state and serving large numbers of Latinx and low-income students. The teachers participating in the Washington State project included science, mathematics, environmental science, earth science, agriculture, and special education instructors. In this section, we describe how the high school special education teacher in the project, Denise, gained an understanding of ArcGIS Online during the summer PD experiences, planned and implemented an activity with her 12th-grade students, and reflected on her use of technology with her special needs students. Denise teaches in a large comprehensive high school; her students are children with intellectual and learning disabilities who have limited experience using technology in the classroom. Denise stated that she wanted to expose the students to authentic experiences using technology and expand their awareness of their capabilities in using technology to solve problems. She also had a goal of connecting her students to the school to increase their feeling of belonging.

Prior to the implementation of this activity, the teachers involved in this project met at the university campus during the summer to learn about designing and implementing SESI investigations to be integrated into their curriculum during the upcoming academic year. The focus of this PD was to introduce teachers to using ArcGIS Online through experiences as students themselves, followed by planning time where, partnering with faculty, they designed SESI activities and investigations that could be embedded into their existing curriculum. Teachers were also encouraged to pilot their lessons with the whole group during the final day of the PD. The expertise of faculty on geospatial activities and the teachers' knowledge about their curriculum and students provided a strong collaboration on which to develop the SESI investigations.

Denise had no prior experience using the geospatial tools introduced in the PD and spent time during the sessions planning for lessons where she would introduce her students to navigating and accessing data within various layers on a map. For example, as a sample activity in the PD, Denise created a lesson that involved students navigating a map of the contiguous United States that included one data layer of over three hundred civil war battles and another layer color-coding the states' allegiance. Denise also spent time with the other teachers learning how to manipulate maps in ArcGIS Online, collecting data with the ArcGIS Field Maps app, and designing an ArcGIS StoryMap. From her summer experiences, Denise gained the knowledge and confidence to start a project with her students when the school year began. She stated that she wanted to engage her students in a problem involving authentic data collection using a data collector app that would connect the students to their local environment and school context.

Trash Is a Problem

Denise started the new school year with a plan to engage her students in an authentic project that involved mapping and data collection. She drew on everything that she had learned over the summer and applied it to her preparation for the new school year. Her class included eight 12th-grade students, two with intellectual disabilities, three with learning disabilities, and three with health impairments. Denise initiated the project by asking students to identify issues within the school that they believed could be investigated. The students talked about various issues, but many of them shared experiences about how they regularly see an excessive amount of garbage on the school grounds; Denise viewed this as a topic that could be used for data collection and mapping in ArcGIS Online. Engaging the students in an authentic problem and pushing them to propose solutions were part of her stated goals.

Using ArcGIS Online, Denise created a feature layer that would allow the students to collect data regarding the trash on campus using Esri's ArcGIS Field Maps app. To do this, Denise used an ArcGIS StoryMap created by the University team members that guided them in creating feature layers using ArcGIS Online. The students were able to use their mobile devices to download the app and record data on the trash they found around campus (e.g., type of trash) and include a photo. Denise created a second feature layer that allowed students to collect data on the school grounds' garbage cans. Students mapped where the garbage cans were located and the cleanliness of the container. Students only collected data outside on the school's grounds, and they did not collect data from inside garbage sites. The class used the map to discuss where garbage was located, the type of garbage found, and its proximity to garbage cans (see Figure 2). Students were genuinely surprised to see how much garbage there was on school grounds when they viewed a map depicting locations where trash had been spotted. One student said, "we were shocked when we saw all the garbage on the map," and "it made me disgusted to see all that garbage." The students came to some conclusions about why garbage was located in specific locations

Figure 2. Trash is a Problem map
Note: Map displays two data layers (a) trash data collected before Trash Awareness Week (purple), and (b) garbage cans on campus (blue)

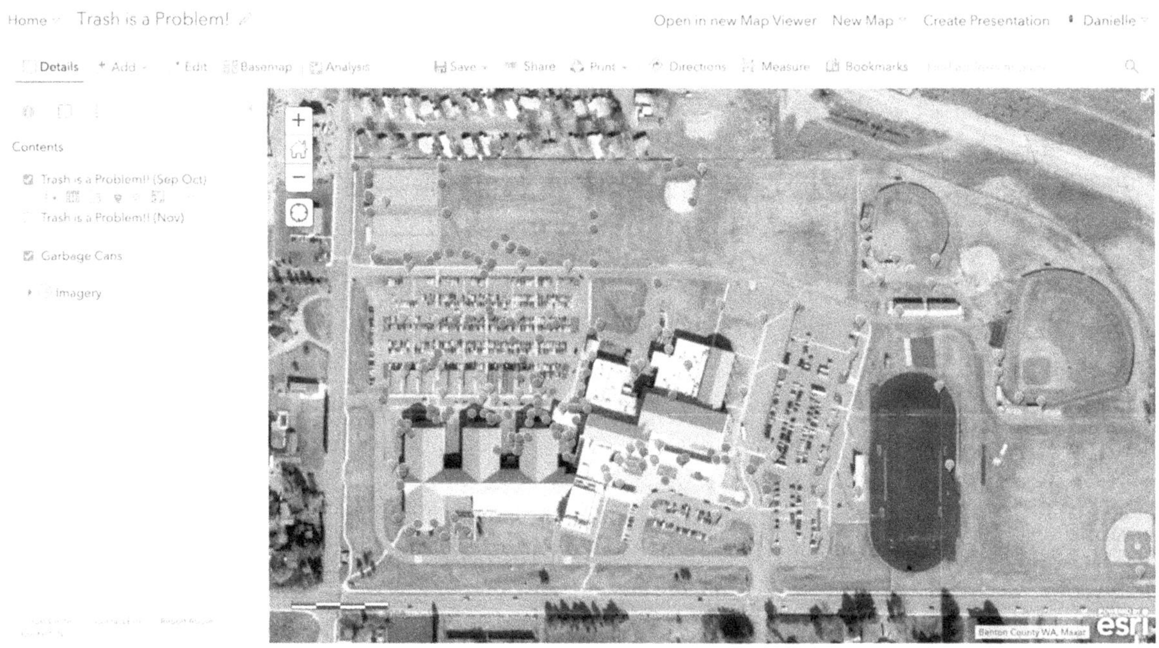

(e.g., where students ate lunches) and why some garbage cans were not used (e.g., "they were disgusting"). Using the data they had collected themselves to make inferences and draw conclusions is a critical piece of this lesson. Students were highly engaged because they had direct experience with the data used to draw conclusions.

Their data provided entry into a whole class brainstorming sessions, discussing the various ways to reduce trash around the school. For example, students proposed different sites for garbage cans, suggested cleaner garbage cans, and suggested creating a plan to bring awareness of the trash problem through "Trash Awareness Week." During this week, the students designed posters and displayed them in the school cafeteria (Figure 3), wrote notices about trash on school grounds, read the student-created announcements over the intercom during the school's daily announcements, and created coupons for HERO points (i.e., school incentive points) to hand out to students they saw disposing of trash in the proper locations. All of these activities were envisioned by students as they brainstormed ways to inform the entire school that trash on school grounds is a significant problem. When students were asked about their participation in the Trash Awareness Week activities, they said they loved making posters and telling the story, and they hoped it made people see that trash is a concern on campus. Through observations and discussions with the students involved, they expressed a sense of ownership over this project and the intervention strategies they implemented. Participation in these activities provided a sense of purpose for students who may not always feel a strong connection to their school. They saw their posters in the cafeteria and heard their announcements over the intercom. One student commented about being too shy to make the public, all-school announcement, but the student talked about it proudly after doing it.

Data on garbage location and type were again collected after the whole school Trash Awareness Week to determine if there were any changes in the amount of trash on campus (see Figure 4). Students looked

Figure 3. Student-created posters displayed in the school's cafeteria during Trash Awareness Week

at maps and compared data collected before and after the awareness interventions during Trash Awareness Week. The students concluded that there was less trash in some areas of the school that seemed to be problem areas (e.g., right outside of the cafeteria). However, the students also noted the change in weather, suggesting that the cold weather impacted the trash because more students were eating indoors. Esri's ArcGIS Online and the ArcGIS Field Maps app allowed the students in Denise's class to collect data and use geospatial technologies to identify areas that amassed the most trash. They discussed student body movement on the campus grounds and how this affected the accumulation of trash. These conclusions may not have been easily identifiable without using the geospatial technologies used in this project.

Reflection

As Denise reflected on her integration of geospatial technology into her special education classroom and the impact this project had on her students, she stated that the gains her students made in conducting authentic data collection and then using the data to make inferences, draw conclusions, and propose changes were of the utmost importance to her. She said that her students certainly gained experiences in using the geospatial technology but having this use embedded in a real-world, local context problem was most important. The students Denise worked with experienced seeing the data they collected on the map of the whole campus and then used those data points to make inferences about the main campus trash issue. These skills are essential in many discipline-based curriculum contexts but may not often

Figure 4. Trash is a Problem II map
Note. There are three data layers: trash collected before intervention (purple), trash collected after intervention (orange), and garbage cans on campus (blue)

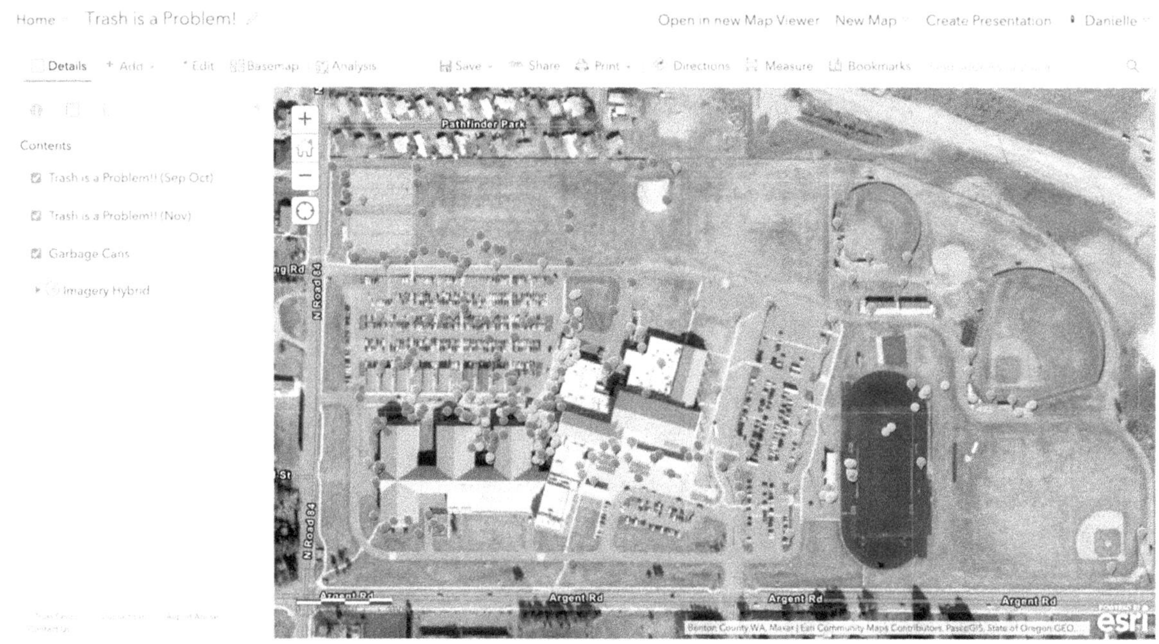

be available to the students in a special education classroom. Denise created a project that strengthened students' technology skills, improved their connections with their campus environment, and increased their confidence in communicating with other students and administrators.

CASE THREE: ECOSYSTEM RESTORATION IN DELAWARE

At the Philadelphia, Pennsylvania and Wilmington, Delaware area sites, the University team is working with six teachers from two schools: a Philadelphia, PA environmental science magnet high school and a Wilmington, DE comprehensive high school. The teachers (including a special education teacher) teach a variety of science and social studies classes, comprised of 9th through 12th grade students. One of the teachers is in his first year of teaching, while others are more experienced. Two teachers have worked with GIS technology in the past, with the remaining four being new to the practice. The hybrid, curriculum-linked PD began with the teachers during the summer of 2020. Due to the COVID pandemic, all of the work with teachers and their students was virtual from the summer of 2020 through the rest of the 2021 academic year.

Because of the shift to online instruction, the university team elected to focus on supporting teachers as much as they could, making their work as easy as possible under the trying circumstances. One adaptation was to reduce the frequency of outdoor data collection by students and instead emphasize exploring online maps in the context of their curriculum. In year two, with the return of students to the physical classroom–albeit with continued COVID restrictions–students were able to engage more fully in SESI investigations. In data collection, students download ArcGIS maps created by the teachers and researchers to the freely available Esri Field Maps app on either their mobile phone or a classroom GPS-enabled iPad. Then, students go outdoors to collect georeferenced data points on their maps, which then can be viewed by the class as a whole. This outdoor data collection step is essential to the SESI process of place-based education as students gather data about real world objects in their environment, which often includes a photo. The students are then able to analyze data they themselves have gathered. The focus of this case study is Sean, a trained and experienced Biology teacher, who also teaches Environmental Science and Earth & Space Science courses. Because of his experience and characteristics, Sean was the first teacher in our group to implement a SESI project with his students.

Classroom Investigations to Set the Scene

Sean was a participating teacher in our precursor National Science Foundation grant in a different high school, so he entered the project with more experience with our hybrid, curriculum-linked PD model and using ArcGIS Online and the related apps, such as ArcGIS Field Maps and Survey123, in the classroom (Carrigan et al, 2019; Hammond et al, 2018; Hammond et al, 2019). In fact, Sean has reached the point where he often creates SESI investigations and lessons for his students on his own, with little assistance needed from our research and development team. He is a teacher who is unafraid of trying new things in support of his curriculum and has the creative control to take his curriculum in new directions.

In the second year of the grant-funded project, Sean implemented a variety of ArcGIS-based investigations in his Environmental Science classes. Some of these activities were adaptations from the precursor grant, such as the Built Environment Scavenger Hunt, Trees & Ecological Services, and Urban Heat Islands (UHI). In the Built Environment Scavenger Hunt, students explore their school grounds for examples of human infrastructure (Figure 5) and code examples by service provided (see https://eli.lehigh.edu/sesi/instructional-sequence/built-environment-scavenger-hunt). For the Trees and Ecological Services investigation, students explore their school property in order to locate, measure and identify trees and examine their impacts on the community (Carrigan et al., 2019). Lastly, in the Urban Heat Island investigation, students measure the temperature of a variety of outdoor surfaces (e.g., asphalt, concrete, grass) in sun and shade at different times of the day. Students learn that human-made surfaces often absorb and retain a large amount of heat from the sun, resulting in urban areas often being hotter than surrounding suburban and rural communities, which usually have more shade trees and natural areas (https://eli.lehigh.edu/sesi/instructional-sequence/urban-heat-islands). Other SESI investigations were new creations from the current grant, such as an investigation of community waste management and sustainable growth and development. In addition to the geospatial SESI investigations, Sean also included instruction on habitat fragmentation, animal migration, native and invasive plants, pollinators, and human impact on the natural environment. This sequence of instruction culminated in an Ecosystem Restoration project.

Ecosystem Restoration Project

A SESI project is different from a SESI investigation in that students are expected to draw data from previous lessons, synthesize a solution to a problem, and demonstrate geospatial thinking and reasoning skills. For the Ecosystem Restoration project, students were charged with developing a proposal suggesting three changes to the school property that would enhance its natural ecosystem, with the potential of

Figure 5. Sample Built Environment Scavenger Hunt data

school district funding for the most highly rated proposal(s). For each of the three proposed changes, students were to identify the environmental issue(s) and how their change would address the issue(s). In addition, they were instructed to rank order their changes and select the best two for potential funding. They used the data they had gathered in the prior SESI investigations to support their proposals. Students' proposed changes included consolidating athletic practice fields to allow for surrounding forest recovery, planting native plants and creating a pond in a central courtyard to draw pollinators, and installing bird houses and feeders on school grounds in order to attract native species and educate other students about them. To illustrate and support their proposals, students developed maps in ArcGIS Online to show where their changes would take place and also created either a slide presentation or an ArcGIS StoryMap to explain their intended solutions and anticipated impacts. Using these materials, they gave an oral presentation to make their case. Completed student presentation files were also distributed to school administrators. This project will be done with new students in year three of the program, with school administration being invited to the presentations.

Figure 6 shows a sample student map, which the student then embedded into a StoryMap. As directed, this student suggests three potential areas to address by planting native flowers, shrubs, and trees. Then, in the related StoryMap (https://storymaps.arcgis.com/stories/5d074af7a38a49dfb380227 44378083c), the student discussed why he ultimately chose only two of the areas (southeast play field and immediately west of the school) for restoration, as the area directly southeast of the school has too much student foot traffic to and from the bus loading area.

An important feature of SESI projects is the ability of the teacher to assess students' geospatial thinking and reasoning skills. In our precursor grant, we developed a reliable and valid rubric for that purpose, which we adapted for the Ecosystem Restoration project evaluation. Table 1 shows the components used to assess each student's geospatial analysis and geospatial reasoning skills. In this first implementation of a SESI project in our current work, we worked closely with Sean to refine the rubric to accurately

Figure 6. Sample student map for the Ecosystem Restoration project, showing areas of proposed changes

assess students' geospatial thinking and reasoning skills. As we enter year three of our four-year grant, we expect all of our teachers to implement at least one SESI project in their classes. We will then be able to use the refined rubric to ascertain what geospatial thinking and reasoning skills students are able to demonstrate.

Table 1. Skills assessed in SESI projects

Category	Skill
Geospatial Analysis	• Use GIS to manage, display, query, and analyze geospatial data. • Use geospatial analysis to process geospatial data for the purpose of making calculations, and inferences about space, geospatial patterns, and geospatial relationships. • Use geospatial data analysis in which geospatial relationships such as distance, direction, and topologic relationships (for example, adjacency, connectivity, and overlap) are particularly relevant.
Geospatial Reasoning	• Use inductive and deductive reasoning to analyze, synthesize, compare, and interpret information. • Use logic and reasoning to identify relative strengths and weaknesses of conclusions, or approaches to problems.

Reflection

The SESI projects allow students to synthesize the data they have gathered throughout the school year in order to make a case to have an effect on their school community. This can be an impactful experience for students. High school students can often feel powerless in their school community and their community at large as they progress to young adulthood. Presenting their ideas about making changes in their environment, with the potential for their ideas to be implemented, can be a very powerful learning experience. In our previous work at another high school, students completed a SESI project where they presented their proposals to plant trees on school property. With support from the school and city, the winning proposal was implemented, with the student author and classmates participating in planting the trees in front of the school. These projects can help students be seen as true members of the community and can empower them to continue in such work

CONCLUSION

Implications and Recommendations

From the work with teachers at three different geographic sites, each with unique school cultures, it is evident that the SESI curriculum approach on which they based their PD helped high school teachers develop practical teaching strategies that integrated a variety of geospatial digital technologies into their teaching and learning activities. The technologies used in the three cases presented in this chapter provided the teachers with the ability to help students engage with authentic content and collect authentic data in the field. The geospatial tools highlighted in the chapter included iPads with embedded GPS without a data plan, ArcGIS Field Maps for data collection, and ArcGIS online for data analysis. These tools were selected specifically for their ability to engage learners with outdoor data collection either around their school grounds or in locations in students' communities. The collected data was then analyzed to examine geospatial relationships and patterns in the collected data. Using the SESI curriculum approach, the three cases illustrate how geospatial thinking and reasoning skills were promoted. Further, each case dealt with an authentic socio-environmental issue that was inherent to the students' immediate environment. The River Project is an example of a first step into SESI, whereas the Campus Trash Project is further along a continuum toward project implementation. The Ecosystem Restoration Project is an example of a SESI project. The Ecosystem Restoration Project and the Campus Trash Project enabled students to engage in authentic decision-making to make recommendations to enhance their school environment. These place-based learning activities made the SESI investigations and projects meaningful and relevant to the learners.

Implementing SESI investigations and projects requires preparation and planning. There are no costs involved for schools to use ArcGIS Field Maps and ArcGIS Online. To use ArcGIS Field Maps for data collection, either a cell phone with a telecommunication plan is needed or an iPad with a built-in GPS without a telecommunication plan. As part of the grant, we funded the purchase of a classroom set of iPads for each school, along with durable cases and a charging cart. These will remain in the schools after the grant is completed. During the CoVID-19 pandemic, we found that almost all the high school students had access to a cell phone that could be used for data collection. While ArcGIS Online is also freely available to schools, it involves setting up an organizational account for a school. At the classroom level, individual student logins and passwords need to be set up and managed. Quite often, some students forget their login passwords, and it is important that teachers keep a good organizational record system to access student passwords when this occurs. We have provided that support for teachers in the first two years of the grant but are now scaffolding teachers to promote sustainability of this work.

As teachers consider using SESI-style investigations and projects with students, we recommend coordinating set-up with a school or school district's technology support staff. Such support staff may serve as invaluable resources for installing applications on school computers and iPads, ensuring network connectivity on the devices, and maintaining software and hardware updates on school equipment. Over time, applications such as ArcGIS Field Maps may become updated or even overhauled completely, and ArcGIS Online may undergo interface changes and updated analysis tools. School technology staff can serve as helpful resources to work with teachers to become familiar with these changes when software and Web application updates occur.

There are many resources available to help teachers get started with their own geospatial investigations. Examples of SESI investigations and projects that have been implemented in high school classrooms are available at https://eli.lehigh.edu/sesi. For teachers who are not yet ready for outdoor data collection with their students, Esri maintains a library of GeoInquiries at https://www.esri.com/en-us/industries/k-12-education/geoinquiries. GeoInquiries are short, structured, and guided data-based learning activities that use dynamic maps to investigate geospatial patterns and relationships among specific data. In addition, Esri also provides ArcGIS StoryMaps (https://storymaps.arcgis.com/stories), which enable students and teachers to select from a suite of templates to craft a digital story with maps and text. In the Ecosystem Restoration Project case, the classroom teacher had each student develop a StoryMap to highlight their recommendations for enhancing the ecosystem on the school property.

The three cases presented in this chapter highlight how teachers used SESI investigations with their students. The cases demonstrate how place-based, socio-environmental investigations using geospatial technologies actively engage learners with authentic data collection and analysis to explore geospatial patterns and relationships. The students were involved with geospatial thinking and reasoning skills. In two of the cases, students were involved in making recommendations to enhance their local environment. The hybrid form of PD was effective for promoting enhanced geospatial pedagogical knowledge of the teachers resulting in powerful learning activities for the students.

ACKNOWLEDGMENT

This material is based upon work supported by the National Science Foundation under grant #DRLs-1949400, 1949393, and 1949388. Any opinions, findings, and conclusions or recommendations expressed in this material are those of the authors and do not necessarily reflect the views of the National Science Foundation.

REFERENCES

Aikenhead, G. S. (2005). Science-based occupations and the science curriculum: Concepts of evidence. *Science Education*, *89*(2), 242–275. doi:10.1002ce.20046

Bednarz, S. W., & van der Schee, J. (2006). Europe and the United States: The implementation of geographic information systems in secondary education in two contexts. *Technology, Pedagogy and Education*, *15*(2), 191–205. doi:10.1080/14759390600769573

Bodzin, A. (2011). The implementation of a geospatial information technology (GIT)-supported land use change curriculum with urban middle school learners to promote spatial thinking. *Journal of Research in Science Teaching*, *48*(3), 281–300. doi:10.1002/tea.20409

Bodzin, A., & Cirruci, L. (2009). Integrating geospatial technologies to examine urban land use change: A design partnership. *The Journal of Geography*, *108*(4-5), 186–197. doi:10.1080/00221340903344920

Bodzin, A., Fu, Q., Kulo, V., & Peffer, T. (2014). Examining the enactment of a geospatial curriculum design approach on students' geospatial thinking and reasoning. *Journal of Science Education and Technology*, *23*(4), 562–574. doi:10.100710956-014-9488-6

Bodzin, A., Fu, Q., Peffer, T., & Kulo, V. (2013). Developing energy literacy in U.S. middle level students using the geospatial curriculum approach. *International Journal of Science Education*, *35*(9), 1561–1589. doi:10.1080/09500693.2013.769139

Bodzin, A., Peffer, T., & Kulo, V. (2012). The efficacy of educative curriculum materials to support geospatial science pedagogical content knowledge. *Journal of Technology and Teacher Education*, *20*(4), 361–386.

Bourke, T., & Lidstone, J. (2015). Mapping geographical knowledge and skills needed for pre-service teachers in teacher education. *SAGE Open*, *5*(1). doi:10.1177/2158244015577668

Carrigan, J., Bodzin, A. M., Hammond, T. C., Rutzmoser, S., Popejoy, K., Farina, W., Hanson, I., Salter, S., Anastasio, D., Kangas, S., Holland, B., & Sahagian, D. (2019). Investigating urban trees. Exploring the impact of trees around our school with geospatial technologies. *Science Teacher (Normal, Ill.)*, *86*(8), 27–35.

Chin, P., Munby, H., Hutchinson, N. L., Taylor, J., & Clark, F. (2004). Where's the science? Understanding the form and function of workplace science. In E. Scanlon, P. Murphy, J. Thomas, & E. Whitelegg (Eds.), *Reconsidering science learning* (pp. 118–134). Routledge Falmer. doi:10.4324/9780203464021_chapter_2.6

DiBiase, D., Corbin, T., Fox, T., Francica, J., Green, K., Jackson, J., Jeffress, G., Jones, B., Jones, B., Mennis, J., Schuckman, K., & Smith, C. (2010). The new geospatial technology competency model: Bringing workforce needs into focus. *URISA Journal*, *22*(2), 55–72.

Goodchild, M. F., & Janelle, D. G. (2010). Toward critical spatial thinking in the social sciences and humanities. *GeoJournal*, *75*(1), 3–13. doi:10.100710708-010-9340-3 PMID:20454588

Hammond, T. C., Bodzin, A., Anastasio, D., Holland, B., Popejoy, K., & Sahagian, D. (2019). Shoulder-to-shoulder: Teacher professional development and curriculum design and development for geospatial technology integration with science and social studies teachers. *Contemporary Issues in Technology & Teacher Education*, *19*(2). https://www.citejournal.org/volume-19/issue-2-19/current-practice/shoulder-to-shoulder-teacher-professional-development-and-curriculum-design-and-development-for-geospatial-technology-integration-with-science-and-social-studies-teachers/

Hammond, T. C., Bodzin, A., Anastasio, D., Holland, B., Popejoy, K., Sahagian, D., Rutzmoser, S., Carrigan, J., & Farina, W. (2018). "You know you can do this, right?": Developing geospatial technological pedagogical content knowledge (GS-TPACK) and enhancing teachers' cartographic behaviors with Socio-Environmental Science Investigations (SESI). *Cartography and Geographic Information Science*, *45*(4), 305–318. doi:10.1080/15230406.2017.1419440

Huynh, N. T., & Sharpe, B. (2013). An assessment instrument to measure geospatial thinking expertise. *The Journal of Geography*, *112*(1), 3–17. doi:10.1080/00221341.2012.682227

National Research Council. (2006). *Learning to think spatially: GIS as a support system in K-12 education*. National Academies Press.

National Research Council. (2011). *Expanding underrepresented minority participation: America's science and technology talent at the crossroads*. National Academies Press.

National Science Board. (2015). *Revisiting the STEM workforce*. National Science Foundation.

NGSS Lead States. (2013). *Next Generation Science Standards. For States, By States*. The National Academies Press.

Penuel, W., Fishman, B., Yamaguchi, R., & Gallagher, L. (2007). What makes professional development effective? Strategies that foster curriculum implementation. *American Educational Research Journal*, *44*(4), 921–995. doi:10.3102/0002831207308221

Sadler, T. D., Amirshokoohi, A., Kazempour, M., & Allspaw, K. M. (2006). Socioscience and ethics in science classrooms: Teacher perspectives and strategies. *Journal of Research in Science Teaching*, *43*(4), 353–376. doi:10.1002/tea.20142

Semken, S. (2005). Sense of place and place-based introductory geoscience teaching for American Indian and Alaska Native undergraduates. *Journal of Geoscience Education*, *53*(2), 149–157. doi:10.5408/1089-9995-53.2.149

Semken, S., Ward, E. G., Moosavi, S., & Chinn, P. W. U. (2017). Placed-based education in geoscience: Theory, research, practice, and assessment. *Journal of Geoscience Education*, *65*(4), 542–562. doi:10.5408/17-276.1

Sobel, D. (2004). *Place-Based Education: Connecting Classrooms and Communities*. Orion Society.

Zeidler, D. L., & Nichols, B. H. (2009). Socioscientific Issues: Theory and Practice. *Journal of Elementary Science Education*, *21*(2), 49–58. doi:10.1007/BF03173684

Chapter 7
Science Practices Innovation Notebook (SPIN):
Cultivating Data Practice Skills Through Computational Thinking and Self-Regulated Learning

Erin E. Peters-Burton
https://orcid.org/0000-0003-1233-9119
George Mason University, USA

Peter J. Rich
https://orcid.org/0000-0002-0039-966X
Brigham Young University, USA

Timothy Cleary
Rutgers University, USA

Anastasia Kitsantas
George Mason University, USA

ABSTRACT

This chapter describes the development, characteristics, and applicability of a web-based interactive notebook, the Science Practices Innovation Notebook (SPIN), for use by high school students and teachers during science investigations. SPIN integrates data practices, computational thinking (CT), and self-regulated learning (SRL) principles to support high school students as they engage in science investigations. SPIN uses CT tactics and underlying SRL principles to optimize student progression through five core data practices. SPIN also incorporates loop mechanisms that foster communication, prompting, and feedback among individual students, their teachers, and peers. SPIN gives teachers data about student progress, communication, and learning analytics that assists in providing high quality feedback to students. This chapter will describe how the theoretical underpinnings of SPIN reinforce each other, how collaborative interactions between teachers and researchers led to SPIN development and design, and how SPIN implements support for student data practices in an online setting.

INTRODUCTION

With reports indicating that U.S. students are falling behind international peers in science proficiency,

DOI: 10.4018/978-1-6684-5585-2.ch007

particularly for students from racial and ethnic underrepresented groups (NCES, 2013), there is a need to focus attention on established and emerging constructs, such as students' academic self-regulated learning (SRL; Peters-Burton et al., 2018), computational thinking (CT; Weintrop et al., 2016), and the use of technology as leverage points for analyzing and optimizing learning processes (Azevedo et al., 2009). In the context of science education, scholars have argued that effective instruction should provide SRL supports that empower and enable students to problem solve and overcome barriers and challenges during learning (Millar & Osborne, 1998; Peters & Kitsantas, 2010; Schraw et al., 2006). One of the most promising features of SRL supports is the ease with which they can be embedded in technological platforms and their effectiveness in enhancing student self-awareness, the quality of their strategic thinking, and their confidence to learn in science contexts (Cleary et al., 2017; Peters-Burton & Kysar-Mattietti, 2017). A critical issue, however, is that SRL interventions and practices are not routinely incorporated into course curricula, and although teachers find value in providing SRL supports to their students, they often struggle to independently infuse these ideas into their teaching (Cleary & Zimmerman, 2006; Grigal et al., 2003; Wehmeyer et al., 2000).

Educators have also begun to emphasize the importance of CT skills in STEM contexts (NRC, 2010; 2011). CT is an approach to solving problems and designing systems that requires students to think recursively, reformulate perceptions of a problem, model relevant aspects of problems, and use abstraction and decomposition in tackling complex problems (Wing, 2006). From our perspective, CT skills can be a useful complement to traditional science pedagogy given their potential for enhancing students' data practice skills (e.g., data collection, data analysis) when engaged in scientific investigations (Weintrop et al., 2016). Given that data analysis practices are often overly prescribed or missing in high school classrooms, many students do not receive adequate opportunities to hone their skills in breaking down problems into smaller components (i.e., decomposition), looking for patterns (i.e., pattern recognition), and automating analysis or transferring these skills to other problems (i.e., automation). Unless adequate support is provided to guide these instructional efforts, the integration of CT instruction into established curricula can be overwhelming for both teachers and students. In this chapter, we provide a rationale for the development of an electronic performance support system called the Science Practices Innovation Notebook (SPIN). We also detail its design, key features, underlying conceptual framework, and applicability for teachers and students. We end the chapter with a brief overview of practice implications for educators.

Issues, Problems, and Trends

In the 21st century, technological advances have spearheaded important innovations in educational contexts. Technological tools can aid teachers in integrating important scientific practices with instruction, such as CT in data practices, while simultaneously providing opportunities for student SRL. Learning technologies, when used in conjunction with personalized feedback from the classroom teacher, can provide students immediate feedback and support that has the potential to enhance learning progress (Kitsantas & Dabbagh, 2010). When learning technologies take into account the needs of educators , however, they have the potential to assist educators in facilitating SRL skills in classroom contexts (Berglas-Shapiro et al., 2017). In this chapter, we focus on SPIN, an electronic notebook or workstation that supports students' strategic and regulatory engagement in science investigations. Through integrating CT practices and SRL scaffolds and by leveraging learning analytics and other assessment data,

SPIN provides a unique platform for students to think computationally and strategically as they progress through data practices in science investigations.

The field of education has witnessed a proliferation of various learning technologies to support student success in school. For example, Artificial Intelligence (AI) tutoring systems, such as Operation ARIES, DeepTutor, and MetaTutor (Azevedo et al., 2009; Graesser et al., 2004; Millis et al., 2011) use AI agents to interact with and guide students in their learning and to promote the development of students' metacognitive (e.g., planning, reflection) and SRL skills (Berglas-Shapiro et al., 2017). MetaTutor is an online learning system that uses dialogue and prompts provided by AI agents to engage students in SRL practices as they learn important concepts in biology. The system scaffolds learners' cognitive and metacognitive abilities to help them become more goal-directed in their execution of learning activities. Although research shows that MetaTutor offers great promise for enhancing students' knowledge and use of SRL practices as well as their content knowledge in biology (Azevedo et al., 2009), it may not generate information about SRL in ways that could be of *immediate benefit* to students and teachers during the learning process. Although not an AI technology, SPIN addresses these gaps by providing students additional guidance or redirection when necessary.

Other learning technologies, such as Virtual Reality (VR), inclusive of Mixed-reality (MR), Extended reality (ER), Khan Academy, Academic Earth, and mobile-app daily diaries, are distinctive because they emphasize self-paced student learning opportunities. Thus, rather than attempting to replace traditional instructional formats, self-paced learning technologies often serve as a supplemental instructional tool (Broadbent et al., 2020; Ludlow, 2015; Putri, 2021). Supplemental, self-paced technologies serve as invaluable tools for facilitating SRL strategy use for in-person or synchronous virtual instruction, but they tend to be less effective when used in isolation (Broadbent et al., 2020). Thus, high levels of teacher engagement and involvement is often needed (Broadbent et al., 2020). Many of these types of programs also do not provide adequate personalized feedback or individualized supports foy students to optimize their SRL processes and overall success (Araka et al., 2020). SPIN is innovative in that its entire developmental progression was conducted in close collaboration and partnership with high school educators to blend learning technology and authentic classroom lessons in science.

The use of "big data" and learning analytics to guide instructional support and feedback is a topic of increasing interest among learning and cognitive scientists (Ferguson, 2012). Learning analytics reflects the collection and analysis of data about learners and their environments in order to better understand and enhance student outcomes (Lee et al., 2020). In principle, learning analytics have the potential to provide teachers with valuable information about student learning and behavior that can subsequently be used to inform how teachers can optimally engage, instruct, and support students in the classroom. Learning analytics, which include task completion time, button clicks, or the pattern of cursor movement of a learner, operate behind the scenes of the learning program without disruption to student learning (Araka et al., 2020; Kitsantas et al., 2019). Many learning technologies (e.g., MetaTutor) emphasize the importance of learning analytics and utilize traces about student behaviors. Relative to self-report measures, learning analytics are desirable because they are objective and reflect the actual behaviors and actions that students exhibit within a particular context or learning situation. SPIN utilizes learning analytics by collecting data about teacher lesson design, SRL prompting designs of the teacher, teacher and student communications, and student processes and outputs.

RATIONALE FOR THE DEVELOPMENT OF SPIN

SPIN offers promise as a learning technology because of its uniqueness in application to the field of science education, its integration of CT, SRL, and data practices, and its flexible structural design. In essence, SPIN provides an electronic workspace for students to engage in teacher-led STEM-related activities while leveraging CT practices and SRL-related supports and scaffolds. The development of SPIN was guided by researcher-practitioner partnership principles (Penuel & Gallagher, 2017), and thus took into account teacher perspectives and authentic course materials while engaging in a collaborative and iterative design process (Broadbent et al., 2020). SPIN also provides learning analytics to evaluate students' process of learning, and provides opportunities for students to interact with teachers and peers in real-time during task completion. SPIN is directly relevant to science education given that it embodies the intent behind recent calls for STEM subjects to closely mirror authentic professional STEM practices (NGSS Lead States, 2013) and to leverage a growing initiative to use electronic research notebooks as part of science investigations (Rubacha et al., 2011). Finally, SPIN offers teachers a highly flexible and customizable infrastructure that enables them to embed their own lesson plans within the system and to determine when and what prompting students receive. When teachers are afforded the opportunity to use their own content area expertise within an electronic performance support system that allows them to select the desired regulatory prompts, CT practices, and data practices, it is likely that their teaching experiences and the learning experiences of students will be enhanced.

Design and Key Features of SPIN for Teachers and Students

To help the reader understand SPIN as an electronic performance support system, we describe the system and its various features and then provide an in-depth explanation of the conceptual framework underlying its design.

SPIN was developed as part of a collaborative partnership between researchers and 20 high school science teachers. As conveyed in Figure 1, teachers, educational psychologists, and science education researchers contributed their specific expertise on planning and implementing science lessons using data practices, supporting student SRL, and teacher professional development. Our researcher-practitioner partnership focused on developing SPIN, a web-based science notebook that (a) focuses science investigations on data practices, (b) supports CT by providing computational tools to process data, (c) enables students to maintain a complete record of their work on activities and tasks, (d) supports students' SRL through a series of customizable prompts, (e) promotes social collaboration and co-regulation among students, and (f) gathers analytics on student SRL behaviors and thought processes for teacher instructional use. Although SPIN contains science investigations as lessons, it is not intended to replace the teacher; rather, it serves as an interactive, reciprocal support system for both students and teachers in the context of data practices.

SPIN Features for Students

When students open a science investigation in SPIN, they proceed sequentially through seven tabs to complete the investigation: Introduction, Create Data, Collect Data, Prepare Data, Visualize Data, Analyze Data, and Submit (see Figure 2). The Introduction page gives students background information on the phenomenon they are studying and introduces them to the research question they will investigate.

Figure 1. Logic model for SPIN development

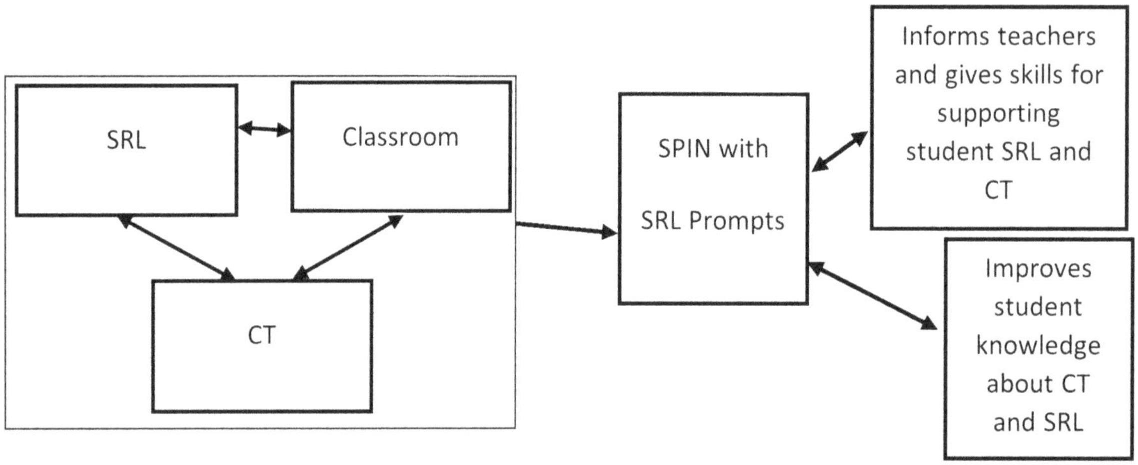

The next five data pages are structured by an organizational structure called GAT-P, which stands for Goals, Aims, Tactics and Prompts. The GAT-P framework organizes students' investigation from a higher level (e.g., goal of a data practice) to granular CT tactics that students might employ to optimize both CT and data practices. When a student opens any data page, a box emerges with the *Goal* (i.e., for a data practice) and a more specific plan statement for what students are to accomplish on the page. By hovering a cursor over the CT practices on a data practice page, students gather information about the *Aim* of the focus skill. Each data practice page also includes *Tactics* which help students accomplish the *Aim* (i.e., for CT skills). Students can access specific tactics by hovering over the *Prompt*.

Figure 2. Screenshot of SPIN Create Page

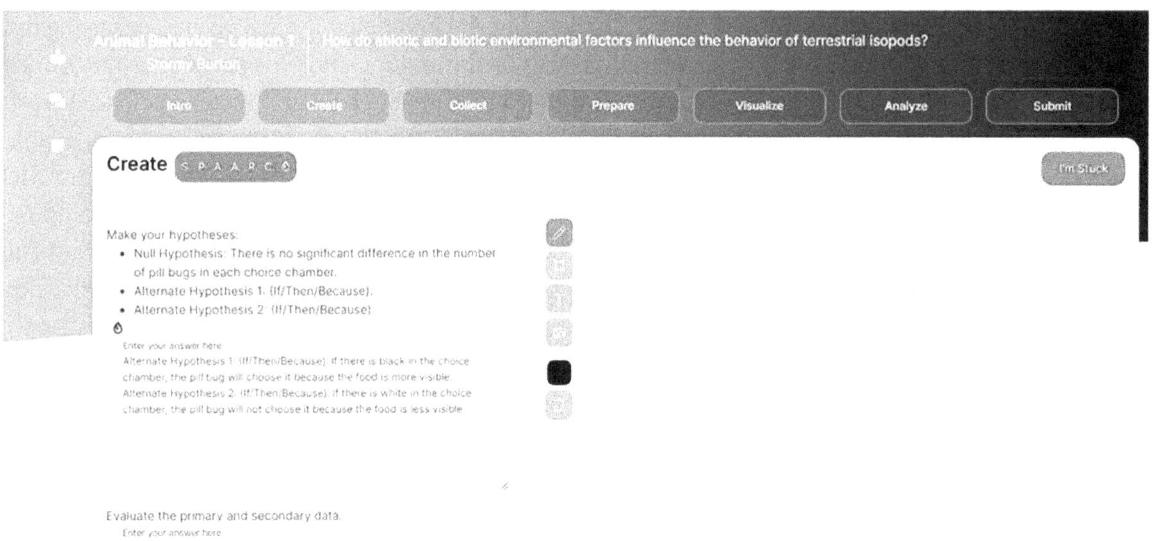

The *Create page* (Figure 2), as well as the five other data practices pages, are presented in a two-sided format. The left side of the notebook displays instructions and questions, while the right side of the notebook displays data practice workspaces. For example, the right side of the *Create page* has two parts: a whiteboard to draw or copy a picture and label the scientific phenomenon of interest and a table to input variables for data collection. To be able to move to the next page, Collect, students must accomplish the Create goal of identifying variables in the scientific phenomenon in the investigation and input variables that are relevant to the research question to be transferred to the *Collect page*.

The right side of the *Collect page* displays a data table with the variables identified from the *Create page*, and the left side has guiding questions about data collection so that students can monitor the quality of data collection. Once data is collected, students can move to the *Prepare page* where the goal is to become familiar with, clean, and prepare data for analysis. On the right side of the page, students have access to tools to automate data preparation, such as a box-and-whisker plot to identify outliers and quartiles as well as central tendency and variance information such as mean, median, mode, and range or standard deviation of variables, as appropriate. Students record this information as responses to teacher questions on the left side of the page. Once the students respond to all of the questions on the *Prepare page*, SPIN allows students to open the *Visualize* and *Analyze pages* at the same time, as students may want to toggle between the two.

Identical to the *Create*, *Collect*, and *Prepare pages*, the left side of the *Visualize* and *Analyze pages* are dedicated to teacher questions and student responses, and the right side of the pages are working spaces for the data practices. The *Visualize* working space features a graph choice flowchart in the form of a dropdown menu, which guides students to select an appropriate graph format based on their responses. The *Analyze page* enables students to perform statistical tests such as ANOVA, post hoc tests, linear regression and other tests that are commonly used in a high school setting (see Figure 3). The final page, *Submit*, gives students a chancc to see the information that was collected from all of the pages for the final investigation report and includes a notes section for students to write comments, observations, or other notes to their teacher. When the student clicks *Submit*, the investigation is locked until the teacher provides feedback to the student on their work. Although students need to complete a new data practice before proceeding to another, they are free to navigate back.

Students can also communicate to teachers, other individual students, and student groups through the *Communication Hub* on SPIN. Students can direct messages to any other student or group in the class, at the teacher's discretion, or they can direct messages to the teacher. SPIN features an "I'm Stuck" button that will open a direct message to the teacher (i.e., if the teacher is logged on to SPIN) or send an email (i.e., if the teacher is not logged on to SPIN).

There are several tools embedded into SPIN to support students' CT and SRL. Buttons labeled Decomposition, Pattern Recognition, Abstraction, and Algorithmic Thinking appear on every page. If students are not clear about the CT practices they should engage with on the data practice page, they can scroll over the live buttons to receive a message about the Aim of the practice. CT practices that are not supportive for the data practice in any particular investigation are grayed out (i.e., based on teacher judgment), so students are specifically directed to focus only on relevant CT practices.

Another student support tool is the *SPAARC promp*t. The acronym SPAARC represents a general strategic approach for organizing and linking CT skills and data practices:

- **Sketch:** tactics that involve creating some type of visual display to help students understand the demands of the lesson, the nature of the phenomenon, and/or the key data or results of the lesson.

Figure 3. Screenshot of SPIN Analyze page with dropdown choices and SPAARC prompt viewed

- **Play:** tactics that encourage students to physically manipulate materials (model or lab equipment), to engage in informal trial and error, and/or to explore in informal ways regarding the nature of a given phenomenon.
- **Ask:** tactics that involve students engaging in self-questioning to guide their thinking about the lab activities, materials, or overall phenomenon being studied.
- **Automate:** tactics that entail the use of one or more automation tools (such as a simulation or simulated model) to optimize or create efficiencies (e.g., reduce time, effort, skills required of a learner) as part of the lesson.
- **Review:** tactics that involve examining or looking over existing lab materials, data, or results germane to the given lesson that are either provided by others or developed by the student (e.g., self-monitored information).
- **Collaborate:** tactics that involve engaging with and utilizing social supports, such as classmates, teachers, or others, to verify, clarify or expand thinking.

Students who need help with the data practice but do not want to message the teacher can scroll over the SPAARC icon in SPIN to receive a prompt on how best to proceed (see flame icon on Figure 3). Another support to encourage students to persist in the investigation is the progress bar. Each of the seven tabs (e.g., Introduction, Create, Collect) entails a progress bar so that students can easily keep track of their rate of completion.

SPIN Features for Teachers

Three years of collaboration on SPIN with 20 classroom teachers has resulted in a flexible, interactive notebook that blends authentic lessons in a user-friendly way. To begin using lessons in SPIN, teachers set up a class name and input student names and emails. Teachers then add an investigation from the

SPIN Global Marketplace or create their own investigation by inputting the instructions and questions for display on the left side of each notebook page. The SPIN Global Marketplace provides teacher-created investigations in biology, chemistry, Earth science, and physics, but which are fully editable. Teachers can edit any part of what the students see on the left side of the page, including the CT buttons available, the instructions and questions, and the SPAARC prompts. The right side of the page is not editable as it is the workspace that has tasks that are consistent across the data practices. In addition to adding text, teachers can add text questions, multiple choice questions, and multiple answer questions. Teachers can decide which questions will be output on the Submit page, so they can reduce clutter in final products.

In addition to receiving direct messages from individual students and groups in the *Communication Hub*, teachers can send out announcements to all students in a class through the investigation. Teachers can monitor student progress by going to their investigation page. This page displays student progress by percent completed and enables teachers to click on a student's name to go into their notebook in real time to see the work that the student is doing. The SPIN screen will display the student's name, the title of the investigation, and the page as the student is working. This feature is particularly helpful to teachers when they are teaching in a hybrid mode. That is, teachers can simply click on a student's name and see the student's work at that point in time, whether they are face-to-face or remote.

Learning Analytics From SPIN

Teachers and educational researchers can access learning analytics on SPIN to assess student learning processes and outcomes. Teachers can access three types of learning analytics related to student behaviors. Teachers can query the *changes* that students make on SPIN, such as when they move backwards to another tab rather than forwards, or when they change an answer to a question, a graph, or an analysis choice. This type of analytics conveys when students have corrected their work as well as the quality of their corrections. Teachers can also query learning analytics for *help-seeking* behaviors of students, such as when they clicked the "I'm stuck" button or messaged the teacher or another group throughout an investigation and the nature of their difficulties. Teachers can also find out through the learning analytics which SPAARC prompts were used and how long the student hovered over the icon. These usage statistics can help teachers identify productive and less productive learning behaviors.

In addition to student learning analytics, information is also gathered about teacher behaviors. As discussed earlier, teachers can edit a lesson from the Global Marketplace or create their own lesson. From the learning analytics, researchers can identify changes that teachers made to a Global Marketplace investigation or to the SPAARC prompts, which can indicate the types of CT practices that teachers use to support data practices. SPIN also gathers learning analytics about when teachers view student work in progress or send messages to students to support them. Not only will the learning analytics help teachers develop a better understanding of how students proceed through data practices in an investigation, but they can also help educational researchers examine important issues regarding the design and implementation of data practices in science lessons.

CONCEPTUAL FRAMEWORK OF SPIN

SPIN is a technological workspace that integrates teacher-generated science investigations (i.e., biology, chemistry, earth science, and physics content area lessons) within a structure of data practices (e.g., data creation, data preparation), CT (e.g., decomposition, pattern recognition), and SRL principles (e.g., self-monitoring, self-evaluation). Although SPIN pulls from theory and research across these areas, the promotion and cultivation of students' SRL skills represents the central, integrative concept. SPIN is grounded in several theoretical perspectives (i.e., social cognitive theory, self-regulation theory) as well as emerging areas of research across educational, technological, and psychological domains (Bandura, 1997, Peters-Burton et al., 2018; Zimmerman 2000). Broadly speaking, social cognitive theory emphasizes the reciprocal relations among personal (e.g., beliefs), behavioral (e.g., strategy use), and contextual factors (e.g., technology, teacher or peer collaboration) and underscores that students' regulatory processes emerge and develop within these contexts (Bandura, 1997). Thus, how learners think and act can influence the environments in which they learn, but those environments can also impact students' learning and SRL processes.

Two models emanating from social-cognitive theory are central in our conceptualization of SPIN as an SRL-supporting system: a three-phase model of SRL (i.e., forethought, performance, self-reflection) and a four-level model of SRL development (Zimmerman, 2000). Social cognitive theorists define SRL as a goal-directed, cyclical process through which individuals deploy specific learning tactics and strategies to attain personal goals (Cleary, 2018; Zimmerman, 2000). Based on Zimmerman's (2000) three-phase cyclical model, sophisticated self-regulated learners set clear and tangible goals and plans (i.e., forethought), effectively use powerful strategies to enhance learning (i.e., performance phase), and understand when and how to adapt their strategic approaches to learning when encountering challenges (i.e., self-reflection phase). SRL is considered a type of cyclical feedback loop in that individuals use self-monitored information or feedback from external sources (e.g., teachers, peers, technology) to refine or adapt their strategic approaches to learning. Figure 4 illustrates the reciprocal and dynamic interaction between contextualized sources (i.e., SPIN features, peers, and teachers) and students' engagement in cyclical, regulatory thinking and action. In essence, SPIN creates opportunities for students, their peers, and teachers to interact in meaningful ways to optimize student SRL and learning. It is through exposure to the various SPIN features that individual students develop more effective goals and plans while learning, generate useful feedback, and refine and adapt their use of CT and data practice skills.

Another conceptual model that underlies the operation of SPIN is the four-level model of SRL and strategic skill development. Broadly speaking, this model explains the developmental process through which individuals become strategic and regulated learners. It emphasizes the gradual progression from reliance on social or external influences (i.e., observation and emulation levels) in the early phases of learning to a more self-directed and internal process over time (i.e., self-control and self-regulation levels; Zimmerman, 2000). Relative to SPIN, this model explains how teacher feedback and peer collaboration (i.e., social influences) and the various SPIN features and prompts (i.e., technological influences) predominate during the early stages of learning but can be reduced over time as students master their skills. Thus, when students initially engage in science investigations using SPIN, teachers will structure SPIN lessons with high frequencies and access to prompts or scaffolds about CT and data practices (i.e., *observation level*). They will also emphasize a feedback-oriented learning environment to help students practice and adapt their use of CT and data practices tactics (i.e., *emulation level*). As students become more skilled in using these skills, teachers can customize SPIN to reduce the level or types of scaffolds and feedback (i.e., *self-control level*) so that students eventually become more autonomous and self-directed learners (i.e., *self-regulated level*; Cleary et al., 2020; Cleary & Zimmerman, 2004).

To understand the conceptual foundation of SPIN, one must also recognize the close and nuanced link between CT practices (i.e., decomposition, pattern recognition, abstraction, and algorithmic think-

Figure 4. Reciprocal Interactions between Contextual Factors and Student SRL Processes

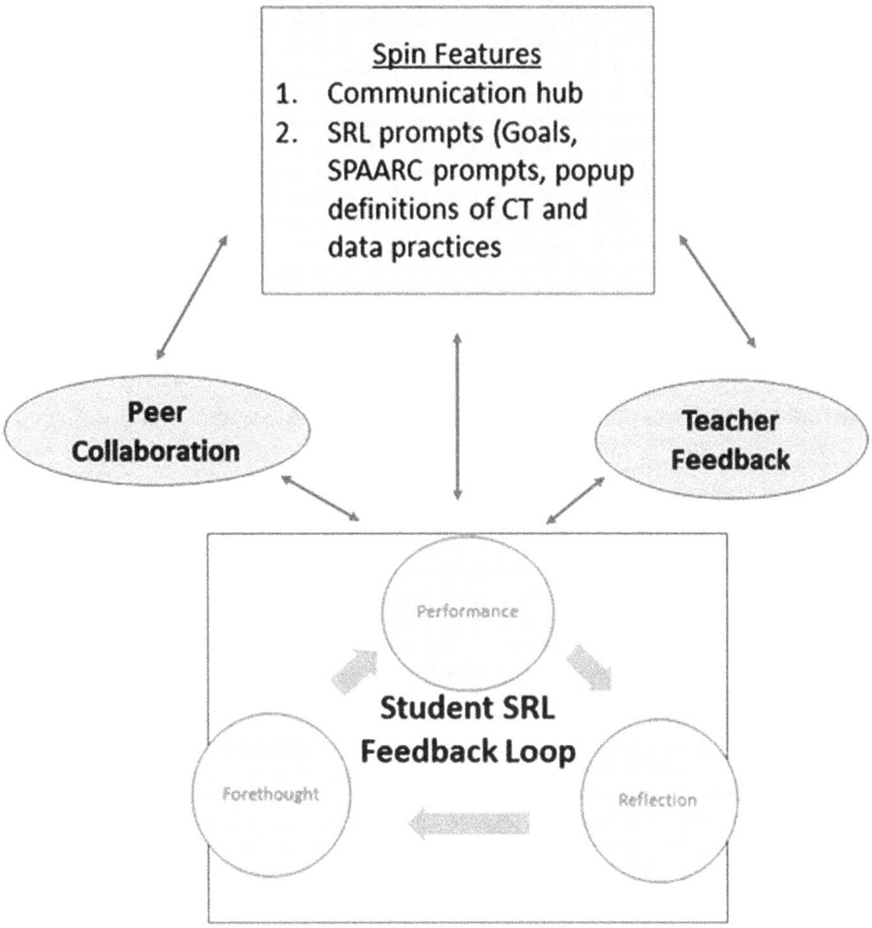

ing) and SRL. Peters-Burton et al. (2018) expanded on Wing's concept (2006) of CT by discussing this construct as a series of goal-directed processes that enable individuals to think recursively, reformulate perceptions of a problem, model relevant aspects of problems, and to draw conclusions about how to create automated solutions to problems. As mentioned previously, we use the acronym GAT-P (Goals, Aims, Tactics, and Prompts) to explicitly represent the integration of CT, SRL, and data practices within the SPIN technological structure (see Figure 5). Framed in a backwards design, through customizable prompts (P), students will learn to use one or more tactics (T) to reach the aims (A) of one or more target CT skills. It is through effective engagement and regulatory use of CT skills (i.e., decomposition, abstraction, pattern recognition, and algorithmic thinking) that students will be able to reach the target goal (G) for the five data practices (i.e., creating, collecting, preparing, visualizing, and analyzing). The SPIN system promotes regulatory actions and thinking of students as they engage in the GAT-P process.

In sum, SPIN emphasizes the use of customizable prompts based on the SPAARC strategy. Students can use these SPAARC prompts to optimize their CT skills, which in turn, can enhance their data practices (see Figure 5). As a flexible part of the system, teachers can make decisions about the type and number of CT skills to target for each specific lesson. Thus, the GAT-P reflects the goal-directed, strategic process

Figure 5. Conceptual link between SPIN prompts, SPAARC, CT practices, and data practices

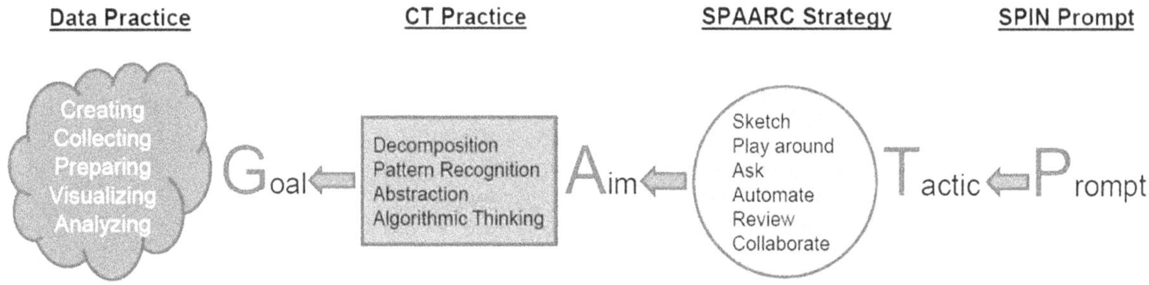

embedded within SPIN that can be modified and customized to address the teachers' specific content areas of interest and lesson plans and to most effectively address students' learning needs.

Implications

Twenty-first century science education looks different than it did in the past. Modern tools enable science educators to teach in ways that increasingly mimic scientific inquiry practiced by researchers. These tools enable the ability to work with data in ways that increase scope, accuracy, collection, and analysis of data. At the same time, there are increasing calls to bring CT into science education (Park & Green, 2019), as demonstrated in The Next Generation Science Standards (NGSS Lead States, 2013). SPIN is a tool that has been created to help answer the call for more authentic scientific inquiry in science education. Specifically, there are four key implications that come from its creation, which we outline below.

1. SPIN provides a tool for considering data practices inherent in scientific investigations from conception to presentation. To our knowledge, no such tool currently exists, so this provides fertile ground for more fully understanding how students work with data. Specifically, SPIN guides teachers to break down scientific inquiry by engaging with data in individual phases of data practices. Making these practices more explicit has the potential to make students more aware of how data can be created, collected, prepared, analyzed and visualized in every inquiry. Each of these data practices is a key part of every scientific inquiry. However, one stage or another is often missing from classroom experiences (e.g., starting with already collected or prepared data), possibly leading students to misunderstand the nature of data in research. Data practices are at the heart of scientific inquiry. Explicitly using this structure makes data practices explicit, whereas they may often be otherwise overlooked or taken for granted.
2. SPIN scaffolds scientific inquiry by incorporating CT throughout these data practices. Thus, rather than CT being "one more thing" for teachers to think about, SPIN depicts CT as a tool for scaffolding inquiry, encouraging students to better break down complex problems, identify patterns throughout the process (not just during analysis, which is when pattern identification is most obvious), abstract the key elements of their projects, and learn to think algorithmically about their data with the potential for creating generalizable solutions from their scientific inquiries.
3. SPIN encourages teachers to be strategic and intentional in enhancing students' strategic and regulated behaviors by using SPIN features, such as selectively using SPAARC prompts to guide SRL processes and to enable students to better manage their own learning. SPAARC tactics are

grounded in the GAT-P framework, which allows for making explicit the Goals and Aims of each phase of scientific inquiry and the approaches for reaching those goals. Furthermore, through built-in learning analytics, SPIN will enable us to better understand which SRL practices are more common amongst successful and unsuccessful students; knowledge that could prove helpful for enhancing or optimizing the key scaffolds for student learning.

4. The convergence of SRL + CT + data practices can create a synergistic experience that enables teachers to integrate modern scientific practices with established theory linked to student success. Little is known about how science teachers actually integrate CT and SRL into existing lessons. The learning analytics built into SPIN can provide researchers and educators a way to identify learning patterns in relation to student behaviors while dealing with data in computational ways. Identifying these learning patterns could potentially lead to other tools, models, or frameworks that might guide our understanding of how students successfully and unsuccessfully think computationally to create and interpret scientific data.

CONCLUSION

There are increasing calls to integrate CT and SRL into existing science practices. SPIN is a technological and data-based tool that provides explicit CT and SRL scaffolds and prompts that focus student attention on a progression of data practices central to scientific inquiry. Teachers can create investigations within the SPIN system and use the existing CT and SRL prompts to enhance students' metacognitive practices that direct their learning. As students engage with each data practice, they are encouraged to think computationally by decomposing the goal of each data practice, identifying patterns in the data, abstracting the most relevant data to their question, and creating algorithms to explain and explore data. In this way, CT and SRL integration facilitates scientific inquiry rather than becoming "one more thing" for teachers to implement.

REFERENCES

Araka, E., Maina, E., Gitonga, R., & Oboko, R. (2020). Research trends in measurement and intervention tools for self-regulated learning for e-learning environments—Systematic review (2008–2018). *Research and Practice in Technology Enhanced Learning*, *15*(6), 1–21. doi:10.118641039-020-00129-5

Azevedo, R., Witherspoon, A., Graesser, A., McNamara, D., Chauncey, A., Siler, E., . . . Lintean, M. (2009). MetaTutor: Analyzing self-regulated learning in a tutoring system for biology. In Artificial intelligence in education (pp. 635-637). IOS Press.

Bandura, A. (1997). *Self-efficacy: The exercise of control*. Macmillan.

Berglas-Shapiro, T., Eylon, B.-S., & Scherz, Z. (2017). A technology-enhanced intervention for self-regulated learning in science. *Teachers College Record*, *119*(13), 1–26. doi:10.1177/016146811711901301

Broadbent, J., Panadero, E., & Fuller-Tyszkiewicz, M. (2020). Effects of mobile-app learning diaries vs online training on specific self-regulated learning components. *Educational Technology Research and Development*, *68*(5), 2351–2372. doi:10.100711423-020-09781-6

Cleary, T. J. (2018). *The self-regulated learning guide: Teaching students to think in the language of strategies*. Routledge. doi:10.4324/9781315693378

Cleary, T. J., Gregory, A., Kitsantas, A., Slemp, J., & Panish, D. (2020). Implementation of self- regulated learning-focused interventions in schools: The intersection of sociocultural factors and social validity principles. In G. A. D. Liem & D. M.McInerney (Eds.), Promoting motivation and learning in contexts: Sociocultural perspectives on educational interventions (pp. 143-172). Information Age Publishing.

Cleary, T. J., Velardi, B., & Schnaidman, B. (2017). Effects of the Self-Regulation Empowerment Program (SREP) on middle school students' strategic skills, self-efficacy, and mathematics achievement. *Journal of School Psychology*, *64*, 28–42. doi:10.1016/j.jsp.2017.04.004 PMID:28735606

Cleary, T. J., & Zimmerman, B. J. (2004). Self-regulation empowerment program: A school-based program to enhance self-regulated and self-motivated cycles of student learning. *Psychology in the Schools*, *41*(5), 537–550. doi:10.1002/pits.10177

Cleary, T. J., & Zimmerman, B. J. (2006). Teachers' perceived usefulness of strategy microanalytic assessment information. *Psychology in the Schools*, *43*(2), 149–155. doi:10.1002/pits.20141

Ferguson, R. (2012). Learning analytics: Drivers, developments and challenges. *International Journal of Technology Enhanced Learning*, *4*(5/6), 304–317. doi:10.1504/IJTEL.2012.051816

Graesser, A. C., Lu, S., Jackson, G. T., Mitchell, H., Ventura, M., Olney, A., & Louwerse, M. M. (2004). AutoTutor: A tutor with dialogue in natural language. *Behavior Research Methods, Instruments, & Computers*, *36*(2), 180–193. doi:10.3758/BF03195563 PMID:15354683

Grigal, M., Neubart, D. A., Moon, S. M., & Graham, S. (2003). Self-determination for students with disabilities: Views of parents and teachers. *Exceptional Children*, *70*(1), 97–112. doi:10.1177/001440290307000106

Kitsantas, A., Baylor, A. L., & Hiller, S. E. (2019). Intelligent system support to optimize performance: Augmenting cognitive capacity and supporting self-regulation of critical thinking skills in decision making. *Cognitive Systems Research*, *58*, 387–397. doi:10.1016/j.cogsys.2019.09.003

Kitsantas, A., & Dabbagh, N. (2010). *Learning to Learn with Integrative LearningTechnologies (ILT): A Practical Guide for Academic Success*. Information Age Publishing.

Lee, L.-K., Cheung, S. K. S., & Kwok, L.-F. (2020). Learning analytics: Current trends and innovative practices. *Journal of Computers in Education*, *7*(1), 1–6. doi:10.100740692-020-00155-8

Ludlow, B. (2015). Virtual reality: Emerging applications and future directions. *Rural Special Education Quarterly*, *34*(3), 3–10. doi:10.1177/875687051503400302

Millar, R., & Osborne, J. F. (Eds.). (1998). *Beyond 2000: Science education for the future*. King's College London.

Millis, F. C., Butler, H., Wallace, P., Graesser, A., & Halpern, D. (2011). Operation ARIES! A serious game for teaching scientific inquiry. In Serious Games and Edutainment Applications (pp. 169–195). Springer London.

National Center for Educational Statistics (NCES). (2013). *The condition of education 2013*. Washington, DC: National Center for Educational Statistics. Retrieved from https://nces.ed.gov/pubs2013/2013037.pdf

National Research Council (NRC). (2010). Report of a workshop on the scope and nature of computational thinking. Committee for the Workshops on Computational Thinking, Computer Science and Telecommunications Board, Division on Engineering and Physical Sciences. Washington, DC: National Academies Press.

National Research Council (NRC). (2011). Report of a workshop on the pedagogical aspects of computational thinking. Committee for the Workshops on Computational Thinking, Computer Science and Telecommunications Board, Division on Engineering and Physical Sciences. Washington, DC: National Academies Press.

NGSS Lead States. (2013). *Next generation science standards: For states, by states*. The National Academies Press.

Park, Y. S., & Green, J. (2019). Bringing computational thinking into science education. *Korea Science*, *40*(4), 340–352. doi:10.5467/JKESS.2019.40.4.340

Penuel, W. R., & Gallagher, D. J. (2017). *Creating research-practice partnerships in education*. Harvard Education Press.

Peters, E. E., & Kitsantas, A. (2010). The effect of nature of science metacognitive prompts on science students' content and nature of science knowledge, metacognition, and self-regulatory efficacy. *Journal of School Science and Math*, *110*(8), 382–396. doi:10.1111/j.1949-8594.2010.00050.x

Peters-Burton, E. E., Cleary, T. J., & Kitsantas, A. (2018). Computational thinking in the context of science and engineering practices: A self-regulated learning approach. In D. Sampson, D. Ifenthaler, J.M. Spector & P. Isaias (Eds.), Digital technologies: Sustainable innovations for improving teaching and learning (pp. 223-240). Springer. doi:10.1007/978-3-319-73417-0_13

Peters-Burton, E. E., & Kysar-Mattietti, G. (2017). Building student self-awareness of learning to enhance diversity in the sciences. In D. D. Liston & R. Rahimi (Eds.), *Promoting social justice through the scholarship of teaching and learning* (pp. 270–296). Indiana University Press. doi:10.2307/j.ctt2005zc2.18

Putri, A. A. (2021). The effectiveness of Khan Academy as a science learning support to improve student's mastery of skills: Literature review. *Journal of Environmental and Science Education*, *1*(2), 52–56. doi:10.15294/jese.v1i2.50370

Rubacha, M., Rattan, A., & Hosselet, S. (2011). A review of electronic laboratory notebooks available in the market today. *Journal of Laboratory Automation*, *16*(1), 90–98. doi:10.1016/j.jala.2009.01.002 PMID:21609689

Schraw, G., Crippen, K. J., & Hartley, K. (2006). Promoting self-regulation in science education: Metacognition as part of a broader perspective on learning. *Research in Science Education*, *36*(1-2), 111–139. doi:10.100711165-005-3917-8

Wehmeyer, M. L., Agran, M., & Hughes, C. A. (2000). National survey of teachers' promotion of self-determination and student-directed learning. *The Journal of Special Education*, *34*(2), 58–68. doi:10.1177/002246690003400201

Weintrop, D., Beheshti, E., Horn, M., Orton, K., Jona, K., Trouille, L., & Wilensky, U. (2016). Defining computational thinking for mathematics and science classrooms. *Journal of Science Education and Technology*, *25*(1), 127–147. doi:10.100710956-015-9581-5

Wing, J. M. (2006). Computational thinking. *Communications of the ACM*, *49*(3), 33–35. doi:10.1145/1118178.1118215

Zimmerman, B. J. (2000). Attaining self-regulation: A social-cognitive perspective. In M. Boekaerts, P. Pintrich, & M. Zeidner (Eds.), *Handbook of self-regulation* (pp. 13–39). Academic Press. doi:10.1016/B978-012109890-2/50031-7

Chapter 8
Harnessing the Digital Science Education Revolution: Smartphone Sensors as Teaching Tools

Rebecca E. Vieyra
https://orcid.org/0000-0001-7720-6558
Vieyra Software, USA

Colleen Megowan-Romanowicz
https://orcid.org/0000-0001-5614-2145
American Modeling Teachers Association, USA

Daniel J. O'Brien
Georgetown University, USA

Chrystian Vieyra
Vieyra Software, USA

Mina C. Johnson-Glenberg
Arizona State University, USA

ABSTRACT

Students regularly struggle with visualization and modeling of physical phenomena. Incorporating smartphones into teaching strategies has the potential to improve learning outcomes while simultaneously augmenting students' digital literacy skills in an applied context. This chapter discusses two smartphone applications developed by the authors and the accompanying research efforts in understanding how students might learn with this technology. The first app pairs augmented reality (AR) with smartphone magnetometers to visualize three-dimensional magnetic fields in space. The second uses light detection and ranging (LiDAR) technology available on modern iPhone models to plot position-time and velocity-time graphs based on users' motions. Each was developed with the support of students, teachers, software developers, and educational psychologists. In this chapter, the authors share their perspectives and other recommendations for the use and development of similar technologies to improve learning.

INTRODUCTION

Pressure is increasing on U.S. educators to match the growth of digitization in the scientific endeavor.

DOI: 10.4018/978-1-6684-5585-2.ch008

In 2017, the National Science Foundation (NSF) identified digital literacies within its *10 Big Ideas* for prioritizing federal investment. Among these ideas are *The Future of Work at the Human-Tech Frontier* and *Harnessing the Data Revolution*. Both these priority areas emphasize the need for more effectively bridging classroom learning with workforce readiness. Further, a variety of educational initiatives promote learners' development and use of models, analysis, and interpretation of data (National Research Council, 2012a; NGSS Lead States, 2013), which researchers have identified as a pathway to achieving digital literacy (France, 2018; Hallström & Schönborn, 2019). However, few classrooms are equipped with technologies that prepare students for digital workforce scenarios described by the NSF.

To meet the need for incorporating workforce-relevant technology into science classrooms, teachers and students can now turn to the everyday devices in their pockets—smartphones. Globally, 6.4 billion people use smartphones (Statista, 2022); they are available even in regions of high poverty and low resources and are the most widespread type of mobile device even in developing countries (Pew Research Center, 2019). Within the U.S., 85% of adults ages 18+ own a smartphone (Pew Research Center, 2021), and more than half of parents say that they gave their children their own phone as early as ages 9-11 (Pew Research Center, 2020). By virtue of their widespread availability, smartphones hold potential as game-changing tools for scientific learning, not just in the U.S., but globally as well.

This chapter envisions a future in which educators effectively employ smartphones to help students reach science and digital literacy learning goals. To illustrate this vision, this chapter will explore the general potential of mobile sensors and then describe two specific mobile sensor-based educational projects supported by the NSF: *Magna AR* and *LiDAR Motion*. The authors will share the results of their proto-design guidelines and early-stage research, including suggestions for technology designers and implementers to consider both learning context and instruction scaffolding.

BACKGROUND

The use of digital tools to teach science grew out of educational research on microcomputer-based laboratories (MBLs) (Reddish et al., 1997; Thornton, 1987; Thornton & Sokoloff, 1990). MBLs, produced by companies such as Vernier, PASCO, and Texas Instruments, use specialized software and probeware such as digital motion sensors, force meters, and thermometers to collect data. Research on using these tools demonstrated their pedagogical affordances, including system simplification, real-time data display, automated data collection, increased student control, and easy data display transformation (Anderson & Wall, 2016; Struck & Yerrick, 2010). Further research has focused attention on game-based devices, including those that use embodied motion-capture such as the Wii and mixed reality (Lindgren & Johnson-Glenberg, 2013) and the Hololens (Georgiou & Ioannou, 2019; Walkington et al., 2022). However, MBL technologies are only accessible to students while in the classroom, and game-based technology peripherals like Kinect and head-mounted augmented or video reality displays are rarely available in schools. Further, these technologies are only infrequently used by professional scientists.

In contrast, new smartphone sensors and evolving software can prepare students for real-world problems they will face as part of the future workforce. Research professionals in science (Cartwright, 2016) and engineering (Alexander, 2015) have touted the utility of smartphones for doing their everyday work. Most smartphones include accelerometers, gyroscopes, light meters, magnetometers, and microphones, and many higher-end phones possess sensors as varied as ambient pressure and temperature sensors, UV meters, and more. Research literature details how smartphone data collection can be used for applications ranging from identifying enzymatic reactions (Zarzar et al., 2017) and improving the design of optical gyroscopes (Srivastava et al., 2016) to movement analysis (Ellamil et al., 2016) and aircraft user interfaces design (Avsar, 2015).

An important utility of smartphones within the classroom rests in the flexibility afforded by their suite of sensors and supporting software. Science teachers perennially struggle to budget for classroom technology that becomes obsolete within years as other forms of technology continue to develop. Smartphones are regularly supported by software updates, and researchers have demonstrated the high precision of many of their sensors, including accelerometers and gyroscopes (Almazán et al., 2013; Bittel et al., 2016), LiDAR (Luetzenburg et al., 2021), barometers (Wu et al., 2015), and light sensors (Xu et al., 2015). A few smartphones might therefore replace cabinets filled with carts, color and light sensors, magnetic field sensors, pressure sensors, and accelerometers. Such devices can simplify inventory management and shorten the time teachers spend preparing for labs. Smartphones thus demonstrate the opportunity to make accessible to students what was previously only available to scientists or facilities with expensive, specialized equipment.

Another important aspect of mobile devices is also their potential for taking advantage of body-based learning. Embodied cognition is the theory that bodily experiences influence cognition, and numerous studies demonstrate that activities that incorporate whole body movement, when compared to non-embodied activities, lead to better learning gains across numerous STEM domains including mathematics (Abrahamson, 2020; Alibali & Nathan, 2012), physics (Anastopoulou et al., 2011; Johnson-Glenberg & Megowan-Romanowciz, 2017), astronomy (Lindgren & Moshell, 2011; Lindgren et al., 2016), and geology (Johnson-Glenberg, et al., 2011). Research in other fields have also suggested that movement-based learning techniques result in better learning retention (Mathias et al., 2022; Sadoski & Paivio, 2013). As wireless devices, smartphones are readily carried by students both around the classroom and even outside for whole-body experiences in places like elevators and amusement parks (Vieyra & Vieyra, 2014, 2022; Vieyra, 2018; Vieyra et al., 2015). The work presented in this chapter is grounded in the body's perceptions and actions in the world that the body inhabits (Barsalou, 2008; Duijzer et al. 2019b; Johnson-Glenberg et al., 2014; Johnson-Glenberg, 2018; Kita et al., 2017; Wilson, 2002).

Smartphones have been embraced by science educators for almost a decade, allowing teachers to make use of game-based learning (Gao et al., 2020), smartphone sensors (Monteiro & Martí, 2022; Vieyra et al., 2015, 2020), and digital tools such as audio and video recording, internet browsers, digital graphing calculators, collaborative documents, and much more (González, et al., 2017). Despite the robust literature on their adoption on a teacher-by-teacher basis, the research on their instructional effectiveness as learning tools is more limited. Even so, their use increases significantly in distance learning circumstances (O'Brien, 2021). For example, COVID-19 challenged teachers to provide remote learning opportunities and made it substantially more difficult to use laboratory experiences as what should be "the central approach for teaching and learning science" (National Academies of Science, Engineering, and Mathematics, 2019, Summary).

Within the U.S. in 2019–2020, nearly 55.1 million K-12 students were expected to learn remotely (EdWeek, 2020), and science education undoubtedly suffered. According to a study by the National Academies of Science, Engineering, and Medicine (2020), "only 38% of science teachers reported that students had been engaged in experiments or investigations through remote learning" (p. 1), meaning that students had far fewer opportunities than before to engage in lab experiences. Multiple educational stakeholders addressed this challenge by creating guidelines for remote science teaching (Council of State Science Supervisors, 2020), especially resources for distance teaching and learning of lab courses

(Strubbe & McKagan, 2020), including through the use of smartphones (O'Brien, 2021). The fact that smartphone data collection apps are already viewed as a resource for remote learning and are being used across education and in the workforce makes them an especially appropriate type of technology in which to make further investments in educational research.

TWO CASE STUDIES: MAGNA AR AND LiDAR MOTION

This chapter outlines two case studies of innovative digital technologies for science education that were carried out by the authors. These illustrate two separate projects developed through collaborations among Vieyra Software (developers of the *Physics Toolbox Sensor Suite* smartphone application), the American Modeling Teachers Association (a professional society dedicated to math and science modeling pedagogy), and a group of educators, scientists, and educational researchers. The first case, which describes the app *Magna AR*, makes use of novel augmented reality (AR) frameworks to visualize three-dimensional magnetic fields in space using real data. In the second case, the app *LiDAR Motion* uses light detection and ranging (LiDAR) technology available on modern iPhone models to plot position and time graphs based on user' motions. The authors' experiences with *Magna AR* contributed to the embedded scaffolding built into the design of the *LiDAR Motion* app.

Technology to Teach About Magnetic Fields: Magna AR

Visualizing fields is a critical skill of a technology-enabled workforce. While students have spent more than a century learning about fields in the context of gravity, electricity, and magnetism, the *Framework for K-12 Science Education* (National Research Council, 2012a) has attempted to re-center much of physics education on this foundational—yet highly abstract—concept. In tandem, the ability to work with computers that can sense and visualize invisible data like that of fields is becoming increasingly important for the U.S. workforce (National Academies of Science, Engineering, and Medicine, 2017, pg. 50).

Field visualizations are essential to many occupations. In turn, members of the National Research Council (2013) have asked educational stakeholders to help prepare a "highly trained workforce that is specialized in the knowledge and expertise of magnet design and construction" (164). Understanding the three-dimensional nature of magnetic fields and their interactions with electric fields—not just specialized understandings of a specific tool—is necessary for a well-prepared workforce. Magnetic fields in particular are used in tasks as diverse as mapping geomagnetic fields for coal mining, studying high-energy particles in accelerators, manufacturing hybrid vehicles, conducting biomedical imaging, recycling waste, designing household appliances, studying space weather, and doing the daily work of electricians, plumbers, and ditch diggers. Given the ubiquity of magnetic fields in the environment, individuals in the trades, as well as those with advanced science or engineering degrees, are equally likely to encounter fields in their daily work.

Preliminary research on helping students conceptualize the three-dimensionality of magnetic fields in magnetic resonance imagining (MRI) and heliophysics suggests that using mobile devices with embodied activities that support understandings about three-dimensionality can lead to benefits in understanding over two-dimensional printed images or pure computer simulation (Benli et al., 2015; Lopez & Hamed, 2004; R. Vieyra & C. Vieyra, 2022). This literature supports the findings of later studies that three-dimensional AR visualizations support learning about the concepts of fields generally (Billinghurst &

Duenser, 2012; Dori & Belcher, 2005), and Yoon (2014) posits that augmented reality has attributes that can make invisible, abstract concepts–such complex data patterns–both visible and concrete.

App Description and Design

Magna AR (https://www.magna-ar.net/) is an application for both Android and iOS smartphones that makes use of the device's internal accelerometer, gyroscope, camera, and magnetometer to place three-dimensional visual overlays in space at the point of magnetic field measurement. By moving throughout space, the user can plot the strength and direction of the field. By clicking on the screen, the user anchors a single vector at the location of the smartphone sensor based on the measurement of the magnetometer. The vectors stay in place as the user moves around and changes perspective. Moving the phone in space is an embodied act. By observing how students manipulate their phones, observers can get a sense of how the user is predicting the nature of a field, including its location and points of interest (such as at the poles, where magnetic field strength and direction change dramatically in a small space). Figure 1 illustrates a variety of magnetic field visualizations in two and three dimensions. With *Magna AR*, users can clearly distinguish the north and south poles of magnets, differences in field strength, and 3-D patterns of field lines around highly complex arrangements of magnets.

User Testing and Feedback

To test the utility of this new *Magna AR* technology, the authors carried out early-stage user experience testing with a variety of ed-tech experts, physicists, and physics educators. Relying on the principles of embodied cognition, the authors observed that the use of the technology supported four elements of field conceptualization for novice physics learners as well as experienced physicists. First, most users relied on classical emission theory (Smith, 2018), expecting the smartphone to visualize field vectors in places that they targeted with their camera, but without moving the smartphone sensor through the space of the desired visualization. The tool supported the users' ability to self-correct and refute this erroneous theory. Second, the tool helped users to understand the pervasive nature and directionality of the Earth's magnetic field, with even experienced physicists expressing surprise that the environmental magnetic field is not horizontal to the surface of the Earth. Third, users were often surprised by the sharp decrease in the strength of magnetic fields produced by permanent magnets with respect to the sensor's distance from it, perhaps incorrectly assuming that the Earth's environmental magnetic field was substantially weaker than it is. As such, users' first experiences were to identify the boundaries at which they could visualize a field around a magnet. Finally, the authors observed that users began to express themselves using conceptual language, more reflective of their improved intuition than a reliance on propositional understandings such as equations or symmetry for well-known scenarios.

Beyond potential conceptual affordances, the authors also wanted to explore teachers' perceptions about the practical utility of the *Magna-AR* app. Introductory physics teachers who used the application as part of a focus group shared their experiences through a multimedia portfolio documenting their explorations. Focus group participants noted that the app provided a definite improvement over the use of iron filings to demonstrate field lines—an experience that includes placing a paper on top of a bar magnet and shaking iron filings on top of it to see how they orient to the magnet's field. Participants noted that the app provided a combination of magnitude, direction, and depth information, which is otherwise impossible without a commercial sensor. Regarding the app's field visualization potential,

Figure 1. Magna AR Visualizations of Magnetic Field Lines Around a (a) Button Magnet, (b) Stack of Button Magnets, (c) Series of Magnets

(a)

(b)

(c)

one teacher reflected on the possibility of using fields as a foundation on which they can teach many physical principles:

I have been thinking about a way to make fields a central topic in physics. I think that it might be possible to bridge between mechanics and electricity and magnetism by creating a unit that studies the similarities and differences between gravitational and electromagnetic fields. ... The concept of energy being stored in these fields...would be really beneficial to students.

The app inspired this teacher to consider a decidedly non-traditional approach to teaching physics, one which is more in line with both the views of professional physicists and the vision of the *Framework for K-12 Science Education* (National Research Council, 2012a): placing fields as the foundation of understanding. He saw the opportunity to illustrate Faraday's insights (how magnetic fields curve around current-carrying wires), something widely studied in physics, but typically only theoretically visualized. Never before has there been such an accessible tool for this application as that which can be found in students' pockets.

Testing the Technology in Classroom Learning Environments

This section describes how several of the authors explored how *Magna AR* might be used in classroom environments. The purpose of this exploratory investigation was both to note students' interaction patterns with the technology and to determine how it might have influenced their conceptualization of fields. The authors piloted interventions with high school physics students over three days in two different classrooms: a 9th-grade 'Physics First' classroom and a traditional sequence high school physics class for 11th and 12th-grade students.

Vignette 1: Self-Led Learning With Magna AR

During the first intervention, students worked in pairs with a provided smartphone and were supported by a self-paced, guided online tutorial prepared by the classroom teacher. This experience allowed the authors to observe how a novice teacher chose both to implement the *Magna AR* technology and to passively observe students' reactions to the activity.

The classroom teacher first introduced students to magnetic fields and the use of the app through a multi-part online tutorial spread over three class periods. First, students watched a video that demonstrated how the app worked. They then transitioned to a series of computer-mediated tasks, investigating Earth's environmental magnetic field, fields around magnets in various configurations, and the superposition of magnetic fields. For each task, students followed instructions about how to visualize the field around a magnet, captured the image, and answered a series of questions.

Students in this intervention displayed cooperation, but limited collaboration. Although the classroom teacher directed students to work collaboratively, typically only a single student in each pair used the smartphone with *Magna AR*. Occasionally, the 'observer partner' prompted the device user to do something different or to do something over again to get a better image. In a couple of cases, students asked their partner to give them the phone (and in both cases, the second user was able to produce better images than their partner). A single student decided to download the app onto his own device so that he and his partner could explore the field simultaneously. This one-to-one ratio led to a much more collaborative exploration for this pair of students, better images, and deeper conversation when arriving at a consensus about how to respond to the follow-up questions.

This intervention permitted a somewhat limited opportunity for deducing students' conceptual understandings of magnetic fields. For most pairs, it appeared to the researcher that students' goals were to get finished—to answer all the questions and to capture the required images so they could submit their assignment and spend the remaining class time chatting with their friends or doing homework. Throughout the intervention, the researchers observed very little extemporaneous exploration. Other than the pair of students who used the teacher-provided device and who downloaded the app onto their own smartphones, students rarely compared their field images with those of other student pairs.

Vignette 2: Teacher-Mediated Learning With Magna AR

For the second intervention, the authors worked with a teacher who was substantially more experienced in supporting students' open inquiry and classroom discourse because of her training in Modeling Instruction (Hestenes, 1987; Wells, 1995). To better scaffold student inquiry, the authors worked with the classroom teacher to prepare teacher-mediated exercises, with special emphasis on building comfort with the use of the *Magna AR* app.

On the first day, the teacher asked small groups of students to prepare a whiteboard with what they already knew about fields, followed by an introduction to the *Magna AR* app. Having just completed a unit on electric fields, most groups mentioned charges, electric field vectors, and forces, and drew parallels to what they knew about magnetic and gravitational fields. After watching a video tutorial about the functionality of the *Magna AR* app, students were given the option to either use a teacher-provided smartphone or download the app onto their own devices. Notably, all students chose to download the app onto their personal smartphones.

Working in pairs, students engaged in a guided exploration of Earth's background field and subsequently looked at the fields around magnets in various orientations. Motivated by a challenge to produce the best-in-show screen capture of a magnetic field, students engaged in conversation between individuals and among groups. Students remade images repeatedly until they produced an image that they felt was share-worthy. Compared with the first intervention, the authors observed substantially more student talk, collaboration, and engagement with the app itself. Students in this second intervention were much more likely to take magnetic field data measurements and fill their images with field vectors. However, it should be noted that the images they took, while full of vectors, did not always give a complete picture of the magnetic field they were attempting to visualize. It was not clear that students understood the dipole nature of the magnetic field. Most imaged the field near only one pole of the object they were exploring.

Over the following two days, students worked through most of a structured lesson prepared by the authors: *Modeling Planetary Magnetism* (available from magna-ar.net/lesson-ideas). This lesson asked students to apply their understanding of the Earth's localized magnetic field in the context of a global perspective. To do so, students created visualizations of a magnetic field around a "model Earth" crafted from a small ball into which magnets were placed to represent north and south magnetic field sources at the geographic poles (Figure 2). Students then considered the orientation of the magnetic field at various locations around the Earth from various perspectives.

At the end of the intervention, the teacher again placed students in small groups and asked them to illustrate their understanding of fields on whiteboards. Interestingly, even after working through this activity in pairs, most groups' field pictures were of a single magnetic pole. And on their post-activity whiteboards, only one group of students mentioned poles, although they incorrectly wrote that the field pointed inward from each pole toward the center of the earth.

Conclusions About Magna AR in Educational Contexts

The authors' observations of these two interventions reveal several tentative early findings, including the importance of teacher scaffolding of inquiry explorations and students' sense of ownership of the technology. Despite being so-called "digital natives," students sometimes struggled to produce uniform magnetic field visualizations that would support a holistic understanding of a dipole magnetic field. Students engaged in more authentic inquiry in the second intervention, in which the teacher provided some grounding in using the technology to explore magnetic fields, while also generally promoting an atmosphere that valued open inquiry and knowledge synthesis.

It is clear that the *Magna AR* tool, which involves both novel visualization and foreign science concepts, cannot simply be picked up by students and learned on their own, even with detailed tutorials used

Figure 2. Magna AR Visualization of The Field of a Model Earth Composed of a Ball Containing Small Button Magnets, Showing Field Lines Emanating From the North Magnetic Pole

alone or with a peer. Just-in-time support is needed from teachers. And, before the app is introduced, simple demonstrations should be done of magnetic field effects, (e.g., magnets pulling and pushing on each other, pulling on magnetic materials such as paper clips, or the attractive effect on a ball bearing rolling past a stationary magnet that bends its path). Additionally, the authors observed that students who used their personal smartphones were substantially more engaged in inquiry and rich conversation about the magnetic field concept. This preference deserves study to see if it applies more broadly to classroom technology use.

One important caveat applies to both testing conditions: magnetic fields are rarely taught as a single or three-day exploration in a high school physics class. They are more typically the subject of a two-week instructional unit, whether or not they are given a rigorous mathematical treatment. As such, the authors did not expect conceptual mastery. What they observed during these two interventions provided useful guidance for introducing the app to students and troubleshooting common problems that can arise in the classroom, allowed them to see the constraints and affordances of various tasks and learning environments, and suggested additional tasks and activities that could be used to develop the magnetic field concept more fully.

Technology to Teach About Movement: *LiDAR Motion*

The authors' second case study explores the use of augmented reality (AR) and the motion-sensing capabilities of another smartphone app, *LiDAR Motion* (lidar-motion.net/). Building on the experiences of classroom use of *Magna AR*, including the perceived affordances of embodied cognition and the need for just-in-time scaffolds for students, the authors elected to create a gamified, data-verified tool to teach about data modeling.

Data modeling with graphs is foundational to science learning and careers. Both data modeling and visualization are paramount to understanding trends and graphs, skills that underlie most STEM fields. A pivotal study from the National Research Council (1991) emphasized that "mathematical modeling [has] an importance to economic competitiveness that is very large relative to the recognition given to these activities by the academic mathematical sciences community" (p. 89). Learners especially need to develop skills associated with quantitative data visualization literacy (DVL). DVL encompasses the ability to analyze and interpret visual data (Börner et al., 2019), including rates of change and graphing, something that is important for all learners in this digital age.

Students demonstrate persistent challenges in coordinating graphical representations with spatial and temporal information about objects in motion. They regularly struggle to connect graphs to both physical concepts and real-world events (Beichner, 1994; Duijzer et al., 2019a, 2019b; McDermott et al., 1987). Students often interpret graphs as pictures (Clement, 1985), such as by assuming that an upward-sloping line on a position-time graph suggests that an object is moving up a hill. Graphs with negative velocities (including downward sloping lines on position graphs and curves of any type in the bottom quadrants of velocity graphs) are especially difficult for students to understand (Goldberg & Anderson, 1989), as is possibly ascribed to students' conflation of position and velocity (Madsen et al., 2012; Trowbridge & McDermott, 1980). However, more recent research suggests that some of these difficulties may originate from a combination of limitations in mathematical or scientific understanding (Boote, 2014) with misunderstandings about physical context (Wemyss & van Kampen, 2013), illustrating the importance of a multidisciplinary, multimodal approach to learning about graphical representations.

Despite having detailed taxonomies of student struggles in learning about these concepts, many teachers are uncertain of how to enact pedagogies that address these struggles—a challenge that might be overcome with a tool that takes advantage of embodied cognition. Researchers have recently turned their attention to a variety of technologies to support student learning difficulties, including motion trackers (Anastopoulou et al., 2011), Kinect for Xbox One (Anderson & Wall, 2016; Johnson-Glenberg, et al., 2015; Johnson-Glenberg & Megowan-Romanowciz, 2017) and situated multimedia art learning laboratory (SMALLab) (Birchfield & Johnson-Glenberg, 2010; Johnson-Glenberg et al., 2009, 2014). Similarly, smartphones might be able to provide the kind of scaffolding, embodied experiences, and real-time visualizations that teachers need to help students better understand representations of motion.

App Description

LiDAR Motion, an application for Apple iOS devices—i.e., iPads and iPhones—was developed to address the gap in accessible technologies for teaching motion modeling. The app utilizes data from a device's LiDAR scanner to plot relative motion as the user moves in real-time. This scanner (a) generates an array of laser beams, (b) utilizes a highly sensitive photodetector to identify reflected laser pulses, and (c) subsequently calculates the pulses' time of flight to determine the smartphone's distance from an object.

As users approach or move away from this object, the *LiDAR Motion* app simultaneously plots relative position and/or velocity information on a graph for the user. A gamified set of tasks challenges users to match their motion to the shape of a predetermined curve by moving their body in particular directions at specific speeds (Figure 3), helping them to understand the meaning of slopes and y-intercepts on graphs.

Building on extant knowledge of students' learning challenges, the authors considered multiple mediating factors in their design. The app is unique in that it displays representations on the very screen that is being held and viewed. Research has affirmed that the degree of bodily involvement has a significant influence on understanding motion, especially when this bodily engagement correlates with the generation of representations such as graphs and motion maps (Duijzer et al., 2019a, 2019b). Ultimately, results demonstrated that "learning environments making use of students' own motion immediately linked to its representation were most effective in terms of learning outcomes" (Duijzer et al., 2019b, 597). The

Figure 3. Screenshots Show How the Various Challenge Stages of the Lidar Motion App's Graph Match Movement to and From the White Reticle in the Top Half of the Image, Modeling Variations in (a) Slope, (b) y-Intercept, and (c) Curvature to Their Physical Analogs—Velocity, Initial Position, and Acceleration

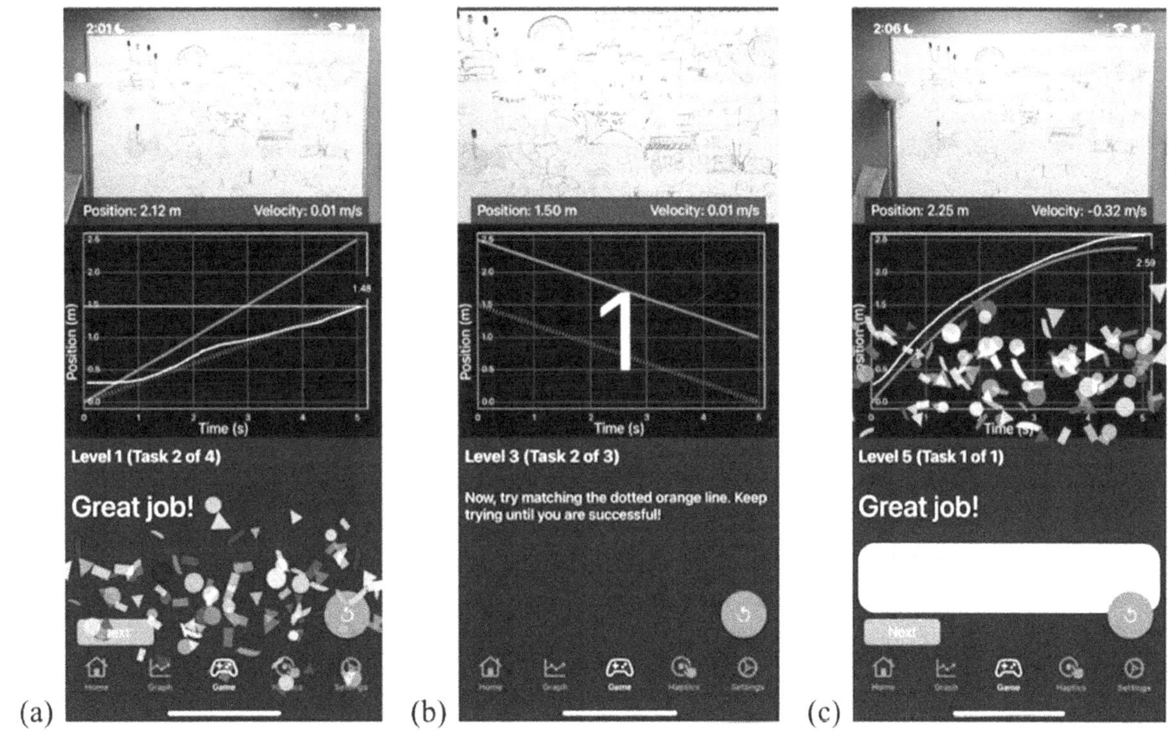

guiding factors used for designing the *LiDAR Motion* app are compiled in Table 1 alongside the resulting design considerations.

This new technology places in students' hands tools that we already know work—e.g., replacing sonic motion rangers with the smartphone, and traditional cart-and-track setups to model constant velocity or accelerated motion with their own bodies. The *LiDAR Motion* function matches these tools in range (*LiDAR Motion* app: 0.25–5 m, compared to the PASCO sonic motion detector: 0.15–4 m) and accuracy (resolution to the nearest mm). Moreover, it improves current technology accessibility: making learning available to all with access to such a smartphone. Learners can also take advantage of embodiment, as they use the app while holding the smartphone in front of them.

Table 1. Eight mediating factors for supporting embodied learning of motion graphing and the associated design considerations of the LiDAR motion visualizer application

Mediating Factors (From Duijzer, et al., 2019b)	**Design Considerations of the *LiDAR Motion* App**
1. Real-world context 2. Multimodality 3. Linking motion to graph 4. Multiple representations 5. Semiotics 6. Student control 7. Attention capturing 8. Cognitive conflict	1. Allows use of own mobile device in a home environment 2. Accounts for sight, imagining, motor actions, and haptic feedback 3. Incorporates whole-body movement and real-time graph overlays 4. Displays of position-time graphs and velocity-time graphs (including slope and y-intercept) 5. Makes use of multiple symbols (graphical, numerical) 6. Encourages users' free exploration and agency to choose target and movements 7. Gamifies the experience for real-time feedback and motivation 8. Embeds productive failure based on common student difficulties

User Testing and Feedback

LiDAR Motion's development and implementation were largely driven by user feedback about its functionality, interfacing, experience, and applications, including from focus groups of undergraduate and graduate students at a large university. Students who evaluated the app noted that it encouraged a more accessible style of learning. One student touched on how she was taught high school-level kinematics through the lens of a driving car, despite her not having driven a car until college. Numerous students expressed appreciation for the self-evaluation questions that followed challenge tasks (Figure 4) and the positive reinforcement they received after successfully completing the tasks (in the form of digital confetti, see Figure 3). They stated that the questions helped them associate tasks with physical concepts from class, while the reinforcement gave them a feeling of accomplishment in having mastered the concepts.

Lastly, these students' comments mirrored the beliefs the authors had about the app's ability to teach through embodiment as introduced above. For example, contrasting with cart-and-track learning, one noted

[w]hen you [emphasis added] are the thing that is moving slower, it's easier to understand," to which another student added, "we learned quicker than you would in a lecture setting because you're doing it.

Figure 4. Screenshots of Free-Response and Multiple-Choice Questions that Followed Motion Tasks in the LiDAR Motion App

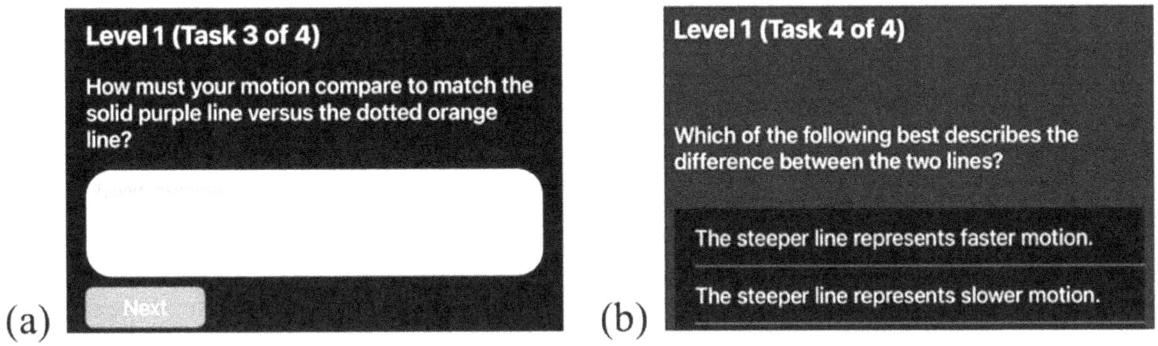

Conclusions About LiDAR Motion in Educational Contexts

LiDAR Motion enables users to "walk the walk," embodying the motion represented by lines and curves on a graph. By pairing gamified challenges with response questions, the app provides scaffolding, embodied experiences, and real-time visualizations and feedback to drive users' learning. While focus group participants had overall positive experiences with the app, full studies with high school and college classes will evaluate students' learning outcomes with the app as compared with existing technology.

SOLUTIONS AND RECOMMENDATIONS

Early experiences with content and education experts suggest that both *Magna AR* and *LiDAR Motion* are technologies with the potential for enhancing the learning experience. However, as was observed in both case studies, the means by which the technology and learning experience are enacted is just as important as the technology's features themselves. Individuals interested in exploring either of these technologies or using them with learners may freely download the apps, embedded within *Physics Toolbox Sensor Suite*, from the Apple App Store (for LiDAR Motion and *Magna AR*) or Google Play stores (for *Magna AR* only).

Magna AR proved to be a tool with few embedded scaffolds, making it difficult even for highly experienced teachers to support a full classroom of students. The highly abstract nature of magnetic fields, for which introductory physics students have limited grounding and researchers have limited understanding of student difficulties, as well as the free-form use of the tool as a sensor to map three-dimensional space, makes magnetism a difficult concept to scaffold (both within and beyond the technology). These challenges are further compounded by the fact that magnetic fields, as opposed to electromagnetic interactions, have not historically been a central component of the introductory physics curriculum in the U.S. The app's ongoing development might be better supported by investing additional efforts in studying student's prior knowledge of magnetic fields, crafting scaffolded activities with structured teacher facilitation, or even developing a data-verified gamified experience.

In comparison, the authors were able to design *LiDAR Motion* founded on well-documented student struggles. Unlike with magnetic fields, introductory physics students have often been exposed to studies of motion and graphical representations—even if they present naïve or even erroneous prior ideas about them. Building on this research, the authors were more carefully able to structure app scaffolds that support the building or modification of prior conceptualizations. Although *LiDAR Motion* has not yet been used in whole-class learning contexts, early experiences with student focus groups suggest that this new, gamified approach lends itself more easily to motivate students and offer just-in-time support than *Magna AR* on its own.

These two case studies, which illustrate new capabilities for visualizing and understanding scientific concepts, albeit with different levels of effectiveness, suggest a number of recommendations for others who might want to develop technology to be used to support science learning. First, it is important to consider the learning context in which the technology will be used, carefully matching it to user background and experience, including both teachers' and students' comfort and motivation levels. A critical part of this process is testing the technology in a large classroom, where one-on-one support is not as likely as in a research laboratory that might engage just one or two students at a time. Truly scalable solutions must meet the demands of a classroom filled with 25-30 students and the real time limitations imposed by school schedules. Second, it is necessary to consider what just-in-time scaffolding and motivational reinforcement should look like from both within the technology and by the instructors themselves. Technology use should always be a part of a unit of instruction, drawing on students' prior knowledge and helping them to achieve an essential learning goal. To do so, it might be necessary to support teachers' implementation of the educational technology through professional growth opportunities, such as workshops, detailed teacher guides, or coaching by the developers themselves.

FUTURE RESEARCH DIRECTIONS

The lessons learned from these two cases illustrate the importance of relying on discipline-based education research (National Research Council, 2012b) to create design principles that account for both technological and pedagogical scaffolds. Realizing that the strength of research varies considerably across topics within STEM disciplines, this also implies the need to better understand the psychology of student learning about different topics. With this in mind, the authors have begun to develop a diagnostic instrument for magnetic fields, the MAGnetic kNowledge Assessment (MAGNA); in contrast, multiple validated instruments already exist for assessing students' comprehension of motion graphs.

Additionally, educational designers need evidence-based guidelines for how to design using emerging technologies that will soon be in the classrooms (Johnson-Glenberg, 2018). Some for AR (Dunleavy, 2014) and VR have begun to merge (Johnson-Glenberg, 2019) but more are needed. For *Magna AR* and *LiDAR Motion* specifically, the authors have begun to explore the balance between providing digital tutorials and encouraging initial open play, managing visual and cognitive overload on individual app screens, and addressing very app- or topic-specific issues, such as the precision level for data verified-challenges (such as mapping body motion to a graph).

All of the aforementioned opportunities beg the research community to answer fundamental questions about learning: Do students learn more or different with smartphone technologies such as *Magna AR* and *LiDAR Motion*? How do these technologies impact student motivation, especially when devices are loaned or personal? What explicit connections to the workforce can this technology help educators to make?

CONCLUSION

Assuming technologies such as those presented in this chapter can be burnished enough to be useful to the everyday teacher and the everyday learning, other opportunities arise for possible research on their linkages to real-world digital literacies. It is now possible to speculate as to what types of digital literacy learning goals can be addressed with smartphones beyond fundamental science concepts, including specific skills desired by future employers. For example, magnetic field visualizations might be useful to visualize current-carrying wiring behind drywall, or a LiDAR-enabled smartphone could be attached to a robot for a mock remote reconnaissance mission. Using smartphones, students might be able to use sensors to contribute to crowdsourced research efforts with professionals, such as monitoring sound- or light pollution in a community.

Both *Magna AR* and *LiDAR Motion* are the products of a vision crafted among physics educators, software developers, and educational psychologists. The early-stage research presented in this chapter demonstrates the challenges and opportunities for using smartphone sensors to enhance student learning. As augmented reality, video reality, and LiDAR-enabled technology become more prevalent, existing tools sitting in students' pockets can help prepare students for an increasingly digital future.

ACKNOWLEDGMENT

This research was supported by the National Science Foundation grants 1822728 and 2114586; and the National Aeronautics and Space Administration grant NXX16AR36A.

REFERENCES

Abrahamson, D., Nathan, M. J., Williams-Pierce, C., Walkington, C., Ottmar, E., Soto, H., & Alibali, M. W. (2020). The future of embodied design for mathematics teaching and learning. *Frontiers in Education*, *5*(147), 147. doi:10.3389/feduc.2020.00147

Alexander, K. (2015). Instruments, sensors, and apps are turning smartphones into engineering measurement tools. *Mechanical Engineering (New York, N.Y.)*, *137*(9), 42–45. doi:10.1115/1.2015-Sep-3

Alibali, M. W., & Nathan, M. J. (2012). Embodiment in mathematics teaching and learning: Evidence from learners' and teachers' gestures. *Journal of the Learning Sciences*, *21*(2), 247–286. doi:10.1080/10508406.2011.611446

Almazán, J., Bergasa, L. M., Yebes, J. J., Barea, R., & Arroyo, R. (2013). Full auto-calibration of a smartphone on board a vehicle using IMU and GPS embedded sensors. *IEEE Intelligent Vehicles Symposium (IV)*, 1374-1380. 10.1109/IVS.2013.6629658

Anastopoulou, S., Sharples, M., & Baber, C. (2011). An evaluation of multimodal interactions with technology while learning science concepts. *British Journal of Educational Technology*, *42*(2), 266–290. doi:10.1111/j.1467-8535.2009.01017.x

Anderson, J. L., & Wall, S. D. (2016). Kinecting physics: Conceptualization of motion through visualization and embodiment. *Journal of Science Education and Technology*, *25*(2), 161–173. doi:10.100710956-015-9582-4

Avsar, H., Fischer, J., & Rodden, T. (2015). Target size guidelines for interactive displays on the flight deck. *34th Digital Avionics Systems Conference*. 10.1109/DASC.2015.7311400

Barsalou, L. W. (2008). Grounded Cognition. *Annual Review of Psychology*, *59*(1), 617–645. doi:10.1146/annurev.psych.59.103006.093639 PMID:17705682

Beichner, R. J. (1994). Testing student interpretation of kinematics graphs. *American Journal of Physics*, *62*(8), 750–762. doi:10.1119/1.17449

Benli, K. P., Dillman, B., Louelh, R., Poirier-Quinot, M., & Darrasse, L. (2015). Illustrating the quantum approach with an Earth magnetic field MRI. *European Journal of Physics*, *36*, 1–20. doi:10.1088/0143-0807/36/3/035032

Billinghurst, M., & Duenser, A. (2012). Augmented reality in the classroom. *Computer*, *45*(7), 56–63. doi:10.1109/MC.2012.111

Birchfield, D., & Johnson-Glenberg, M. (2010). A next gen interface for embodied learning: SMALLab and the geological layer cake. *International Journal of Gaming and Computer-Mediated Simulations*, *2*(1), 1–10. doi:10.4018/jgcms.2010010105

Bittel, A., Elazzazi, A., & Bittel, D. (2016). Accuracy and precision of an accelerometer-based smartphone app designed to monitor and record angular movement over time. *Telemedicine Journal and e-Health*, *22*(4), 302–309. doi:10.1089/tmj.2015.0063 PMID:26447774

Boote, S. (2014). Assessing and understanding line graph interpretations using a scoring rubric of organized cited factors. *Journal of Science Teacher Education*, *25*(3), 333–354. doi:10.100710972-012-9318-8

Börner, K., Bueckle, A., & Ginda, M. (2019). Data visualization literacy: Definitions, conceptual frameworks, exercises, and assessments. *Proceedings of the National Academy of Sciences of the United States of America*, *116*(6), 1857–1864. doi:10.1073/pnas.1807180116 PMID:30718386

Cartwright, J. (2016). Technology: Smartphone science. *Nature*, *531*(7596), 669–671. doi:10.1038/nj7596-669a PMID:27035011

Clement, J. (1985). Misconceptions in graphing. In L. Streefland (Ed.), *Proceedings of the Ninth Conference of the International Group for the Psychology of Mathematics Education* (pp. 369-375). IGPME.

Council of State Science Supervisors. (2020). *Supporting students' science learning during COVID-19 school closures*. https://stemteachingtools.org/assets/landscapes/Guidance-for-Supporting-Science-Learning-During-COVID-19-School-Closures.pdf

Dori, Y. J., & Belcher, J. (2005). How does technology-enabled active learning affect undergraduate students' understanding of electromagnetism concepts? *Journal of the Learning Sciences*, *14*(2), 243–279. doi:10.120715327809jls1402_3

Duijzer, C., Van den Heuvel-Panheuizen, M., Veldhuis, M., Doorman, M., & Leseman, P. (2019b). Embodied learning environments for graphing motion: A systematic literature review. *Educational Psychology Review*, *31*(3), 597–629. doi:10.100710648-019-09471-7

Duijzer, C., Van den Heuvel-Panhuizen, M., Veldhius, M., & Doorman, M. (2019a). Supporting primary school students' reasoning about motion graphs through physical experiences. *ZDM Mathematics Education*, *51*(6), 899–913. doi:10.100711858-019-01072-6

Dunleavy, M. (2014). Design principles for augmented reality learning. *TechTrends*, *58*(1), 28–34. doi:10.100711528-013-0717-2

EdWeek. (2020). *Map: Coronavirus and school closures in 2019-2020*. https://www.edweek.org/leadership/map-coronavirus-and-school-closures-in-2019-2020/2020/03

Ellamil, M., Berson, J., Wong, J., Buckley, L., & Margulies, D. (2016). One in the dance: Musical correlates of group synchrony in a club environment. *PLoS One*, *11*(10), e0164783. doi:10.1371/journal.pone.0164783 PMID:27764167

France, B. (2018). Modeling in technology education: A route to technological literacy. In M. J. de Vries (Ed.), *Handbook of technology education* (pp. 1–11). Springer. doi:10.1007/978-3-319-44687-5_45

Gao, F., Li, L., & Sun, Y. (2020). A systematic review of mobile game-based learning in STEM education. *Educational Technology Research and Development*, *68*(4), 1791–1827. doi:10.100711423-020-09787-0

Georgiou, Y., & Ioannou, A. (2019). Embodied learning in a digital world: A systematic review of empirical research in K-12 education. In P. Díaz, A. Ioannou, K. K. Bhagat, & J. M. Spector (Eds.), *Learning in a digital world* (pp. 155–177). Springer. doi:10.1007/978-981-13-8265-9_8

Goldberg, F. M., & Anderson, J. H. (1989). Student difficulties with graphical representations of negative velocity. *The Physics Teacher*, *27*(4), 254–260. doi:10.1119/1.2342748

González, M. Á., Martín, M. E., Llamas, C., Martínez, Ó., Vegas, J., Herguedas, M., & Hernández, C. (2017). Teaching and learning physics with smartphones. IGI Global.

Hallström, J., & Schönborn, K. J. (2019). Models and modelling for authentic STEM education: Reinforcing the argument. *International Journal of STEM Education*, *6*(1), 22. doi:10.118640594-019-0178-z

Hestenes, D. (1987). Toward a modeling theory of physics instruction. *American Journal of Physics*, *55*(5), 440–454. doi:10.1119/1.15129

Johnson-Glenberg, M. C. (2018). Immersive VR and education: Embodied design principles that include gesture and hand controls. *Frontiers in Robotics and AI: Hypothesis and Theory*, *5*(81), 1–19. doi:10.3389/frobt.2018.00081 PMID:33500960

Johnson-Glenberg, M. C. (2019). The necessary nine: Design principles for embodied VR and active stem education. In P. Díaz, A. Ioannou, K. K. Bhagat, & J. M. Spector (Eds.), *Learning in a digital world: Perspective on interactive technologies for formal and informal education* (pp. 83–112). Springer Singapore. doi:10.1007/978-981-13-8265-9_5

Johnson-Glenberg, M. C., Birchfield, D., Megowan-Romanowicz, M. C., & Snow, E. L. (2015). If the gear fits, spin it! Embodied education and in-game assessments. *International Journal of Gaming and Computer Based Simulation*, *7*, 40–65.

Johnson-Glenberg, M. C., Birchfield, D., Savvides, P., & Megowan-Romanowicz, C. (2011). Semi-virtual embodied learning-real world STEM assessment. In L. Annetta & S. C. Bronack (Eds.), *Serious educational game assessment* (pp. 241–257). Brill Sense. doi:10.1007/978-94-6091-329-7_15

Johnson-Glenberg, M. C., Birchfield, D. A., Tolentino, L., & Koziupa, T. (2014). Collaborative embodied learning in mixed reality motion-capture environments: Two science studies. *Journal of Educational Psychology*, *106*(1), 86–104. doi:10.1037/a0034008

Johnson-Glenberg, M. C., Birchfield, D. A., & Usyal, S. (2009). SMALLLab: Virtual geology studies using embodied learning with motion, sound, and graphics. *Educational Media International*, *46*(4), 267–280. doi:10.1080/09523980903387555

Johnson-Glenberg, M. C., & Megowan-Romanowicz, C. (2017). Embodied science and mixed reality: How gesture and motion capture affect physics education. *Cognitive Research: Principles and Implications*, *2*(4), 1–28. doi:10.118641235-017-0060-9 PMID:28603770

Kita, S., Alibali, M. W., & Chu, M. (2017). How do gestures influence thinking and speaking? The gesture-for-conceptualization hypothesis. *Psychological Review*, *124*(3), 245–266. doi:10.1037/rev0000059 PMID:28240923

Lindgren, R., & Johnson-Glenberg, M. (2013). Emboldened by embodiment: Six precepts for research on embodied learning and mixed reality. *Educational Researcher*, *42*(8), 445–452. doi:10.3102/0013189X13511661

Lindgren, R., & Moshell, J. M. (2011). *Supporting children's learning with body-based metaphors in a mixed reality environment* [Paper presentation]. *10th International Conference on Interaction Design and Children*, Ann Arbor, MI, United States.

Lindgren, R., Tscholl, M., Wang, S., & Johnson, E. (2016). Enhancing learning and engagement through embodied interaction within a mixed reality simulation. *Computers & Education*, *95*, 174–183. doi:10.1016/j.compedu.2016.01.001

Lopez, R. E., & Hamed, K. (2004). Student interpretations of 2-D and 3-D renderings of the substorm current wedge. *Journal of Atmospheric and Solar-Terrestrial Physics*, *66*(15-16), 1509–1517. doi:10.1016/j.jastp.2004.03.030

Luetzenburg, G., Kroon, A., & Bjørk, A. A. (2021). Evaluation of the Apple iPhone 12 Pro LiDAR for an application in geosciences. *Scientific Reports*, *11*(1), 1–9. doi:10.103841598-021-01763-9 PMID:34782692

Madsen, A. M., Larson, A. M., Loschky, L. C., & Sanjay Rebello, N. (2012). Differences in visual attention between those who correctly and incorrectly answer physics problems. *Physical Review Special Topics. Physics Education Research*, *8*(1), 010122. doi:10.1103/PhysRevSTPER.8.010122

Mathias, B., Andrä, C., Schwager, A., Macedonia, M., & von Kriegstein, K. (2022). Twelve- and fourteen-year-old school children differentially benefit from sensorimotor- and multisensory-enriched vocabulary training. *Educational Psychology Review*, *34*(3), 1739–1770. doi:10.100710648-021-09648-z

McDermott, L. C., Rosenquist, M. L., & van Zee, E. H. (1987). Student difficulties in connecting graphs and physics: Examples from kinematics. *American Journal of Physics*, *55*(6), 503–513. doi:10.1119/1.15104

Monteiro, M., & Martí, A. (2022). Resource letter MDS-1: Mobile devices and sensors for physics teaching. *American Journal of Physics*, *90*(5), 328–343. doi:10.1119/5.0073317

National Academies of Sciences, Engineering, and Medicine. (2017). *Information technology and the U.S. workforce: Where are we and where do we go from here?* National Academies Press.

National Academies of Sciences, Engineering, and Medicine. (2019). *Science and engineering for grades 6-12: Investigations and design at the center.* The National Academies Press.

National Academies of Sciences, Engineering, and Medicine. (2020). *Teaching K-12 science and engineering during a crisis*. National Academies Press.

National Research Council. (1991). *Mathematical sciences, technology, and economic competitiveness.* National Academies Press. doi:10.17226/1786

National Research Council. (2012a). *A framework for K-12 science education*. National Academies Press.

National Research Council. (2012b). *Discipline-based education research: Understanding and improving learning in undergraduate science and engineering*. National Academies Press.

National Research Council. (2013). *High magnetic field science and its application in the United States: Current status and future directions*. National Academies Press.

National Science Foundation. (2017). *NSF's 10 big ideas*. https://www.nsf.gov/news/special_reports/big_ideas/

NGSS Lead States. (2013). *Next generation science standards: For states, by states*. National Academies Press.

O'Brien, D. (2021). A guide for incorporating e-teaching of physics in a post-COVID world. *American Journal of Physics*, *89*(4), 403–412. doi:10.1119/10.0002437

Pew Research Center. (2019). *Mobile connectivity in emerging economies*. https://www.pewresearch.org/internet/2019/03/07/mobile-connectivity-in-emerging-economies/

Pew Research Center. (2020). *Parenting children in the age of screens*. https://www.pewresearch.org/internet/2020/07/28/parenting-children-in-the-age-of-screens/

Pew Research Center. (2021). *Mobile fact sheet*. https://www.pewresearch.org/internet/fact-sheet/mobile/

Reddish, E. F., Saul, J. M., & Steinberg, R. N. (1997). On the effectiveness of active-engagement microcomputer-based laboratories. *American Journal of Physics*, *65*(1), 45–54. doi:10.1119/1.18498

Sadoski, M., & Paivio, A. (2013). *Imagery and text: A dual coding theory of reading and writing*. Routledge. doi:10.4324/9780203801932

Smith, A. (2018). Greek optics. In A. Jones & L. Taub (Eds.), *The Cambridge history of science* (pp. 413–427). Cambridge University Press. doi:10.1017/9780511980145.023

Srivastava, S., Rao, D. S., & Nandakumar, H. (2016). Novel optical gyroscope: Proof of principle demonstration and future scope. *Scientific Reports*, *6*(34634). Advance online publication. doi:10.1038rep34634 PMID:27694987

Statista. (2022). *Number of smartphone subscriptions worldwide from 2016 to 2027*. https://www.statista.com/statistics/330695/number-of-smartphone-users-worldwide/

Strubbe, L., & McKagan, S. (2020, 25 March). *I suddenly have to move my lab course online! What should I do?* PhysPort. https://www.physport.org/recommendations/Entry.cfm?ID=119927

Struck, W., & Yerrick, R. (2010). The effect of data acquisition-probeware and digital video analysis on accurate graphical representation of kinetics in a high school physics class. *Journal of Science Education and Technology*, *19*(2), 199–211. doi:10.100710956-009-9194-y

Thornton, R. K. (1987). Tools for scientific thinking-microcomputer-based laboratories for physics teaching. *Physics Education*, *22*(4), 230–238. doi:10.1088/0031-9120/22/4/005

Thornton, R. K., & Sokoloff, D. R. (1990). Learning motion concepts using real-time microcomputer-based laboratory tools. *American Journal of Physics*, *58*(9), 858–867. doi:10.1119/1.16350

Trowbridge, D. E., & McDermott, L. C. (1980). Investigation of student understanding of the concept of velocity in one dimension. *American Journal of Physics*, *48*(12), 1020–1028. doi:10.1119/1.12298

Vieyra, R. (2018). Teaching students metacognition through discipline-based research and technology. In T. Spuck, J. Leigh, T. Rust, & R. Dou (Eds.), *Best practices in STEM education: Innovative approaches from Einstein Fellow alumni* (2nd ed., pp. 503–528). Peter Lang, Inc.

Vieyra, R., & Vieyra, C. (2014). Analyzing forces on amusement park rides with mobile devices. *The Physics Teacher*, *52*(3), 149–151. doi:10.1119/1.4865516

Vieyra, R., & Vieyra, C. (2022). Immersive learning experiences in augmented reality (AR): Visualizing and interacting with magnetic fields. In S. Macrine & J. Fugate (Eds.), Movement matters: How embodied cognition informs teaching and learning. MIT Press.

Vieyra, R., Vieyra, C., Marti, A., Monteiro, M., & Jeanjacquot, P. (2015). Turn your smartphone into a science laboratory. *Science Teacher (Normal, Ill.)*, *82*(9), 32–40. doi:10.2505/4/tst15_082_09_32

Vieyra, R., Vieyra, C., Pendrill, A.-M., & Xu, B. (2020). Gamified physics challenges for teachers and the public. *Physics Education*, *55*(4), 1–7. doi:10.1088/1361-6552/ab8779

Walkington, C., Nathan, M. J., Hunnicutt, J., Washington, J., & Holcomb-Webb, K. (2022). Learning geometry through collaborative, embodied explorations with augmented reality holograms. In C. Chinn, E. Tan, C. Chan, & Y. Kali (Eds.) P*roceedings of the 16th International Conference of the Learning Sciences – ICLS 2022* (pp. 1992-1993). International Society of the Learning Sciences.

Wells, M., Hestenes, D., & Swackhamer, G. (1995). A modeling method for high school physics instruction. *The Physics Teacher*, *63*(7), 606–619.

Wemyss, T., & van Kampen, P. (2013). Categorization of first-year university students' interpretations of numerical linear distance-time graphs. *Physical Review Special Topics. Physics Education Research*, *9*(1), 010107. doi:10.1103/PhysRevSTPER.9.010107

Wilson, M. (2002). Six views of embodied cognition. *Psychonomic Bulletin & Review*, *9*(4), 625–636. doi:10.3758/BF03196322 PMID:12613670

Wu, M., Pathak, P. H., & Mohapatra, P. (2015). Monitoring building door events using barometer sensor in smartphones. *Proceedings of the 2015 ACM International Joint Conference on Pervasive and Ubiquitous Computing*, 319-323. 10.1145/2750858.2804257

Xu, Q., Zheng, R., & Hranilovic, S. (2015). IDyLL: Indoor localization using inertial and light sensors on smartphones. *Proceedings of the 2015 ACM International Joint Conference on Pervasive and Ubiquitous Computing*, 307-318. 10.1145/2750858.2807540

Yoon, S., & Wang, J. (2014). Making the invisible visible in science museums through augmented reality devices. *TechTrends*, *58*(1), 49–55. doi:10.100711528-013-0720-7

Zarzar, L., Kalow, J., Xinping, H., Walish, J., & Swager, T. (2017). Optical visualization and quantification of enzyme activity using dynamic droplet lenses. *Proceedings of the National Academy of Sciences of the United States of America*, *114*(15), 3821–3825. doi:10.1073/pnas.1618807114 PMID:28348236

ADDITIONAL READING

Birchfield, D., & Megowan-Romanowicz, C. (2009). Earth science learning in SMALLab: A design experiment for mixed reality. *International Journal of Computer-Supported Collaborative Learning*, *4*(4), 403–421. doi:10.100711412-009-9074-8

Johnson-Glenberg, M. C. (2018). Immersive VR and education: Embodied design principles that include gesture and hand controls. *Frontiers in Robotics and AI*, *5*, 81. doi:10.3389/frobt.2018.00081 PMID:33500960

Johnson-Glenberg, M. C. (2022). Evaluating embodied immersive STEM VR: Quality of Education in Virtual Reality Rubric (QUIVRR). In S. L. Macrine & J. Fugate (Eds.), *Movement matters: How embodied cognition informs teaching and learning*. The MIT Press.

Johnson-Glenberg, M. C., Birchfield, D., Megowan-Romanowicz, C., Tolentino, L., & Martinez, C. (2009). Embodied games, next gen interfaces, and assessment of high school physics. *International Journal of Learning and Media*, *1*(2). Advance online publication. doi:10.1162/ijlm.2009.0017

Johnson-Glenberg, M. C., Megowan-Romanowicz, C., Birchfield, D. A., & Savio-Ramos, C. (2016). Effects of embodied learning and digital platform on the retention of physics content: Centripetal force. *Frontiers in Psychology*, *7*, 1819. doi:10.3389/fpsyg.2016.01819 PMID:27933009

Tolentino, L., Birchfield, D., Megowan-Romanowicz, C., Johnson-Glenberg, M. C., Kelliher, A., & Martinez, C. (2009). Teaching and learning in the mixed-reality science classroom. *Journal of Science Education and Technology*, *18*(6), 501–517. doi:10.100710956-009-9166-2

Vieyra, R. (2018). Teaching students metacognition through discipline-based research and technology. In T. Spuck, J. Leigh, T. Rust, & R. Dou (Eds.), *Best practices in STEM education: Innovative approaches from Einstein Fellow alumni* (2nd ed.). Peter Lang, Inc.

KEY TERMS AND DEFINITIONS

Augmented Reality: An interactive technology that overlays computer-generated images atop the real-world environment.

Embodiment: In cognitive psychology, a theory that espouses that the body and cognition are inextricably linked and that learning can be both supported by and take place through bodily experiences.

LiDAR: "Light Detection and Ranging," a remote sensing technology that measures the distance to an object or surface by calculating the time a light pulse takes to reach and reflect back from said object.

Magnetic Field: The space near a magnetic material or moving charge in which magnetic forces act.

Mobile Devices: A portable form of technology that typically operates through its connection to the internet and/or other sensor-based networks, and includes smartphones and tablets.

Modeling: The creation, refinement, and/or application of representations that illustrate principles or relationships within a system.

Vector: A quantity having both direction and magnitude, often represented by an arrow whose length corresponds to its magnitude and whose orientation in space corresponds to its direction.

Section 4

General Principles for the Effective Integration of Technology Into the Science Classroom

Chapter 9
Identifying In-Service Teachers' Perceptions of Developing 21st Century Skills Through Science Education Using TPACK-21 Framework

Salma Ali
Texas A&M University, USA

James Hernandez
University of Texas, USA

ABSTRACT

In this digital age, science teachers are continuously confronted with new emerging opportunities and challenges as a result of the rapid rate of discoveries in science and technology and information technology as well. Science educators, therefore, emphasize restructuring the education system to prepare 21st century learners to face complex challenges and prepare them to avail the opportunities at their fullest. This present study aimed to identify in-service science teachers' perceptions of developing 21st century skills through science education by employing an updated TPACK-21 framework. A mixed-methods study, using a survey-based questionnaire and focus group interviews with selected teachers, was undertaken to collect teachers' perspectives. The study revealed that when technology is used by teachers, it is often paired with specific content and pedagogically very different from what teachers are used to performing. The results highlight the need to use technology for innovation and to renovate contemporary teaching practices for 21st century learning.

INTRODUCTION

The development of a knowledge-based society and the impact of globalization have led educators,

DOI: 10.4018/978-1-6684-5585-2.ch009

policymakers, and stakeholders to argue that adopting twenty-first-century learning skills in education is essential and technology is central to the development of those skills. 21st-century skills such as collaborating, problem-solving, creative and innovative thinking, and taking advantage of information and communication technology (ICT) applications have received significant attention in recent years (Valtonen et al., 2017). To bring these skills to the classrooms, teachers must know the content knowledge and various pedagogical strategies to take advantage of educational technology and support the development of students' 21st-century skills (Voogt et al., 2013). Researchers and scholars believe that science education can play an important role in developing and promoting some of those skills in students (Jang & Tsai, 2013; Voogt et al., 2013) through using a variety of educational technologies or digital tools such as interactive Probe ware, mind-mapping tools, whiteboards, geospatial technology, etc. (Trautmann & MaKinster, 2010). However, researchers have not explored teachers' perceptions of developing 21st-century skills in science classrooms through integration technology. Although, researchers have conducted studies to examine in-service teachers' TPACK (Technology Pedagogy and Science Content Knowledge) enactment and development in science classrooms (Hechter, 2012; Jang & Chen, 2010; Phillips, 2017; Lin et al., 2013; Adam, 2017). However, exploring teachers' understanding and perception of developing 21st-century skills in students while teaching science by employing an updated TPACK-21 framework (Valtonen et al., 2017) has remained unexplored. Therefore, to address this gap, this mixed-method approach study aims to explore in-service teachers' perception of developing 21st-century skills through science education by employing the TPACK-21 framework. This study uses an exploratory mixed-method design to allow us to triangulate quantitative data which is coming from the questionnaire survey and qualitative data from teachers' focus group interviews to provide a more robust understanding of the phenomenon (Denzin & Lincoln, 2006).

Theoretical Background of TPACK -21 Framework

Technology Pedagogy and Content Knowledge (TPACK) is one of the widely used frameworks for technology integration in teaching concepts or subject matter (Phillips, 2017; Warr, 2019). It has its roots in work done in 1986 by Shulman who realized that teachers needed to understand the overlap of those two areas (Content and Pedagogy) and the pedagogical practices specific to their intended subject area such as science education. Mishra and Koehler (2006) added technology as the third component and emphasizes the interactions between and among content, pedagogy, and technology, arguing that for effective integration of technology into the classroom, teachers' experiences with a technology need to be discipline-specific (Mishra & Koehler, 2006).

This framework contains three core types of knowledge: content knowledge (CK), pedagogy knowledge (PK), and technology knowledge (TK), and four interrelated types of knowledge between the three core types of knowledge: PCK, technological content knowledge (TCK), technological pedagogical knowledge (TPK), and technological pedagogical content knowledge (TPCK). The outer circle of the framework described the contextual knowledge (XK) which describes classroom setting, schools/ districts, teachers' knowledge of their students' prior knowledge and learning difficulties, and teachers' workplace (Mishra, 1990; Greene & Jones, 2020). This knowledge of TPACK influences teachers' construction of educational content in any context.

To understand how the TPACK framework is useful in promoting twenty-first-century skills while teaching science or science concepts, Valtonen et al., (2017) developed a TPACK-21 measurement instrument aligning with 21st-century skills. The original TPACK framework (Mishra & Koehler, 2006;

Figure 1. TPACK diagram
(Mishra & Koehler, 2006)

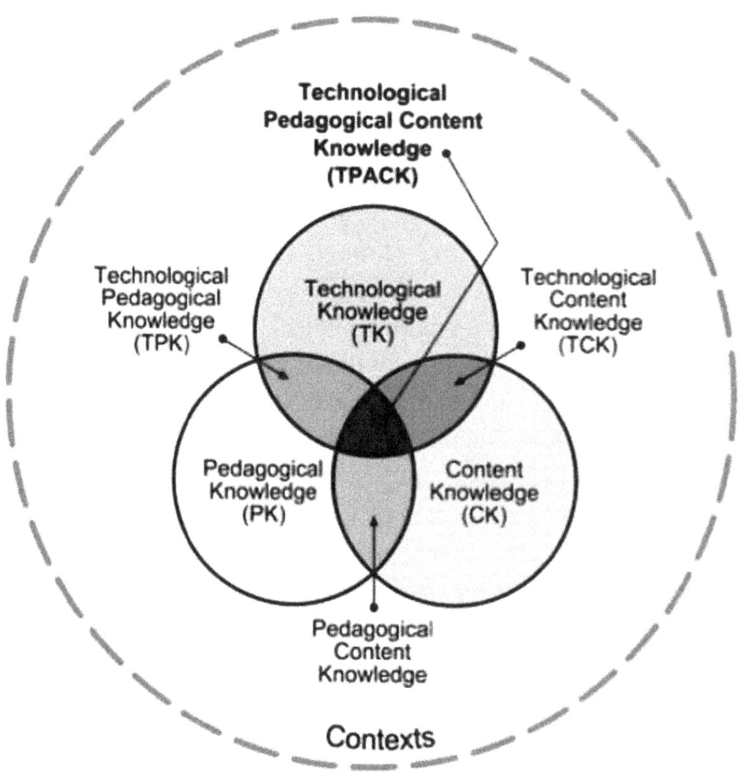

Valtonen et al., 2017) aligned the content (subject matter), 21st-century pedagogies, and technology aspects of twenty-first-century skills. This provides a comprehensive framework for studying and supporting in-service teachers' TPACK development aligning with 21st-century skills while teaching the concepts. This chapter follows the TPACK-21 framework to evaluate how science education can play an important role in promoting high-order thinking skills.

LITERATURE REVIEW

In this digital age, science teachers are continuously confronted with new emerging opportunities and challenges because of the rapid rate of discoveries in science and technology and information technology as well (Darling-Hammond et al., 2020). Many education researchers believe science as the subject matter can play an important role in preparing students to avail themselves of educational opportunities, and prepare them to face the complex challenges of today's time (Jang & Tsai, 2013; Save, 2020). This belief is because science plays an important role in daily human life, students can connect science concepts with their own lives, and can systematically understand the environment (Turiman et al., 2012). Whether the content is natural science, biology, chemistry, or physical science, students are able to learn the concepts in a meaningful way based on the constructivist approach (Cakir, 2008). Several types of research revealed that science education improves students' abilities to transfer their knowledge to problem in daily life. According to (Turiman et al., 2012) teaching science through problem-based learning (PBL) is found to encourage the development of students' concepts. Their findings show that students will better understand the practical application of concepts through the transfer of knowledge to problems in daily life. There is a consensus that teachers must provide students with learning content in ways that support the student's development of 21st-century skills (Rotherham & Willingham, 2009). Research shows that problem-solving and social skills are not practiced separately from subject knowledge (Murgatroyd, 2010). Thinking skills and working skills are best learned together in their natural context (Rotherham & Willingham, 2009; Silva, 2009). 21st-century skills are emphasized in different countries' national educational goals. However, in terms of actual teaching practices, the role of such skills is much weaker (Voogt et al., 2013).

Several studies have been conducted regarding exploring the relationship between TPACK and 21st-century skills in the process of teaching and learning in the classroom showing positive relationships between TPACK and fostering 21st-century skills in schools (Rotherham & Willingham, 2009; Silva, 2009; Turiman et al., 2012; Shafie et al., 2019). These studies also revealed that the integration of technology and 21st-century skills has made it complicated for teachers to educate students to reach national goals. For example, the research about challenges to integrating technology while promoting 21st-century skills (Handal et al., 2014). Teachers' level of TPACK and understanding of 21st-century skills (Walker, 2017; Fontanilla, 2016), development and validation of TPACK measurement instruments (Valtonen et al., 2017), and integration of 21st-century skills in teaching and learning (shape et al., 2019). Fontanilla (2016) compared beginner and experienced teachers' readiness to promote 21st-century skills in a techno-oriented environment and found a significant negative correlation between years of teaching and promoting soft skills in a technologically-oriented environment. Walker (2017) found that elementary teachers are not confident with the construction of TPACK while teaching those soft skills.

Significantly, different variables have been examined to evaluate the implementation of the TPACK framework in science classrooms such as teachers' beliefs about self-efficacy (Lin et al., 2013; Helppolainen & Aksela, 2020), science teachers' gender and TPACK (Lin et. al., 2012), skills of integrating technology into learning and teaching (Jang, 2010). Helppolainen & Aksela (2020) compared chemistry teachers' skills, knowledge, and beliefs on using technology in the classroom to other science teachers. Their findings revealed how chemistry teachers have positive beliefs about the use of ICT in teaching and learning settings and they reported good basic ICT skills in general, but they pointed to the lack of skills that are needed to integrate technology in chemistry teaching. The results of their study indicated their deficiency of TCK and TPK. Other frameworks, like the Technological Pedagogical Science Knowledge (TPASK), are based on three knowledge domains: pedagogical science knowledge (a science education of PCK), technological science knowledge (a science education of TCK), and technological pedagogical knowledge (TPK) (Jimoyiannis, 2010). However, the role of types of knowledge in science education is still in development.

Importance of Learning 21st-Century Skills

Development and advancement in technology in the fourth industrial revolution (4IR) changed the economic, political and social systems of the world (Shafie et al., 2019) and also changed the education scenario. These changes come along with challenges such as socioeconomic inequality, climate change, globalization, unemployment, and cultural diversity (Shafie et al., 2019). Therefore, Educators need to prepare students to avail themselves the opportunities, and to face and overcome these complex challenges too (Goradia, 2018). Technology advancement impacts students' learning in the classroom which is different from previous generations. For example, students are dependent on technology and always look for practical applications (Lemley et al., 2014), as they are surrounded by technology.

Educators, therefore, emphasize restructuring the education system to prepare 21st-century learners to face complex challenges successfully. Scholars and researchers demanded that high-order thinking skills include "creativity, critical thinking, collaboration and lifelong learning" (Goradia, 2018, p. 47; Acedo & Hughes, 2014) in the student curriculum where students learn those soft skills while learning the concepts or subject matter. On the other side, teachers are aware of the need to change their ways of teaching in today's classrooms (Shafie et al., 2019) and several educators and teachers adopted the pedagogies that develop higher-order thinking skills and support deeper learning through collaboration (Acedo & Hughes, 2014). For example, Goradia (2018) revealed that inquiry-based learning, problem-based learning and constructivist learning are highly rated and demanded because these pedagogies enhance deeper learning as opposed to superficial learning and promote higher-order thinking skills.

21st-Century Skills in Science Education

Based on the opportunities and challenges of today's current digital era, the framework for 21st-century learning identifies the skills and knowledge that we students need to acquire to be successful in the 21st century. Those soft skills are categorized as learning and innovative skills, information skills, media and technological skills, and life and career skills (Goradia, 2018). This study emphasizes learning and innovative skills that students develop through science education in schools and classrooms. These learning and innovation skills are briefly described below.

- Reflective thinking – Allows students to review concepts and reflect and think about their studying, learning, and skills
- Problem-solving – Allows students to be able to solve previously unknown tasks and problems in an effective and timely manner.
- Creative thinking – Allows students to make use of one's skills and to combine different sources of information to create something new
- Critical thinking – Allows students to process large amounts of information, analyze the information, and to compare different sources of information
- Communication and collaboration- Allows students to work together (small or large groups) efficiently to contribute by understanding their roles.
- Information and communications technology (ICT) – A wide arrange of different devices and tools used to disperse the information, such as computers, tablets, smartphones, etc., as well as web-based applications and software, social media services (e.g. blogs, Facebook, YouTube, WhatsApp, Instagram) and online learning environments (e.g. Moodle, Office365).

Goradia (2018) revealed that inquiry-based learning, problem-based learning and constructivist learning in science education are highly rated and demanded because these pedagogies enhance deeper learning as opposed to superficial learning and promote higher-order thinking skills. Glowatz and O'Brien (2017) highlighted the discussion forums or group discussions in the classrooms as the highly rated pedagogies and this finding is similar to that found in research conducted by Goradia (2018) where in the science classroom students' engagement appeared as the driving influences for using various technologies.

These scholars have revealed that 21st-century pedagogies help teachers to promote high-order skills in teaching the concepts and subject matter such as science education.

PURPOSE OF THE CURRENT STUDY AND RESEARCH QUESTIONS

This study was conducted to identify in-service science teachers' perception of developing 21st century skills through science education by employing an updated TPACK-21 model. The data obtained in the study were analyzed to understand how updated TPACK models help Science teachers to examine the development of 21st-century skills in the classroom and to explore the challenges and the confidence of developing those skills in the secondary science education context.

The study was designed to answer the following questions:

1. To what extent does the TPACK-21 model effectively examine teachers' perception of developing 21st-century skills in students in the science education classroom?
2. Does the TPACK of secondary school science teachers differ according to gender and teaching experience?
3. What are the challenges teachers face in developing 21st-century skills in students through teaching science (content)?

Research Design

We used a mixed methods approach to the qualitative findings about science teachers' perception of developing 21st-century skills through science education by employing an updated TPACK-21 model should support the results of the quantitative findings. The homogenous results which are derived from the two sets of data (quantitative analysis and qualitative analysis) give us a depth and breadth view of teachers' perspectives on the development of 21st-century skills in the science education classroom and exploring the challenges and the confidence around them. All data were collected in two phases. In the first phase, quantitative data was collected from a population of in-service teachers to identify teachers' perceptions of developing 21st-century skills through Science education using an updated TPACK-21. The descriptive data furnished in this phase of the study were used to guide the purposeful sampling of participants for the second face of the study. In the second phase, the focus-group interview was conducted with six science teachers to get meaningful and in-depth data for research questions. These findings show the disparities among in-service teachers regarding the aspect of developing 21st century skills through Science education and discuss the challenges and confidence in promoting 21st century skills through science education.

We intended to combine the qualitative data which comes from focus-group interviews and quantitative data which comes with the six item Likert-type scale survey questionnaire. In this way, each method confirms the finding from the other method (Johnson & Onwuegbuzie, 2004).

Participants and the Context

A total of forty male and female middle and high school science teachers were recruited for this study (N=40). All participants are certified science teachers to teach grades 7th to 12th grades. 68% of participants

Table 1. Demographics of study participant

		Male	Female	Total
Age	22 to 30 years old	6	11	17
	30 to 40 years old	5	8	13
	40 plus	4	6	10
Years of Teaching	0 to 3	3	7	10
	4 to 6	4	6	10
	7 to 10	4	7	11
	11 plus	4	5	9
Ethnicity	White, Non-Hispanic	7	10	17
	Black, Non-Hispanic	2	5	7
	Hispanic	4	7	11
	Asian or Pacific Islander	2	3	5
Total Participants				40

were female science teachers. The number of years of teaching experience is separated into four categories (see Table 1), one to three years of teaching experience, four to six years of teaching, seven to ten years, and ten or more years of teaching experience. In phase 1, the data collection was conducted using an online TPACK-21 questionnaire (a sample is available in the next section). For phase 2, six teachers are selected based on their average TPACK score and their number of years teaching science content.

In phase 1, the data collection was conducted using an online Technological pedagogical content knowledge for twenty-first-century skills (TPACK-21) questionnaire a (sample is available in the next section). TPACK-21 includes seven TPACK framework areas: pedagogical knowledge (PK; 7 items), content knowledge (CK; 4 items), technology knowledge (TK; 4 items), pedagogical content knowledge (PCK; 6 items), technological pedagogical knowledge (TPK; 6 items), technological content knowledge (TCK; 4 items) and technological pedagogical content knowledge (TPACK; 7 items).

Questionnaire

The science teachers assessed their knowledge of TPACK areas using a 6-point Likert-type scale (1 = I need a lot of additional knowledge about the topic; 6 = I have strong knowledge about the topic). Mean and the standard deviation is used to examine teachers' perception of developing 21st-century skills through science education. Then the Pearson Correlation (r) test was used to compare the components of TPACK-21 where teachers express their confidence or challenge. For example, teachers believe that the TPK (Technology pedagogy knowledge) component is helpful in developing the 21st century while teaching science, however, the TCK (technology content) component is less useful in developing those skills with teaching science subjects.

In phase two, data was collected to gather information to answer the third research question guiding this study. The focus-group interview was semi-structured and guided by open-ended questions based on disparities among teachers' responses to surveys. Six participants were selected for the focus group. Two teachers received a high score in the survey, two teachers received an average score, and two teach-

Figure 2.

Report

Gender		CKScore	PKScore	TKScore	PCKScore	TPKScore	TCKScore	TPACKScore
Male	Mean	3.6000	3.9600	4.2000	3.7111	4.3000	3.5889	3.7933
	N	15	15	15	15	15	15	15
	Std. Deviation	.92967	.40848	.71047	.69141	.32244	.70953	.38999
Female	Mean	3.7800	3.9200	4.0533	3.8813	4.2333	3.7333	3.8000
	N	25	25	25	25	25	25	25
	Std. Deviation	.93630	.36515	.78575	.60988	.40540	.60093	.35385
Total	Mean	3.7125	3.9350	4.1083	3.8175	4.2583	3.6792	3.7975
	N	40	40	40	40	40	40	40
	Std. Deviation	.92603	.37727	.75254	.63833	.37354	.63872	.36285

ers received a low score on the survey. Teachers who scored low on the TPACK-21 questionnaire were able to discuss the challenges and struggles they face in developing 21st-century skills in students. In Phase 2, data were analyzed using an a priori thematic analysis method. The a priori themes reflected the variables that have been identified in the interview as the challenge or confidence factors. The interview was audiotaped after obtaining informed consent from the participants. The Focus group interview was transcribed into the text through Nvivo software for qualitative analysis. Three major themes were identified from teachers' interviews through Nvivo 1) Positive about technology, 2) technological-pedagogical knowledge (TPK) promotes 21st-century skills, 3) Science concepts highlight 21st-century skills in a technology-oriented environment (TCK).

Results and Findings

Quantitative Findings and Analysis

The quantitative portion of this study consisted of a total of forty participants, with fifteen identifying as males and twenty-five identifying as females. The number of years teaching science instruction for the participants ranged from one year to ten or more years and averaged four to six years of experience teaching science for the sample with no significant difference between genders. The means for each metric for the entire sample size were: CK (M=3.7, SD=.93), PK (M=3.94, SD=.38), TK (M=4.1, SD=.75), PCK (M=3.8, SD=.64), TPK (M=4.3, SD=.37), TCK (M=3.7, SD=.64), and TPACK (M=3.8, SD=.36). There is no significant difference of mean scores between genders.

Correlations

The Pearson correlation(r) test was performed on the data to identify any relationships between demographic variables in the study and metric variables. For example, correlations were tested between gender and CK scores, gender and years of experience, and any relationships between content knowledge.

Gender and Years of Teaching Experience

Regarding gender, no significant correlations were found between any metric testes. The number of years of teaching, however, was positively correlated with several metrics including CK score (p=.000, r=.890), PCK scores (p=.000, r=.846), PK score (p=.000, r=.590), and TPACK scores (p=.001, r=.506). The weakest positive correlation with the number of years of teaching was TCK score (p=.013, r=.388). TK score (p=.000, r= - .806), however, was negatively correlated with participants' number of years teaching. These results show that years of teaching experience strongly influence the ability of teachers to promote 21st-century skills in science education. The result of the negative correlation with years of teaching experience supports the need for ongoing training in TK throughout a science teacher's career, likely meaning that the more teachers have a command of content knowledge the more teachers are able to promote 21st century skills effectively.

Metric Relationships

The researchers also conducted a Pearson correlational analysis between the different metric scores to find if any specific metrics stood out. This was done to identify if any specific metrics were more important to strengthening TPACK scores in general. In our analysis, CK score was positively correlated with PK (p=.000, r=.635), PCK (p=.000, r= .827), TPACK (p=.001, r= .522), and TCK (p=.007, r=.422). These findings present a positive relationship between Science Education and the development of 21st-century skills, based on teachers' perceptions. TK (p=.000, r= -.782) was negatively correlated with CK, meaning that teachers perceive themselves as incapable of using technologies in the classroom if it involves diminishing their knowledge of science content.

PK was positively correlated with CK (p=.000, r=.635), PCK (p=.000, r= .692), TCK(p=.005, r=.440) and TPACK (p=.034, r=.336). These results indicate that teachers perceived PK to be supported by their CK since it is involved in all positively correlated metrics. In other words, if science teachers have a strong CK, they are much more likely to have strong pedagogical beliefs. TK (p=.038, r= -.330) was negatively correlated with PK meaning that as teachers gained in their PK, their technological skills and knowledge of technology decreased. This may be due to the possibility that science teachers place more emphasis on their pedagogy of science content, rather than the technology that is delivering the content. Teachers may also perceive technology as not required to teach students or that they are better capable using a teacher-centered approach to technology, rather than a more technologically focused one.

TK was negatively correlated with PK (p=.038, r= -.330), CK (p=.000, r= -.782), PCK (p=.000,r= -.661) and TPACK(p=.000,r= -.564), making it negatively correlated with the most metrics. This finding shows that an emphasis on teaching content with new or emerging technologies is not occurring in pre-service teacher classrooms or that preconceived ideas of technology in the classroom are being brought to teacher training environments. The only positive correlation of TK was with TPK (p=.002, r=.481). This supports that teachers who place an emphasis on technology and pedagogy, place less emphasis, or perceive themselves as less content oriented.

Lastly, TCK was positively correlated with PK (p=.005, r=.440), CK (p=.000, r=.635), PCK (p=.005, r=.550), and TPACK (p=.001, r=.522). This finding shows that teachers that are confident in their content knowledge can use technology to support pedagogy and pedagogy to support technology. It seems, however, that knowledge of technology does have an influence on pedagogy and content.

Overall these findings indicate levels of specific knowledge that contribute to overall TPACK scores. Some variables, like years of teaching experience may negatively affect technology scores. The results of

Figure 3.

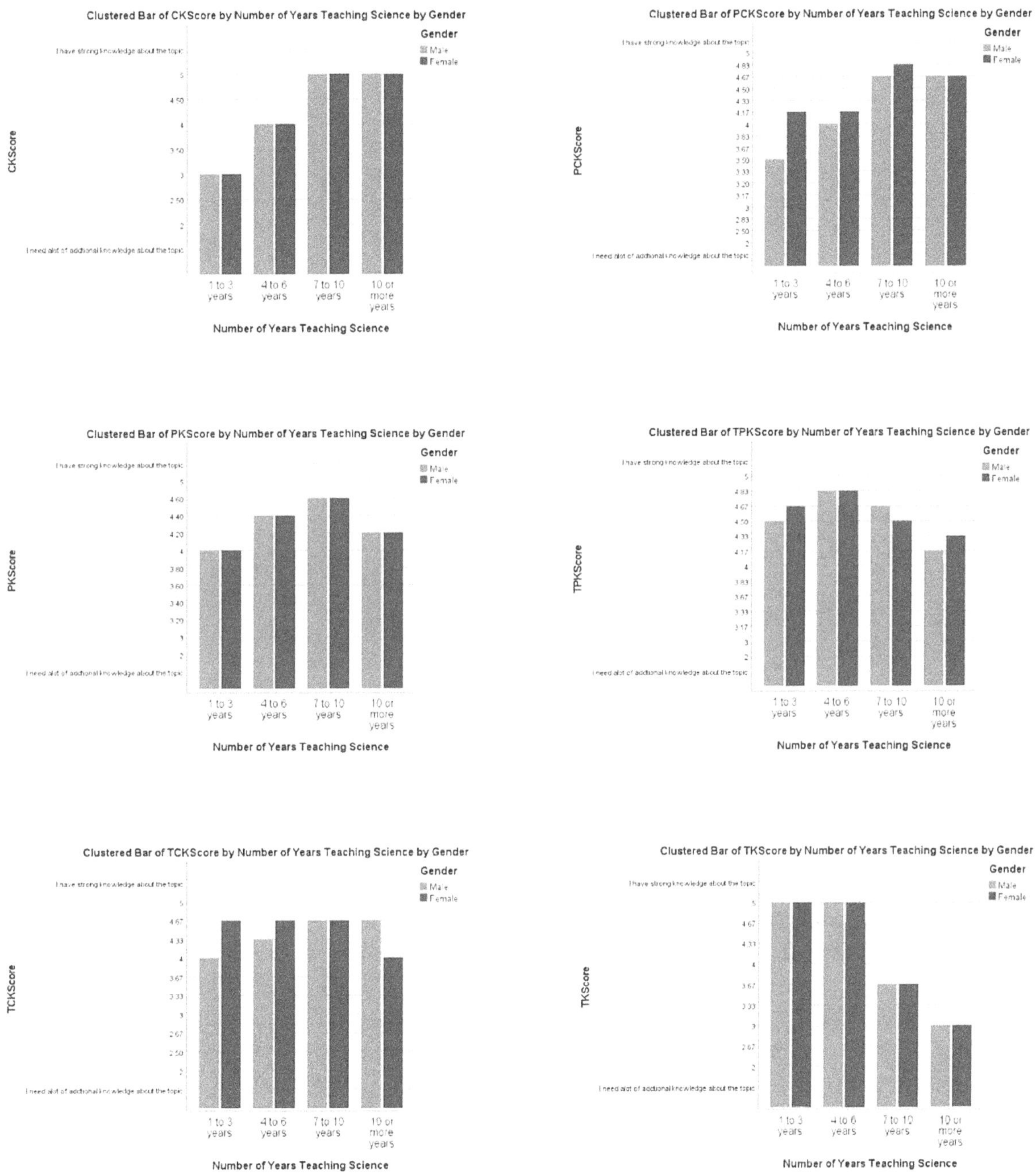

the quantitative analysis in this study suggest that gender has no significant influence on science education knowledge held by teachers. Instead, the number of years' teachers taught science has a strong correlation with many knowledge metrics, suggesting that these are aspects of science education that teachers get the most training on during their pre-service education or that these skills develop naturally during the science teaching process. Lastly, TK was negatively correlated with the most metrics. This shows

Table 2. Participated teachers demographics and list of skills

S.No.	Teaching Experience	Content	Technological Tools	Major 21st Century Skills
1	3 years	Chemistry	Interactive white board W2 tools/blogs	Critical thinking Communication skills Creative thinking
2	5 years	Biology	Interactive white board Interpret multimodal text/W2 tool	Reflective thinking Technological skills
3	6 years	Natural Science	Social network	Creativity
4	7 years	Physics	Video making (YouTube) Apps/Gamifications New software	Technological skills Critical thinking Communication
5	8 years	General Science	Image editors/Video making Presentations (Prezi/Animoto)	Teamwork/Cooperation Creativity
6	10 years	General Science	Mind maps (lucidchart/flowchart) Presentation (Doodle/EyeJot)	Problem solving Communication/coordination

that TK must be further emphasized in teacher education and professional development opportunities as they progress in their teaching roles.

Qualitative Findings and Analysis

A total of six teachers were interviewed via focus group during an online meeting for the qualitative data collection process. The group interview questions aimed were to identify the challenges and confidence in developing 21st - century skills in students through teaching science(content). This qualitative data further illuminates the above survey findings. Teachers are selected based on teaching experience and their average TPACK-21 gain in the survey. Interview was audiotaped after obtaining informed consent from the participants. The Focus group interview was transcribed into the text through Nvivo software for qualitative analysis. Three major themes were identified from teachers' interviews through Nvivo 1) Positive about technology, 2) technological-pedagogies (TPK) promote 21st century skills, 3) Science concepts highlight 21st century skills in a technology-oriented environment (TCK). These themes were common to all the selected participants irrespective of whether they were in high, medium, or low perceived TPACK-21 in the survey questionnaire. Before we dive into the challenges and confidence in developing 21st century skills through science education, we start our conversation with what is the nature of the relationship between science content and 21st century skills? Below are the demographics of six teachers by their teaching experience, what science content they teach and what educational tools they use to promote 21st century skills.

Table 2 shows that teachers utilize technology and digital tools to develop 21st century skills in students such as critical thinking skills, reflective skills, creativity, communication and teamwork when they deliver specific science content. Table 2 also shows connections between the column 'content' and the column 21st century skill. For example, general science teachers feel teaching general science would help to develop problem-solving skills or cooperation or group working skills in the classroom. Similarly, both chemistry teachers mentioned 'creativity' as the developing skills through teaching chemistry concepts. One general science teacher mentioned that

I use online resources with topic-specific content activities. Content comes first in my class, but we need to present it creatively so students can learn the concepts, and the skills. Last week, I used an interactive volcano website during the geology lesson and in the same class, students used google earth to find the locations of volcanoes that help them to learn new technological skills.

Most teachers emphasize the 'C' component in the TPACK-21 framework and explain that what we need to ask ourselves as educators is what today's students need to know or learn. Once we have identified the purpose of the concepts and skills to teach the students, then this framework can help us determine how to achieve them. Teachers mentioned that we can't ignore the content and even we can't teach critical thinning or any skills in a vacuum. These findings corroborate those (Warr, 2019). Overall we found a greater relationship between the content in school as disciplinary knowledge and the 21st century learned skills.

One chemistry teacher stated,

How the content is conceptualized is critical too, in terms of how it is learned. We deliver the content by using the content-specific technology such as websites so students can conceptualize the concepts and engage with the content. Creativity can't occur without some domain.

These findings connect previous research showing problem-solving or social skills cannot be practiced without the content and creativity does not emerge separately from subject knowledge. These skills are best learned together in the natural context (Valtonen et al., 2017). During the conversation, a physics teacher talked about the approach where teachers value the content and the strategies or content-specific strategies as well in the technology-oriented environment. This trans-disciplinary approach (Warr et al., 2019) prepares teachers to think effectively across a range of domains.

One physics teacher brought an example from her lesson to show how students become critical thinkers and learn communication skills through learning physics-specific content.

I always make sure that my students are actively and voluntarily involved in all my learning content. I mostly use Problem-based learning strategy and students in the group solve the case studies through the acquisition of problem solving skills. I provide them detailed rubrics with their assigned roles that help them at every level from understanding the problem to finding the actionable solution.

A biology teacher shares the strategies she has been using for years with high school students. She emphasizes Universal Design Learning (UDL) as a good teaching method in technology-oriented culturally diverse classrooms. She mentioned,

UDL framework provides the students with multiple options to process and present the concepts and content of the subject matter. I understand that students from different backgrounds bring different experiences in the classroom and learn more or less. Therefore, I give them multiple opportunities to learn and present their knowledge. Everytime students pick different options and become more creative.

She mentioned that after learning a new concept using UDL, I always give them 10 minutes of reflective activity. The time allowed for reflection helps them discover more about the concepts of the subject matter and about themselves. Participating teachers mentioned how technology helps them to address their students' needs so in terms of giving them multiple options to learn and process the concepts.

These qualitative findings strengthen the evidence from our quantitative analysis (survey questionnaire), which reported a strong, positive correlation between teachers' overall average TPACK with their Content Knowledge (CK) and Technological Content knowledge (TCK). Teachers' perceived Technological Pedagogical knowledge (M=4.3, SD=.37) and Technological Knowledge (M=4.1, SD=.75) were reported above average in our survey analysis. It is important to note the overall TK for participants in this study varied greatly and was significantly negatively correlated with PK, CK, PCK, and TPACK. This suggests that teachers who understand how to teach within their content area actually lack the necessary skills to utilize technology effectively. However, our participating teachers who had sound pedagogy and command of science content may have a better understanding of promoting and developing 21st century skills in their students, rather than a teacher with strong technological skills. Further, the findings suggest that the more pedagogical knowledge and content knowledge teachers have, the less command they have over technology to promote and develop the 21st century-based skills in their students.

Challenges Related to TPACK-21 in Science Education to Develop 21st-Century Skills

Despite numerous viewpoints regarding teachers' confidence in developing 21st century skills through science education, teachers talk about the barriers and struggles they face in the science classroom. The most prominent factors listed by the participants in the focus group interview include 'Lack of time'. Participating teachers stated that they always need to spend extra time preparing resources and planning activities that promote the skills with the content of the subject matter. This was exemplified in teachers' statements such as, *"I did not have a lot of preparation in selecting appropriate technology to develop the skills in my students"* and participants describe educational technology as the 'fastest vehicle' referring to how technology is changing rapidly. The teacher mentioned, *"every month I learned a new digital tool and wanted to keep updating myself. So I can pick the best option for my students."* This perception reflected technological integration as more than an instructional tool, rather than a learning tool for the students (Ashe & Bibi, 2011). In the end, some teachers think technology is an effective tool to incorporate into their teaching and instruction, but not for students to facilitate their learning in the classroom.

Another main challenge with regard to TPACK in science education, it seems to be in finding ways to facilitate science teachers' recognition of the value of TPACK and how to facilitate their efforts for the development of TPACK knowledge. Some teachers believe that pedagogies play a vital role in developing the skills of students, but some teachers feel that content is everything.

Ashe and Bibi (2011) conclude technology is an effective tool teachers incorporate into their teaching. Still, technology is also an additional tool available to students to facilitate their learning in the classroom. Some Teachers show their confidence in perceived technological and Content Knowledge (TCK), but some teachers are confident with their perceived technology and pedagogical knowledge (TPK), with the mention of how their content delivery and pedagogical practices are guided by technology. Some teachers mentioned they often use digital technologies to merely deliver content they want to teach, whereas others also use digital technology as a transformative tool in their teaching of subject matter. Additionally, their technological practices are influenced by the social-cultural classroom environment. Therefore, having a balanced role of each component in the TPACK can support the development of 21st century skills, and enhance existing classroom practices in science education.

DISCUSSION

The purpose of our study was to identify in-service science teachers' perception of developing 21st century skills through science education by employing an updated TPACK-21 model and to explore their confidence and challenges in developing those skills in secondary students. First, we looked at whether TPACK of secondary school science teachers differs according to gender and teaching experience. Our findings reveal that teachers' gender has no relationship with their TPACK scores. Years of teaching, however, plays a larger role for a teacher regarding their perceptions of TPACK and its individual metrics. The findings suggest that participants in this study have a strong foundation of overall science content and pedagogy, but less technological knowledge as their years of science teaching increase. In line with Aksela (2015), our findings suggest the number of years a teacher has been in the classroom is not a strong indicator of a teacher being technologically informed. In fact, according to our results, a teacher right out of pre-service preparation will be more technologically savvy than a teacher who received their certification many years prior but may lack strong pedagogical and content skills. However, the newer teacher may lack strong pedagogical and content skills. This finding is contrary to the qualitative theme in this study of teachers responding positively to technology, yet it seems the training or focus is not achieved.

These findings are strong evidence to support findings related to teachers with less teaching experience who should be asked to provide professional development to teachers with more years of experience on student-focused ways to use technology in the classroom. On the other hand, teachers with more science teaching experience should be called upon to guide newer teachers with sound pedagogy and content. This finding also supports Goradia's (2018) and Hughes's (2014) assertion of the way technology education must be ongoing with teachers throughout their teaching career, not only during pre-service teaching education. This finding is in line with previous research done by Lemley et al., (2014) which supports findings of how 21st-century students 21st-century about technology while they are a student, but may not keep the same rigor for technology learning while they are in service. Since any results from our and others' studies indicate that years of teaching is also positively correlated with age, teachers with fewer years of experience are likely to be younger and may have more knowledge of current technology, while older teachers with more years of experience may put more focus on PK and CK.

Correlations between metrics also showed interesting findings related to CK. In our analysis, CK score was positively correlated with most survey metrics showing a positive relationship between Science Education and the development of 21st century skills, based on teachers' perceptions. Similar to correlations with years of teaching experience, the results of our study show CK has the most influence on overall scores of knowledge, further suggesting content should be taught with high standards to pre-service teachers. Additionally, our findings corroborate with the qualitative results pertaining to CK coming first, then PK coming second when planning a lesson. TK is often the last piece of knowledge considered when planning a lesson, according to the qualitative analysis and is supported by the quantitative results. If pedagogy or technology is considered first in the curriculum, how would it change the way teachers provide science content? These findings support the need to experiment with other ways to formulate lessons and possibly the way teachers organize their classrooms, possibly around technology or pedagogy, rather than content.

The qualitative analysis revealed teachers often see content as ignored if it is not first in the planning process. However, this way of planning seems to push technological knowledge, according to the quantitative results, to the point where it actually harms the inclusion of new or emerging technology in the classroom. Furthermore, when technology is used by the teachers in the study, it is often paired with specific content and pedagogically very different from what teachers are used to performing. This finding is in line with Ashe and Bibi's (2011) assertion related to how teachers see technology as valuable, but technology is added to the content, not on par.

Lack of planning time was also expressed as a factor of inclusion for technology in the qualitative results. This lack of planning time corroborates the quantitative findings, that technological knowledge is low since it is often the last part of a lesson considered, after pedagogy and content, not allowing for much practice or experimentation with new or emerging technology. Participants also expressed the way that technology changes rapidly and, without enough planning time to include technology, they often get left behind and rely on what they have used in the past or use limited technology during their lessons.

CONCLUSION

To conclude, based on our findings Technological Pedagogical and Content Knowledge (TPACK) can be recognized as a dynamic and integrative knowledge of the content of subject matter (science education) and technology and pedagogy to meaningful integration of 21st century skills in students. Overall, our results show a significant relationship between the TPACK-21 framework and the teaching of 21st century skills among middle and high school science teachers. We conclude TPACK-21 framework helps teachers in recognizing their capacity and ability to work 21st-century educational context and develop the required skills. This updated framework highlights the complex interaction between technology, pedagogy, and science content knowledge with respect to developing 21st century skills. Science teachers are able to identify their strengths and weaknesses by evaluating their knowledge of each of the components of TPACK-21. Additionally, the findings of this study would be beneficial to pre-service teacher education and for the continued professional and personal; development among educators.

Study Limitation

We conducted this study with a small sample. Forty middle and high school science teachers completed the survey and we invited only six teachers participants for the focus group interview. Therefore, the generalization of the findings of these studies is limited and context-based too. Our study captures teachers' perspectives on using various technology, and pedagogies with science content to enhance 21st century learning skills, but it is worthwhile to derive more information from students; feedback, and principal perspectives, and look at other contextual factors.

REFERENCES

Acedo, C., & Hughes, C. (2014). Principles for learning and competencies in the 21st-century curriculum. *Prospects*, *44*(4), 503–525. doi:10.100711125-014-9330-1

Adam, A. (2017). A framework for seeking the connections between technology, pedagogy and culture: A study in the Maldives. *Journal of Open, Flexible, and Distance Learning*, *21*(1), 35–51.

Ashe, D., & Bibi, S. (2011, December). Unpacking TPACK and students' approaches to learning: Applying knowledge in pieces to higher education teaching and learning. In Ascilite (pp. 128-132). Academic Press.

Cakir, M. (2008). Constructivist approaches to learning in science and their implications for science pedagogy: A literature review. *International Journal of Environmental and Science Education*, *3*(4), 193–206.

Creswell, J. W., Hanson, W. E., Clark Plano, V. L., & Morales, A. (2007). Qualitative research designs: Selection and implementation. *The Counseling Psychologist*, *35*(2), 236–264. doi:10.1177/0011000006287390

Darling-Hammond, L., Flook, L., Cook-Harvey, C., Barron, B., & Osher, D. (2020). Implications for the educational practice of the science of learning and development. *Applied Developmental Science*, *24*(2), 97–140. doi:10.1080/10888691.2018.1537791

Denzin, N. K., Lincoln, Y. S., & Giardina, M. D. (2006). Disciplining qualitative research. *International Journal of Qualitative Studies in Education: QSE*, *19*(6), 769–782. doi:10.1080/09518390600975990

Fontanilla, H. S. (2016). *Comparison of beginning teachers' and experienced teachers' readiness to integrate technology as measured by TPACK scores* [Doctoral dissertation]. Brandman University.

Glowatz, M., & O'Brien, O. (2017). Academic engagement and technology: Revisiting the technological, pedagogical and content knowledge framework (TPACK) in higher education (HE)—The academics' perspectives. *IAFOR Journal of Education*, *5*, 133–159.

Goradia, T. (2018). Role of educational technologies utilizing the TPACK framework and 21st century pedagogies: Academics' perspectives. *IAFOR Journal of Education*, *6*(3), 43–61. doi:10.22492/ije.6.3.03

Greene, M. D., & Jones, W. M. (2020). Analyzing contextual levels and applications of technological pedagogical content knowledge (TPACK) in E nglish as a second language s ubject area. *Journal of Educational Technology & Society*, *23*(4), 75–88.

Handal, B., Campbell, C., Cavanagh, M., & Dave, K. (2014, January). Appraising mobile maths apps: The TPACK model. In *Australian Computers in Education Conference* (pp. 251-269). Australian Council for Computers in Education.

Hechter, R. P. (2012). Pre-service teachers' maturing perceptions of a TPACK-framed signature pedagogy in science education. *Computers in the Schools*, *29*(1-2), 53–69. doi:10.1080/07380569.2012.657999

Helppolainen, S., & Aksela, M. (2020). Science teachers' ICT use from a viewpoint of technological pedagogical content knowledge (TPCK). *Teknologia Kemian Opetuksessa*, *1*(1), 2–2.

Jang, S. J., & Chen, K. C. (2010). From PCK to TPACK: Developing a transformative model for pre-service science teachers. *Journal of Science Education and Technology*, *19*(6), 553–564. doi:10.100710956-010-9222-y

Jang, S. J., & Tsai, M. F. (2013). Exploring the TPACK of Taiwanese secondary school science teachers using a new contextualized TPACK model. *Australasian Journal of Educational Technology*, *29*(4). Advance online publication. doi:10.14742/ajet.282

Jimoyiannis, A. (2010). Developing a technological pedagogical content knowledge framework for science education: Implications of a teacher trainers' preparation program. *Proceedings of Informing Science & IT Education Conference*, *597*, 607. doi:10.28945/1277

Johnson, R. B., & Onwuegbuzie, A. J. (2004). Mixed methods research: A research paradigm whose time has come. *Educational Researcher*, *33*(7), 14–26. doi:10.3102/0013189X033007014

Lemley, J. B., Schumacher, G., & Vesey, W. (2014). What learning environments best address 21st-century students' perceived needs at the secondary level of instruction. *NASSP Bulletin*, *98*(2), 101–125. doi:10.1177/0192636514528748

Lin, T. C., Tsai, C. C., Chai, C. S., & Lee, M. H. (2013). Identifying science teachers' perceptions of technological pedagogical and content knowledge (TPACK). *Journal of Science Education and Technology*, *22*(3), 325–336. doi:10.100710956-012-9396-6

Lin, T. C., Tsai, C. C., Chai, C. S., & Lee, M. H. (2013). Identifying science teachers' perceptions of technological pedagogical and content knowledge (TPACK). *Journal of Science Education and Technology*, *22*(3), 325–336. doi:10.100710956-012-9396-6

Mishra, P., & Koehler, M. J. (2006). Technological pedagogical content knowledge: A framework for teacher knowledge. *Teachers College Record*, *108*(6), 1017–1054. doi:10.1111/j.1467-9620.2006.00684.x

Murgatroyd, S. (2010). 'Wicked problems' and the work of the school. *European Journal of Education*, *45*(2), 259–279. doi:10.1111/j.1465-3435.2010.01428.x

Phillips, M. (2017). Processes of practice and identity shaping teachers' TPACK enactment in a community of practice. *Education and Information Technologies*, *22*(4), 1771–1796. doi:10.100710639-016-9512-y

Rotherham, A. J., & Willingham, D. (2009). 21st century skills: The challenges ahead. *Educational Leadership*, *67*(1), 16–21.

Rotherham, A. J., & Willingham, D. (2009). 21st century. *Educational Leadership*, *67*(1), 16–21. PMID:19670694

Savec, V. F. (2020). The opportunities and challenges for ICT in science education. *Teknologia kemian opetuksessa, 1*(1), 1-1.

Shafie, H., Majid, F. A., & Ismail, I. S. (2019). Technological pedagogical content knowledge (TPACK) in teaching 21st century skills in the 21st century classroom. *Asian Journal of University Education*, *15*(3), 24–33. doi:10.24191/ajue.v15i3.7818

Shulman, L. S. (1986). Those who understand: Knowledge growth in teaching. *Educational Researcher*, *15*(2), 4–14. doi:10.3102/0013189X015002004

Silva, E. (2009). Measuring skills for 21st-century learning. *Phi Delta Kappan*, *90*(9), 630–634. doi:10.1177/003172170909000905

Stehle, S. M., & Peters-Burton, E. E. (2019). Developing student 21 st Century skills in selected exemplary inclusive STEM high schools. *International Journal of STEM Education*, *6*(1), 1–15. doi:10.118640594-019-0192-1

Trautmann, N. M., & MaKinster, J. G. (2010). Flexibly adaptive professional development in support of teaching science with geospatial technology. *Journal of Science Teacher Education*, *21*(3), 351–370. doi:10.100710972-009-9181-4

Turiman, P., Omar, J., Daud, A. M., & Osman, K. (2012). Fostering 21st century skills through scientific literacy and science process skills. *Procedia: Social and Behavioral Sciences*, *59*, 110–116. doi:10.1016/j.sbspro.2012.09.253

Valtonen, T., Sointu, E., Kukkonen, J., Kontkanen, S., Lambert, M. C., & Mäkitalo-Siegl, K. (2017). TPACK updated to measure pre-service teachers' twenty-first century skills. *Australasian Journal of Educational Technology*, *33*(3). Advance online publication. doi:10.14742/ajet.3518

Voogt, J., Erstad, O., Dede, C., & Mishra, P. (2013). Challenges to learning and schooling in the digital networked world of the 21st century. *Journal of Computer Assisted Learning*, *29*(5), 403–413. doi:10.1111/jcal.12029

Walker, R. X. (2017). *Assessing teachers' technological, pedagogical, and content knowledge in elementary schools* [Doctoral dissertation]. Wilmington University.

Warr, M., Mishra, P., & Scragg, B. (2019, March). Beyond TPACK: Expanding technology and teacher education to systems and culture. In *Society for Information Technology & Teacher Education International Conference* (pp. 2558-2562). Association for the Advancement of Computing in Education (AACE).

Chapter 10
The Role of Digital Curation in Science Teacher Professional Development

Efrat Dayan
Technion, Israel

Rivka Gadot
Technion, Israel

Dina Tsybulsky
https://orcid.org/0000-0002-2434-8509
Technion, Israel

ABSTRACT

This chapter discusses the goals of digital curation as an educational practice in K-12 science education in the digital age and presents a model for science teachers' professional development through digital curation. The model proposed in this chapter was developed during a multiple-case study with the participation of 44 science teachers. The model outlines a path to enabling teachers to learn and practice curation as a teaching and learning practice and later to direct and apply curation in their classrooms. In addition, it offers guidelines for enabling teachers to practice thinking skills, such as making pedagogical decisions and defining teaching goals, through the digital curation process as a professional learning method: (1) teacher professional learning goals, (2) pedagogical-practical goals, (3) social and cooperative goals, (4) teacher personal goals. This model can be applied to science teachers' professional development and teacher training courses.

INTRODUCTION

An endless flood of information characterizes today's digital age. Information in a wide range of fields is easily accessible from anywhere and at any time, but this avalanche of information requires constant

DOI: 10.4018/978-1-6684-5585-2.ch010

critical thinking and intelligent filtering. Mitchell Kapor (1991) captured this eloquently when he said: "Getting information off the internet is like taking a drink from a fire hydrant."

Digital curation constitutes one way to overcome these challenges. Curation originally related to museums and libraries, and referred to collecting content on a central theme or idea for specific individuals interested in this topic. By extension, digital curation is the process of creating, managing, and using digital materials for a wide variety of activities and content gathering to create a digital collection (Beagrie, 2006). To create the collection, the curator searches information online, chooses the content and its source, filters the information, catalogues it, and then shares the collection (Deschaine & Sharma, 2015). Digital curators do not merely go through a technical process of collecting, but rather distill the meaning of the content (Betts & Payne, 2016) through metacognition and critical thinking (Antonio et al. 2012; Gadot & Tsybulsky, in press).

Educational digital curation refers to creating digital collections for teaching, learning, and assessment purposes. It constitutes one of the core skills for the digital age, which is characterized by masses of information and constant engagement in locating and managing data. It has developed into a new field of research and has led to new practices in educational contexts (Tsybulsky, 2020). Analysis of the decisions that teachers make during the teaching and learning process can contribute to building a model of pedagogical reasoning in the digital age where teachers are called upon to harness their professional judgement when deciding which information to use, for whom, and when (Starkey, 2010).

Curation creates a rich, active environment that provides teachers with the opportunity to build a personal ontology, choose from a rich variety of resources, develop professionally in an open and flexible learning environment, and create a personalized collection (Dayan & Tsybulsky, 2021). Teachers assemble different forms of content such as videos, film stills, short video clips, texts, and others which they use to enhance student learning (Potter, 2012). Through curation, teachers can organize the ideas, sources of information, and digital tools to achieve their educational goals (Trust, 2016). Teachers exercise their professional judgment in terms of the content collected, the teaching goals, as well as the relevance, precision, reliability, and authenticity of the content. These considerations are the foundation of digital curation, since they are the keys to turning content into a collection with value that supports teaching and reflects the curator's goals (Deschaine & Sharma, 2015).

The study described in this chapter examined the teaching and learning goals manifested by science teachers during the digital curation process. The findings are then harnessed to develop a model for K-12 science teachers' professional development and for teacher training courses that integrate digital curation to promote pedagogical reasoning and define teaching goals. The analysis provides unique insights into the value of digital curation as a science teaching practice, and the impact of knowledge and practice on these teachers' decisions with respect to what to do, when, and why.

The first part of this chapter overviews digital curation in general and digital curation in science education in particular, in terms of teachers' professional development and their teaching goals. The second part discusses the use of digital curation in a study of a course for 44 science teachers, and shows how it served the pedagogical goals teachers defined for themselves and improved their teaching.

BACKGROUND

Digital Curation

Curation has been a cultural phenomenon since ancient times, when peoples gathered objects from their surroundings to survive, and subsequently began a learning process that led to the creation of new things based on these items (food, hunting tools, household utensils, and the like). More recent "curation" tended to refer to exhibits in galleries and museums. However, in today's information era, several new concepts have coalesced: *content curation* is the process of gathering relevant information on a particular topic or field, and *digital curation* is defined as using digital media for curation. In other words, there has been a shift from gathering objects to gathering digital information.

The digital curation process comprises five sequential stages (Deschaine & Sharma, 2015): collection, categorization, critiquing, conceptualization, and circulation. Each stage overlaps and leads to the next stage, and during each stage the content is refined in a cyclic reflective process of effective curation. In the first stage, the digital curator collects content and information that seems relevant by comparing new content to that which already exists in the collection. In the categorization stage, similar content is grouped, and the collected content is catalogued, allowing the curator to add additional content in the future. At this stage, the curator's judgment comes into play by deciding why certain content should or should not be included in a specific category. In the third stage, critiquing, the curator's professional judgment and reasoning are engaged to conduct an in-depth assessment of the content in a process of argument and justification. This is a critical stage for refining the collection, since it guarantees that only suitable content will be included. Next, in the conceptualization stage, new connections are made between the artifacts in the collection. In the final stage, circulation, the digital curators can present and disseminate the collection they created, so that it can also serve other community members.

In addition to organizing content, curation is a social way of sharing content. When social media added the option of sharing, it opened up the valuable ability of categorizing content for specific goals. When content is defined as public, the user becomes an expert, and users provide value to one another (Shirky, 2010). Sometimes users who curate and organize content in an interest list can do so better than the network itself and can share their collection. A digital curator acts within a collaborative environment, and the collection is the result of peer-to-peer learning, social communication, knowledge sharing, and reflection. Information sharing between digital curators in similar fields is an essential feature of the curation process, which thus integrates the personal and the social.

Digital Curation in Science Education

The massive growth in digital information, created continuously and in ever-increasing amounts and collected from various media sources and types, has changed the learning process, as the world moves from lacking information to unlimited access to information. Teachers face unique challenges when trying to teach effectively using digital media (Tsybulsky & Levin, 2019), since the primary didactic dilemma is how to integrate technology into the learning environment effectively to create involvement and meaningful learning (Watson & Pecchioni, 2011; Tsybulsky & Levin, 2016). Today, teachers can choose content from a tremendous range of digital resources to mediate their students' learning experience.

Digital curation in teaching characterizes the ways in which elements from digital culture can be combined with the curriculum and teaching materials. Teachers can group a variety of forms of content such as videos, photographs, texts, etc., to provide learning opportunities for their students (Potter, 2012). It is important to encourage students to use digital media to communicate, learn, and collaborate. Using a digital learning environment helps make learning and teaching more effective by increasing motivation, connecting teaching to students' preexisting knowledge, and by providing opportunities to learn regardless of location or time (Kümmel et al., 2020). The digital curation process helps reduce the cognitive load of acquiring new knowledge in a digital environment already saturated with information. Through curation, teachers can organize their ideas, sources of information, and digital tools to help them achieve their learning goals (Trust, 2016).

Teachers need to adopt a new system of pedagogical approaches and skills to prepare students for a life of constant technological development (Hobbs, 2011) that involves analyzing information, examining and assessing information critically, etc. The digital curation process improves learners' knowledge and 21st-century skills and integrates cognitive and metacognitive skills alongside social skills (Forkosh Baruch & Gadot, 2021). Using digital curation can help improve digital literacy and major skills such as critical thinking when engaged in online investigation, knowledge construction, and online navigation and the assessment of credibility. In essence, it promotes the principles of literacy and develops the skills of analysis, assessment, and creation (Mihailidis & Cohen, 2013).

One key science-teaching goal is to help students develop higher-order thinking skills such as critical thinking and decision-making (Zohar & Dori, 2003). Students' transition from knowledge construction to the development of higher-order thinking skills is necessitated by today's changing reality, and it is clear that to achieve this aim, teachers must first acquire and practice these skills themselves (Dori & Herscovitz, 2005). This makes it crucial to integrate teachers' experience with these new learning methods in professional development courses before they use them in the classroom (Kramarski & Kohen, 2017; Kohen et al, 2020).

Changes in society in the digital era of the 21st century are also manifested in the practice of science. Today it is hard to imagine a scientific study that does not implement technology (Tsybulsky & Levin, 2017). Scientists' digital scientific activities include collaborations and communications, data collection and curation, data analysis and sharing, etc. Digital curation is one of the most basic activities scientists engage in and can be considered one of the many science practices of the digital age (Tsybulsky, 2019). This significant change has implications for science education, which provides knowledge and disciplinary practices (Kelly, 2018). According to the Next Generation Science Standards (NGSS, 2013), the expectation in science teaching is to integrate core topics and broad scientific principles with scientific practice and not to separate the content. Thus, it is crucial to find ways in which students can be involved in practices similar to those of the professional science community, while keeping the changes in scientific practices in the digital age in mind. Specifically, new approaches must be developed, such as digital curation, which will enable the use of science practice in science teaching (Tsybulsky, 2020).

Teacher's Professional Development and Goal Setting

Professional learning networks (PLN) are systems of interpersonal connections and resources that support informal learning and consist of two types (Trust, 2012). The first is knowledge acquisition, which involves teachers' acquisition of new knowledge through a variety of online sources of information, such as blogs, articles, etc. Some sites have ways for teachers to follow or register such that they are notified of updated information without needing to search for it. The second type is social connections such as social or group platforms and online communication, where teachers can participate in discussions, ask questions, respond or share their knowledge, support or be supported, and get feedback. Combining these two types of learning allows teachers to obtain practical solutions and experience collaborative learning opportunities, which also shifts the picture from the slow professional development of the individual teacher to constant learning via divided expertise. Many teachers join online teachers' communities and sign up for various blogs and websites to continue improving their professional capacities (Macia & Garcia, 2016; Trust, 2016). Teachers increasingly use the internet to connect to others and improve their professional skills. Teachers build their professional expertise online by connecting to professional networks that provide immediate and effective access to information and connections with experts, colleagues, collaborating opportunities, and practical solutions. These online networks and communities provide an experience of ongoing independent learning connected to field experience, which contrast with traditional professional development activities for teachers such as workshops or official courses. Many teachers see their participation in a network as a significant source of professional development, as well as a place for social and emotional support and the construction and sharing of professional knowledge (Duncan-Howell, 2010, Kelly & Antonio, 2016, Trust & Prestridge, 2021).

Collaborative online platforms provide a space for self and group construction and lead to significant learning (Tour, 2017). Digital curation is a collaborative platform that constitutes a key concept in learning for organizations and individuals. Organizations that engage in quality curation will find themselves able to create value from their knowledge and acquire expertise within the organization. Individuals who adopt the practice of digital curation will be equipped with more personal knowledge than their colleagues (Betts & Payne, 2016). Identifying the goal of the curation process is highly important when curation is implemented in learning; it leads to more efficient decision-making regarding which resources to curate and how and which content best supports learning (Kelly, 2018).

Teachers do not necessarily think about their teaching goals in their daily work. Teacher training programs tend to emphasize teaching practice and activities during lessons rather than the thoughts, knowledge, and concepts underlying classroom activities (Kennedy, 2016). To more precisely design effective teaching practices, it is not enough to analyze teachers' actions. Instead, it is essential to understand not only "what" teachers do and "how" they do, but also "why" they do it, as a basis for understanding teachers' professional knowledge building (Loughran, 2019).

SCIENCE TEACHERS' CURATING AND LEARNING

Digital Curation Courses for Science Teachers

We developed a model of teachers' professional learning for science teaching practices using digital curation. This model served as the basis for the design and implementation of two courses and teaching resources during the 2021-2022 academic year, which were delivered as online professional learning. Forty-four science teachers in post-elementary education volunteered to take part in the courses and the study. The teachers came from various backgrounds and differed in terms of teaching experience, academic background, area of specialization, gender, and ethnicity. Each teacher participated in one 30-hour course, of which 15 hours took place via synchronous meetings on the Zoom platform, and 15 hours involved participants' asynchronous learning. The synchronous sessions were experiential and encouraged collaboration and reflective dialogue about the learning process. The course syllabus, teaching resources, and course assignments were put together by the authors, as were the supervision of the learning and group discussions. The course model for science teachers proposed in this study enables teachers to activate their thinking skills, such as critical pedagogical thinking and setting teaching goals, using the process of digital curation as a method of learning: Which digital content to select? From which sources? How to connect this content to the learning goal? And more.

The courses included four central elements that were integrated dynamically (figure 1):

1. *Theoretical instruction in digital curation* – The teachers were taught the main principles of the digital curation process and educational digital curation in particular. The teachers were then taught the main stages, the goals of curation, and were presented with digital collections that other teachers had made which dealt with a range of topics in science education. In addition, the lecturer discussed the relationships between 21st-century skills and digital curation and then encouraged the participants to talk about the ways they felt these skills could be promoted through digital curation
2. *Familiarity with the digital environment for curation* – The digital curation process was carried out on the "Wakelet" collaborative platform. A pilot study showed it is a suitable environment for users and operations, works on all types of end devices, does not require high technological skills, and allows searching and sharing of digital content all over the world. The participants were introduced to the Wakelet platform and experimented with during two workshop sessions. The teachers created their private accounts, learned about the techno-pedagogical opportunities of using the environment, and practiced creating a new collection by adding items to an existing collection, and defining the visual presentation and the options for sharing. Choosing an easy-to-use learning platform served to focus on the pedagogy of curation rather than the issues involved in teaching the participants, who were not all tech-savvy, to use the technology and later implement it in their classrooms.
3. *Experience with the digital curation process* – During the asynchronous learning hours, each teacher chose a topic in science education and created a digital collection. The teachers defined their collections' pedagogical goals, practiced gathering content items, filtered the content, and exercised pedagogical thinking as to its suitability for the goal that they had defined. They also practiced adding personal comments to digital content, organizing items in a collection, and defining their collections' level of public sharing. In the subsequent synchronous meetings, the teachers presented their collection to their colleagues and discussed their teaching goals, which was followed by a reflective dialogue and feedback from colleagues.
4. *Building a teaching unit that integrates curation* –Two workshops and open discussions were held in which the teachers raised ideas for implementing digital curation in the teaching-learning-assessment process. The teachers suggested a variety of possibilities, such as digital curation as part of scientific inquiry, the use of digital curation for constructive assessment or summative assessment of learning, creating a digital collection as an interesting introduction to a new topic or as a review of a subject previously learned, and others. During the asynchronous learning portion, each teacher planned a teaching unit integrating digital curation as a method of teaching one of the topics learned in class. The teachers explained their pedagogical goals during the planning process, as well as the skills that would need to be used in the lesson they planned, how it would be implemented in the classroom (individual, group, full-class, or a combination), the advantages of a teaching model using curation and the challenges they expected to encounter. In the final synchronous meetings, the teachers presented their plans, received feedback, and fine-tuned their lesson plans in response to the feedback. Roughly half of the teachers implemented digital curation in their classes and shared their experiences.

Figure 1. Model for teachers' professional development course integrating digital curation in science education.

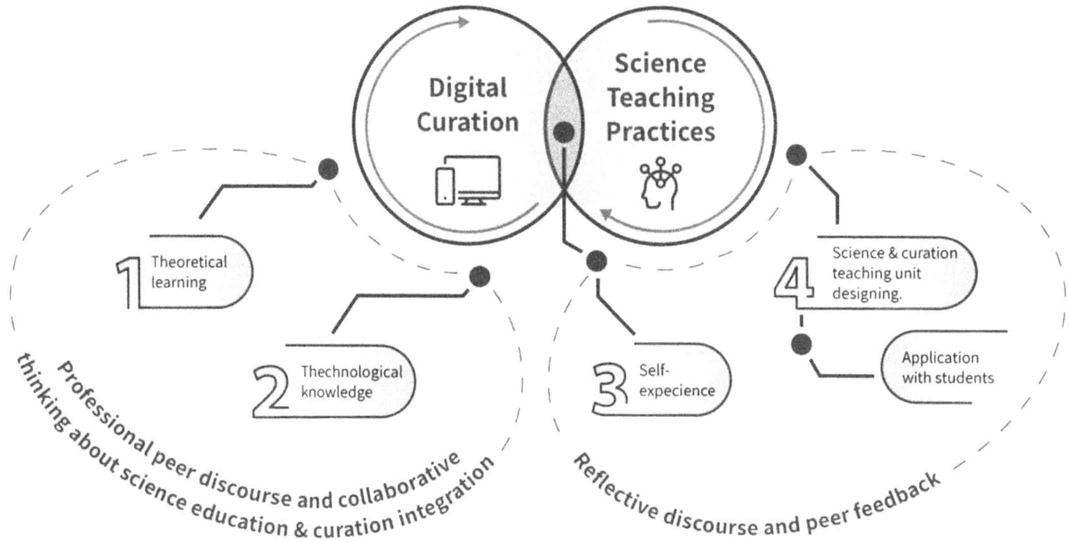

Teachers' Goals Using Digital Curation: Multiple Case Study

The research approach used was that of a multiple-case study. In order to gain an in-depth understanding of the pedagogical goals and methods of teaching involved in integrating curation in science classes, we used multiple qualitative research tools: observation of group discussions and the teachers' presentations of the collections they made during the workshops (n=44), reflective reports written by the teachers about six months after the beginning of the process (n=38); and semi-structured interviews (n=12). Figure 2 describes the timeline of data collection for the two courses as it related to the syllabus.

Figure 2. Data collection flowchart during the course.

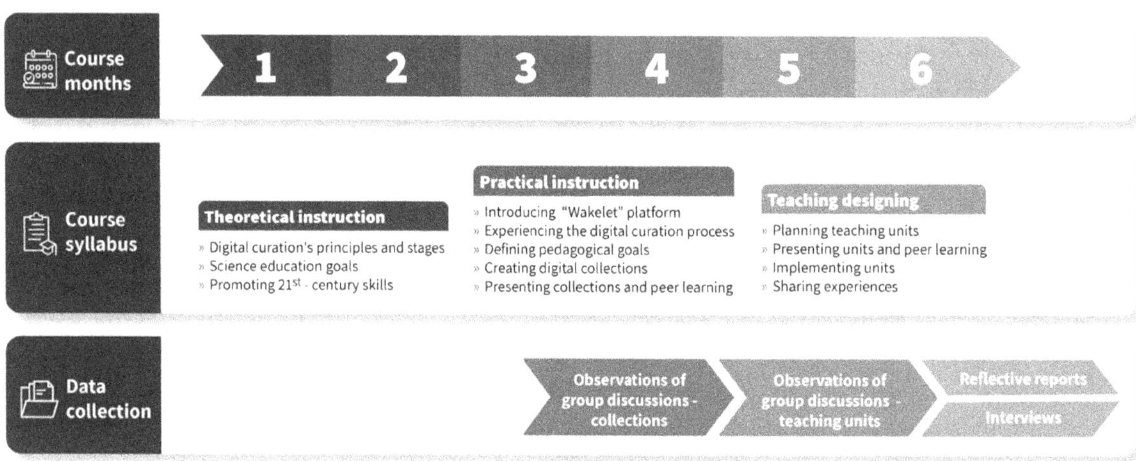

The data analysis included an initial sampling of teachers' general statements about teaching goals, learning, and technological applications in teaching. We conducted a content analysis of the first categories that emerged, and after repeated examination of the definitions of the categories, we arrived at more precise categories and themes. To ensure the reliability of the analysis with respect to labelling and categorizing themes, 25% of the data was analyzed by an additional researcher from the research group who was not part of the current study. Four main goals emerged: Teachers' professional learning goals, teachers' pedagogical and practical goals, Social and cooperative goals and Teachers' personal goals.

Below we describe each theme, and present examples and quotes. For purposes of confidentiality, teachers are identified by number.

1. Teachers' Professional Learning Goals

The teachers defined several goals they hoped to achieve by using digital curation in class and as end products. In this section, we discuss goals relating to cultivating the teachers' knowledge, both in terms of deepening their content knowledge of the subject they teach, and in terms of pedagogical content knowledge pertaining of the topic.

1.1. Enriching Teachers' Content Knowledge of the Subject They Teach

The teachers stated that engaging in collecting materials about a specific topic in science teaching prompted them to read a wider variety of items online and enhanced their knowledge of the subject beyond the content that they teach in class. Teacher 13 defined the goal of increasing content knowledge *before* starting to curate:

I chose materials for myself, to expand the topic beyond what's in the curriculum and the material that is in the textbook. The topic of the digestive system is a field that I felt less connected to at the beginning of the curation process and I wanted to know more about it before teaching the material to my students. I read the material to choose what to add to the collection, and I learned more about the topic (Teacher 13, group discussion).

Teacher 21 also felt that digital curation gave her a better understanding and a broader knowledge base:

It (curation) gives you a broad base. Like if you are learning a chapter and you learn far more than what you are required to, that is how I felt about the collection, when I prepare to teach, it gives me the broadest understanding of this whole topic, in depth (Teacher 21, interview).

Teacher 1 defined a similar learning-related goal *during* the curation process:

A significant advantage to building a collection has to do with the way you learn without feeling like you're learning... It is like a door that opens and leads to another door, it is an excellent opportunity to enrich your knowledge and learning (Teacher 1, group discussion).

Teacher 7 described similar content learning goals at the *end* of the curation process: "And the most important thing was when I felt that I myself was enriching my own knowledge of the topic as a result of making the collection" (Teacher 7, written reflection).

1.2. Enhancing Teachers' Pedagogical Content Knowledge

Beyond increasing their content knowledge, teachers also enhanced their pedagogical content knowledge pertaining to teaching a specific field of science, thus responding to students' critiques of classroom activities. For instance, during one of the group discussions, Teacher 25 said:

The fact that I looked for concepts in itself clarified my thinking about what the students might be thinking, what ideas the students might have and what I need to do in class... There is a very strong element here of learning about the topic that you are creating a digital collection for (Teacher 25, interview).

Some teachers noted that digital curation allows teachers to define their pedagogical goals more precisely and engage in critical thinking as to the choice of items that best suit the collection's structure and goals:

The process of creating the collection made me compare different items with similar content, and then decide which content was better suited to the collection and to the goals of the collection (Teacher 7, written reflection)

I learned that the process of curation requires time and thought... I invested a lot of time in locating content and in making decisions about what content would be included in the collection and what would not... I reviewed each and every resource (Teacher 10, written reflection).

Several teachers mentioned technological pedagogical content knowledge and stated that they upgraded their technological skills and felt at ease integrating digital curation into the learning process. Teacher 41 is one example:

In my work in education over the years, I have tended to stick to older styles of learning, and struggle to bring in new methods of learning ... even though my technological abilities are quite high, I generally struggle to create an environment in which it is easy for me to introduce new tools. So, I was really happy that a big part of the program was preparing a lesson that integrates digital curation, which requires me to use it, and helped let me get over the fixation I had in this area. Now I know that I am able to do it, and I can also include my students in work with new tools (Teacher 41, written reflection).

1.3. Collaborative Learning, Sharing Inspiration and New Ideas

Part of the teachers' professional development involved sharing their collections and a directed review of others' resources, which was a significant part of the course, both during the group discussions and as feedback on the participants' presentations of their own collections:

I learned a lot from others' collections, I loved hearing about other people's different ways of thinking, about what is worth emphasizing. I also loved sharing our experiences, challenges, materials and knowledge and getting the perspectives of the other teachers (Teacher 3, group discussion).

In their written reflections, the teachers described the learning process. They stressed that it was a very significant part of the course: "From a professional perspective the exposure to colleagues' collections may have been the most effective element, because it gave me ideas for future lessons" (Teacher 6, written reflection);

Looking over the collections on different topics introduced me to links from sources I wasn't familiar with, and contributed to my professional development. If I plan lessons on these topics, I can use these sources in my teaching, and I will know where to find them. I was happy to be exposed to more collections, both created by others in the course and those I found in a free search on Wakelet. I found collections and links in collections that were new to me, which I believe will be useful to me in my teaching, both for preparing lessons (as background material for the topic) and for introducing to the students as part of the lesson (Teacher 4, written reflection).

Half a year after the process began, while writing their reflective reports, some teachers still remembered specific examples of new things they had learned from the collaborative work and noted them:

I feel like the collaborative work is valuable in that it really promotes exposure to didactic materials, for example in the field of education and in putting together lesson plans… I thought it was very meaningful, when my classmates were presenting their collections to the class, I was exposed to particularly rich worlds of content in fields that relate to what I teach and collections that are not tangential to my field at all … and for each I learned how rich they are, how interesting, how varied (Teacher 2, written reflection).

The process of creating the collection introduced me to sites and sources that I had not been familiar with previously, and even taught me how fruitful the partnership process can be. When we were asked to search for other collections similar to ours, and to take relevant items from there for our collection, I discovered a collection on "science and technology for all: an in-depth approach" and there I discovered a real digital curator! There were a ton of educational slideshows for middle school and high school that were very, very well done – the kind that could even have served me well when I taught in college, if I had been familiar with the tool and used it then (Teacher 11, written reflection).

1.4. Locating Teaching and Learning Resources

Some teachers defined their goals for creating the collection in advance, and related to a variety of teaching methods and pedagogical creativity when locating educational resources relating to the subject:

Because I wanted to refresh my lesson plan a bit in advance of teaching the topic in 10th grade in the near future. First, the idea was to make the topic as accessible as possible to the students, because we did not know if we would be teaching online or physically in the school. Either way – I felt the need to revise it. That was the principle that guided me. It is hard to describe the satisfaction that comes from *finding new and varied materials, especially when that was my goal for creating the collection (Teacher 5, written reflection).*

Beyond finding new teaching resources, Teacher 5 described how curating expanded her way of thinking:

I think that using digital curation allowed me to push the limits. In other words, although I chose a very specific topic – excretion in humans – adding other topics that were likely to be connected was much simpler. I felt like it broke the mold, it got me out of the box (Teacher 5, interview).

1.5. Creating a Focused Continuum of Instruction

The process of curation and organizing the digital content led some of the teachers to think about the learning sequence of the entire unit, not just the collection of the content items connected to the topic:

It can help focus on lesson preparation, it can give the teacher a kind of timeline for the lesson that's based on digital curation. It gives you a sort of preview of the flow of your learning unit, like points in time. I know that this will be the first lesson, this will be the second, this the third, it is a kind of order (Teacher 24, interview).

Teachers 2 and 7 mentioned that the instructional sequence is important when teachers are weighing their teaching goals:

I think that it would be very effective to prepare a digital collection before any new topic you are supposed to teach, because then the teacher can feel more focused… They know what they are going to teach at each stage of the topic (Teacher 7, group discussion)

I invested a lot of thought into how to build the collection in a didactic way, so that it would be clear to the user, with a structure that is not too complex for the users, but rather draws them deeper into the collection, and look at additional content. It was important to me to start with "teaser" items, to include fairly simple explanations …and to give as many concrete examples as possible for inspiration (Teacher 2, written reflection).

2. Pedagogical: Practical Goals

2.1. Boosting Interest, Experience and Involvement in Learning

One of the important teaching goals of curation that the teachers mentioned was increasing students' interest and curiosity. Teacher 11 wrote:

I thought it would be smart to give them access to a rich digital folder… A folder with learning material presented in a variety of forms… To kindle scientific curiosity in the next generation… The expectation is that while looking for the most relevant things, they will get interested and be curious enough to keep looking, and after all, active curiosity is the basis for innovative research, which is the aim (Teacher 11, written reflection).

The teachers set a goal in advance of looking for items that would get their students out of the rut of their usual learning routine and prompt them to connect to the topic and take a greater interest:

I searched for something on the topic of the cell, something a bit different, non-standard, that would be a kind of interesting introduction to the lesson... When I teach I like to start with something to stimulate their interest. So I downloaded two videos... that were interesting enough and stimulating enough to begin with (Teacher 40, interview)

The students will learn a new technological skill that will make the learning process more experiential, more effective than the regular conventional method which is sometimes boring... The process of curation is very helpful in that it presents the material in a non-routine way, so that students will enjoy learning and their motivation will increase, will be noticeable (Teacher 44, written reflection).

Students who engage in gathering content to create a collection become active in the learning process. As Teacher 29 stated, after trying out curation in class:

What helped them was how to choose and gather material, and phrase questions, and be an integral part of the process of collecting the learning material, including a discussion and debate amongst the students and suggestions for sources and sites (Teacher 29, written reflection).

Teachers who used curation with their students testified that the process elicited more interest and excitement than other teaching methods:

Using curation in class is very easy, flexible, effective, and dynamic, it shakes up the usual class routine, it also boosts the class's aspirations and desire to learn. The students felt like part of the learning experience. There was a group activity, and each student had a specific role within the group. Within this role, the students played the main part and contributed to the learning, there was a positive competition that led to a really perfect lesson and more effective learning (Teacher 31, written reflection).

When students' interest, curiosity, and involvement increase through curation, the learning process becomes more experiential for them. Many teachers mentioned this in their interviews, written reflections, and discussions:

Experiential learning, without a book and notebook, makes for learning that is challenging and interesting... The curious student will enjoy a class that is partly digital more than a class that is just frontal instruction, and that gives me and the students a chance to think outside the box (Teacher 32, interview).

The pedagogical advantages of using curation in class are first of all that it's a different type of learning, it is challenging and novel. The students said that this kind of studying was interesting for them; they enjoyed it and even asked to learn other topics this way. What was meaningful for me about the experience was seeing the students getting excited, and the products they created, and their response in itself, that they were happy to study independently, online, and collaboratively, that gave me a lot of satisfaction (Teacher 28, written reflection).

2.2. Addressing the Differences Between Learners in Heterogeneous Classrooms

Most of the teachers who participated in the course taught in heterogeneous classrooms, in which the students have a range of cognitive abilities, preferred learning styles, and engagement in learning. The process of curation helped teachers to address the needs of different students in one classroom:

My goal was to create a collection that includes materials for me as a teacher to use in class with students, to provide instruction that is suited to each individual student in class, and to the conditions in the classroom (Teacher 13, written reflection).

It is something that basically lets learners connect to what they're learning, the way they are learning... they understand and feel that it is something they are capable of doing... They have a sense of capability, that it is on their level, that they... that it is a match. In terms of the level of difficulty, that it's suitable for them in terms of relevance, the way they prefer to learn, and the level of difficulty (Teacher 25, interview).

Teacher 44 summarized his success in responding to students' various learning needs:

The lesson was fascinating; all the students in class participated and expressed their opinions. Clearly, in this lesson I succeeded in reducing the gap between the strong students and the struggling students (Teacher 44, written reflection).

2.3. Boosting Critical Thinking

In learning activities that involve an online search for information, one of the challenges students face is filtering a large quantity of information effectively. This filtering involves critical thinking and examining the information for relevance. Some teachers felt that curation promotes students' critical thinking:

This way, the learner is required not only to implement the technological skill of searching, but also to demonstrate critical thinking throughout the process of choosing and organizing sources of information. Digital curation integrates not only well-reasoned choices, but also knowledge construction in a deep sense that is achieved gradually, and does not depend on learners' technical abilities, but rather on the understanding they acquire over the course of the process. By building a digital collection the learner internalizes the need to be organized, and the skills of sorting sources of information. In a certain sense it can be said that there is more reflection involved in digital curation tasks or projects than in tasks that only involve searching for and obtaining information (Teacher 6, written reflection).

In an activity built around gathering information the students need to find sources and sites that provide them with the answers they are looking for about heart disease, so they use critical thinking (Teacher 29, written reflection).

The teachers also mentioned learning to assess the reliability of the information as part of the process of critical thinking:

I want the students to learn to upload content that is significant and important, I want them to know how to filter sources of information and learn to distinguish between sources that are good and those that are less good, how to know whether a source is reliable… Filtering sources of information is a crucial thing to practice today, with fake news everywhere (Teacher 34, written reflection).

The students tackled the difficulty of searching for and finding reliable sites, appropriate material, and the need to make decisions about which materials are more suitable and which are less relevant and less suitable (Teacher 28, interview).

2.4. Practicing Digital Skills and Science Practices

The teachers also mentioned that students' digital skills, media, and information literacy are manifested during the digital curation process:

The fact that the collection is generated by the student while learning in the physical world, means that we essentially create an opportunity for meaningful learning that engages higher-order learning skills and thinking skills (investigation, drawing conclusions, constructing and implementing an argument) (Teacher 14, written reflection).

After trying digital curation in class, I understood that today's generation is deeply connected to the world of media, and is very connected to digital technology, and it's impossible to escape the clear and obvious fact that we as teachers need to change the way we teach, and the frontal method of instruction that we've been holding onto so tightly for years, we need to reach the students and awaken the tremendous potential and skills inside of them by coming closer to their real world, which is the media (Teacher 31, group discussion).

Teacher 28 reflected on the potential of curation when it comes to reviewing material and imbuing students with digital skills:

While using digital curation students can practice a lot of skills: digital literacy, critical thinking, independent learning, collaborative learning, information literacy, scientific literacy. The process allows them to practice a large number of skills in a relatively short time (Teacher 28, written reflection).

2.5. Simplifying the Topic of Study Through Visual Representations

Many topics taught in science are difficult for students to understand without visuals. After trying curation themselves, the teachers adopted the idea of scaffolding the learning process as an essential aspect of digital curation:

… Visual ways to illustrate the processes, to give examples of molecular structures. That is why I created a collection of video clips and slideshows for the students' use that can really help them to understand molecular models like the structure of atoms, molecules, cells or understand mechanical things, physical forces or exemplify processes in internal organs of the human body (Teacher 9, written reflection).

Another goal is to implement the process of curation in a way that reflects simplicity and order. It was important to me to make the information accessible in a clear, visual way, accompanied by short, concise explanations of mine that were not too difficult for the user (Teacher 2, group discussion).

After using curation in class with her students, Teacher 33 felt that her students had a better understanding of the topic because of the visual presentation:

I felt like this method made it easier for the students to better understand the material, after I taught them the material in class, and by searching for material and videos and pictures that really gave a visual presentation of the subject the students had a better understanding, and really assimilated the material more (Teacher 33, written reflection).

2.6. Alternative Assessment of Process and Learning Outcomes

Teacher 14 decided to try the digital curation process from the outset with her students as part of a scientific inquiry. She stated that through curation, she was able to make a constructive assessment, which was significant for her and for her students:

Because the process of collecting items for the collection is transparent, and at any point I can see who is adding what, and see from their comments why the participant decided to add that content. I could supervise the process on the spot and point out mistakes or disagreements… The teacher can, at any time, follow the process in real time, respond, guide, and assist – while the students work – it is a step up in our ability as teachers to guide students in making a real investigative process, and to achieve meaningful learning… Ultimately, the collection serves as a product the teacher can assess and give a grade to, and also check and assess the role that each student played in the group work (Teacher 14, interview).

Teachers 27 and 41 also noted that students could engage in assessing their own work or could compare their work to their peers: "They engaged in an ongoing reflection during the whole process, about themselves and their work" (Teacher 27, written reflection);

Beyond the fact that it allows the teacher to follow the class activity in real time, and to observe the processes the students have undergone, curation also allows the students to observe their peers' learning processes (Teacher 41, written reflection).

3. Social and Cooperative Goals

The curation process by nature is also social and collaborative. During the course, sharing occurred between teachers and their classmates, teachers and other online teachers and experts they did not know. Sharing also takes place between teachers and students, or among students. The teachers' goals supported collaboration and social connections.

3.1. Professional Relationships With Fellow Teachers and Subject Experts Online

The teachers mentioned that learning from other teachers or from experts in the field online was quick and easy:

Compiling learning materials – getting inspiration from other teachers' collections, monitoring active collections by colleagues from all around the world; the ability to select the materials that would help me from out of the vast amount of information online, simply because a teacher I follow used a certain item in his collection, which filtered a lot of this information for me (Teacher 2, written reflection).

It is easy to share it with other teachers and for people to share their collections with each other, and that makes it accessible to all of us, in contrast to sharing with other techniques and each time starting to search from scratch, it's right there in the collection with a picture and a brief description, because you can add a picture to every collection and to every (file or slideshow and video), which makes it easier for teachers to find what they are looking for (Teacher 20, interview).

The teachers considered that collaboration was a professional learning process: "Similarly, it was also important to me to follow the collections of people I admire and to learn from them and through them" (Teacher 9, written reflection).

3.2. Collaborative Learning Among Students

Digital curation can be a convenient platform for collaborative learning among students:

Beyond the power of curation as a pedagogical tool that serves as a reference for the student when reviewing material for a certain learning unit, or as a collaborative tool for knowledge sharing for teachers, digital collections have a lot of other pedagogical applications. For example, as a platform to synchronize and coordinate group work on a research project… It's useful for coordinating research work on the topic of organ donation in a group, and as a platform where the students in the work group collect material and edit it as a group – to create a single coherent product (Teacher 14, written reflection)

Collaboration with students – building shared collections with students, giving a class assignment involving the use of this tool. It is simple to use, clear, didactic, and curation can definitely be integrated into class work (Teacher 2, written reflection)

Coordinating material, sources of information, and putting things in a logical order. Thanks to the order and organization, they can get through the program this year. It is a clear, familiar option for finding material, without worrying about 'where did I save that source, where is the picture, etc. (Teacher 34, written reflection).

In addition to ease of use, some teachers found the ability for students to share to be valuable, a pedagogical goal, and a way to teach the students a variety of skills linked to collaborative work:

The socio-cultural interactions promote learning based on a culture of participation, sharing, empathy, mutual respect, and learning about one another. Curation management is based on rules that the group agrees on, and is implemented through practical experience. The activity is conducted in a group in which each student has a certain role, and students share ways of collecting and presenting material with each other, which encourages collaboration and mutual respect (Teacher 16, written reflection).

The collaborative work in curation leads to cooperation among the students, by sharing information and ideas, expanding each student's abilities through interactions in a way that wouldn't have happened through independent work (Teacher 32, written reflection).

One teacher summarized the benefits of collaborative work in a particularly striking way:

The curation process we were introduced to is a different world. Collections that can be shared with teachers across the country, and around the world. Collections that students can create themselves and share with each other, and more... The sky is the limit (Teacher 38, written reflection).

4. Teachers' Personal Goals

Some of the goals that the teachers defined were not specifically pedagogical but rather supported the teachers' pedagogical work in managing and organizing teaching resources and making them available in the short and long term. These goals were expressed in the teachers' reflections after trying out the curation process and completing the course.

4.1. Effective Management of Digital Teaching Resources

The ease of creating a collection, the simplicity of adding comments and organizing it to match the lesson plan. The ability to devise an organized lesson plan with the collection, to organize the ideas ... Even someone who is not great with technology can use this easily (Teacher 3, written reflection).

The next time I need to teach that material, I can go to the collection to use the same materials I already found, and the process will be faster. I am used to saving files on the computer in a very organized way, but today, especially with distance learning, a lot of the learning material isn't in files, it is in links to websites, video clips, etc., and there needs to be a convenient way to save it (Teacher 13, written reflection).

I am a teacher who loves order when it comes to learning topics, so instead of having a lot of files of tests and explanations and worksheets about a certain topic, I can save everything I need about a certain topic in an organized way, in a sequence that I devise for myself... It stays on the computer and does not get lost, and that makes it a lot easier for me (Teacher 32, written reflection).

One of the significant advantages of digital curation is that teachers can react to the items. Teachers who used the 'adding comments' option felt that this was very helpful in managing and retrieving stored items:

I have everything in one place, and I even add things, for example I really like design and I also have collections about design... So that helps me with everything that interests me, everything is organized, and I feel like I am in a library, each folder has its own picture, even if I don't find the page when I copy the link, I can open it and write a note for myself, what's in there, and that usually sets things straight in my head" (Teacher 20, interview). "Curation has a significant advantage over storing items on a personal computer or in the cloud, which is the visibility of the items in the collection. First of all, you 'see' the items because there is a preview. In addition, you can add a written description of the item.

So, I see the item and its content, and I don't need to guess what each item is. Since I started using this platform, I curate all my digital resources on this platform (Teacher 10, interview).

4.2. Long-Term Storage of Teaching Resources

The teachers mentioned the advantages of using curation for storing teaching resources for long periods of time:

The digital curation process helped me organize the materials; instead of saving the links to video clips in my internet browser, email, WhatsApp or on the computer, and then struggling to find specific items, now everything is in one place, with my personal opinion written next to each item, which will help me find all the material I need easily in the future. I really got into this! (Teacher 7, written reflection).

It really allows you to save information for years, without relying on physical storage which might be destroyed or get damaged in the distant future" (Teacher 2, group discussion). "I realized that I can come back to it another year, in my next class, and I will not have to start wracking my brains over 'what did I do with the previous class, what worked, what didn't work.' That was something I thought was awesome, that I can always add (Teacher 22, interview).

4.3. Availability of Teaching Resources From Any Location, at Any Time

The fact that the digital collections are stored online, in the teachers' personal account or a shared account allows them to access the collection they saved any place and at any time. The teachers mentioned this as an additional techno-pedagogical benefit:

I created the collection with the goal of sharing it with my colleague in course development, in a practical way. The platform is collaborative and online, so it is accessible to both of us any time and in any place (Teacher 10, interview).

I realized that instead of saving my files and my collections in folders on my computer and going around with a disk-on-key, I could create collections on this site that I can access from any computer (Teacher 37, written reflection).

It has the advantage that it's accessible from anywhere, so I can go to any computer, and it is there. And if I'm in class or on a trip, and I want to access the collection, I know I will find it there (Teacher 35, interview).

RECOMMENDATIONS AND FUTURE RESEARCH DIRECTIONS

As a pedagogical strategy, digital curation is student-oriented; it enriches students' digital literacy, improves their ability to deal with an abundance of information, and contributes to improving skills that are essential on social media and online communities (Mihalidis & Fromm, 2014). Curation changes the learning process, in a way that is meaningful and relevant for teachers and students. This emphasizes the need to integrate it in teacher training and provide practice with the curation process until a good level of expertise is reached (Forkosh Baruch & Gadot, 2021). More research and additional models for integrating digital curation in teachers' professional development are needed. Digital curation can be integrated into a variety of classes and science lessons. Future research could examine the potential applications of the digital curation process in the classroom and derive other models for students' independent and group learning.

CONCLUSION

This chapter discussed digital curation as a professional learning process for science teachers and is based on a study we conducted in the context of courses we created and designed. Curation creates a rich, active environment that supports teachers' professional development and contributes to refining the curator's knowledge and practices. Curation helps teachers achieve a range of personal, learning-related, social, and professional goals. It provides an opportunity for teachers to build a personal ontology, choose from a rich variety of resources, and expand their own skills within an open and flexible learning environment, and create a personalized collection. Through curation, teachers can organize their ideas and their sources of information and achieve their educational goals.

The model developed here is based on the experiences of 44 science teachers and reflects their learning about curation and its implementation in the classroom. Its prime contribution is to draw attention to the value of digital curation in science education in the digital age. This implementation is likely to lead to innovative teaching practices for today's information era. Digital curation has a high potential for implementation in science teaching in the digital age. It requires curators to use their judgement, higher-order thinking, and critical thinking, connect prior knowledge to new knowledge and organize the collection coherently (Gadot & Tsybulsky, in press). For these reasons, it is crucial for today's teachers, who are equipping students with 21st-century skills, to become good digital curators themselves by recognizing and taking advantage of the digital resources that correspond to the goals set forth in the curriculum and meet students' needs (Deschaine & Sharma, 2015; Forkosh Baruch & Gadot, 2021). These teachers can then provide their students with similar learning opportunities by integrating digital curation into the classroom.

ACKNOWLEDGMENT

This research project was funded in part by the Office of the Chief Scientist, Israel Ministry of Education Research Grant (No. 36/12.2020).

REFERENCES

Antonio, A., Martin, N., & Stagg, A. (2012). Engaging higher education students via digital curation. In *Proceedings of the 29th annual conference of the Australasian society for computers in learning in tertiary education (ASCILITE 2012)* (pp. 1-5). Australasian Society for Computers in Learning in Tertiary Education (ASCILITE).

Beagrie, N. (2006). Digital curation for science, digital libraries, and individuals. *International Journal of Digital Curation*, *1*(1), 3–16.

Betts, B., & Payne, N. (2016). From content to curation. In A. Anderson & B. Betts (Eds.), *Ready, set, curate* (pp. 7–13). Association for Talent Development.

Dayan, E., & Tsybulsky, D. (2021). *Science teachers' perceptions regarding digital curation as a personalized learning activity that promotes professional learning*. In NARST2021 Annual Meeting, Orlando, FL.

Deschaine, M. E., & Sharma, S. A. (2015). The Five Cs of Digital Curation: Supporting Twenty-First-Century Teaching and Learning. *InSight: A Journal of Scholarly Teaching, 10*, 19-24.

Dori, Y. J., & Herscovitz, O. (2005). Case-based long-term professional development of science teachers. *International Journal of Science Education*, *27*(12), 1413–1446. doi:10.1080/09500690500102946

Duncan-Howell, J. (2010). Teachers making connections: Online communities as a source of professional learning. *British Journal of Educational Technology*, *41*(2), 324–340. doi:10.1111/j.1467-8535.2009.00953.x

Forkosh Baruch, A., & Gadot, R. (2021). Social curation experience: Towards authentic learning in preservice teacher training. *Technology, Knowledge and Learning*, *26*(1), 105–122. doi:10.100710758-020-09449-3

Gadot, R., & Tsybulsky, D. (in press). Digital curation as a pedagogical approach for promoting critical thinking. *Journal of Science Education and Technology*.

Kapor, M. (1991). *Internet Quotation Appendix*. Harvard Berkman Klein Center. https://cyber.harvard.edu/archived_content/people/reagle/inet-quotations-19990709.html

Kelly, G. J. (2018). Developing epistemic aims and supports for engaging students in scientific practices. *Science & Education*, *27*(3-4), 245–246. doi:10.100711191-018-9974-y

Kelly, N., & Antonio, A. (2016). Teacher peer support in social network sites. *Teaching and Teacher Education*, *56*, 138–149. doi:10.1016/j.tate.2016.02.007

Kennedy, M. (2016). Parsing the practice of teaching. *Journal of Teacher Education*, *67*(1), 6–17. doi:10.1177/0022487115614617

Kohen, Z., Keller, N., & Koichu, B. (2020). Metacognitive processes in online collaborative problem solving forums: mathematics teachers' dual roles. *Proceedings of the International Commission on Mathematical Instruction (ICMI 25)*, 484-491.

Kramaraski, B., & Kohen, Z. (2017). Promoting preservice teachers' dual self-regulation roles as learners and as teachers: Effects of generic vs. specific prompts. *Metacognition and Learning*, *12*(2), 157–191. doi:10.100711409-016-9164-8

Kümmel, E., Moskaliuk, J., Cress, U., & Kimmerle, J. (2020). Digital learning environments in higher education: A literature review of the role of individual vs. social settings for measuring learning outcomes. *Education Sciences*, *10*(3), 78. doi:10.3390/educsci10030078

Loughran, J. (2019). Pedagogical reasoning: The foundation of the professional knowledge of teaching. *Teachers and Teaching*, *25*(5), 523–535. doi:10.1080/13540602.2019.1633294

Macia, M., & García, I. (2016). Informal online communities and networks as a source of teacher professional development: A review. *Teaching and Teacher Education*, *55*, 291–307. doi:10.1016/j.tate.2016.01.021

Mihailidis, P., & Cohen, J. N. (2013). Exploring curation as a core competency in digital and media literacy education. *Journal of Interactive Media in Education*, *2013*(1), 1–19. doi:10.5334/2013-02

NGSS Lead States. (2013). *Next Generation Science Standards: For states, by states*. National Academies Press.

Shirky, C. (2010). *Cognitive Surplus: How Technology Makes Consumers into Collaborators*. Penguin.

Starkey, L. (2010). Teachers' pedagogical reasoning and action in the digital age. *Teachers and Teaching*, *16*(2), 233–244. doi:10.1080/13540600903478433

Tour, E. (2017). Teachers' self-initiated professional learning through personal learning networks. *Technology, Pedagogy and Education*, *26*(2), 179–192. doi:10.1080/1475939X.2016.1196236

Trust, T. (2012). Professional learning networks designed for teacher learning. *Journal of Digital Learning in Teacher Education*, *28*(4), 133–138. doi:10.1080/21532974.2012.10784693

Trust, T. (2016). New model of teacher learning in an online network. *Journal of Research on Technology in Education*, *48*(4), 290–305. doi:10.1080/15391523.2016.1215169

Trust, T., & Prestridge, S. (2021). The interplay of five elements of influence on educators' PLN actions. *Teaching and Teacher Education*, *97*, 97. doi:10.1016/j.tate.2020.103195

Tsybulsky, D. (2019). Transformations and emerging implementations of scientific practices in the digital age. *Proceedings of the International History, Philosophy, and Science Teaching Conference*, 532-541.

Tsybulsky, D. (2020). Digital curation for promoting personalized learning: A study of secondary-school science students' learning experiences. *Journal of Research on Technology in Education*, *52*(3), 429–440. doi:10.1080/15391523.2020.1728447

Tsybulsky, D., & Levin, I. (2016). SAMR framework for study technology integration in science education. New Perspectives in Science Education Conference Proceedings of the 5th Edition. Florence, Italy: Libreriauniversitaria.it.

Tsybulsky, D., & Levin, I. (2017). Inquiry-based science education and the digital research triad. In *Digital Tools and Solutions for Inquiry-Based STEM Learning* (pp. 140–165). IGI Global. doi:10.4018/978-1-5225-2525-7.ch006

Tsybulsky, D., & Levin, I. (2019). Science teachers' worldviews in the age of the digital revolution: Structural and content analysis. *Teaching and Teacher Education*, *86*, 102921. doi:10.1016/j.tate.2019.102921

Watson, J. A., & Pecchioni, L. L. (2011). Digital natives and digital media in the college classroom: Assignment design and impacts on student learning. *Educational Media International*, *48*(4), 307–320. doi:10.1080/09523987.2011.632278

Zohar, A., & Dori, Y. J. (2003). Higher order thinking skills and low achieving students: Are they mutually exclusive? *Journal of the Learning Sciences*, *12*(2), 145–183. doi:10.1207/S15327809JLS1202_1

Chapter 11
Principled Integration of Technology for Science Learning That Lasts

Kevin Krahenbuhl
Middle Tennessee State University, USA

Jim Rost
Middle Tennessee State University, USA

ABSTRACT

The world of education is caught up in trends that affect all corners of contemporary society. Those who are working in the sciences have to help students learn in this new environment of fast-paced expectations and information overload. The intention of this chapter is to present a set of principles—or axioms—about learning that have direct bearing on instructional utilization of technology in the classroom. This will, in turn, provide a guide to an effective integration of technology that promotes learning that lasts within the sciences. The principles introduced herein balance knowledge of cognition as well as application in K12 science classrooms. The principles range from moderation to when to appropriately utilize technology for enhancing science learning to precision and extension in which emphasis is placed upon in what manner to focus technology integration to enhance science learning and also a principle of timely application that clarifies not just 'when' in terms of content but also in terms of what skills and knowledge are in place among the students.

INTRODUCTION

The world of education today is caught up in trends that are affecting all corners of contemporary society. Among the most relevant of these trends is the shift which has occurred in the past several decades from a world of information scarcity to information overload. In this context, everyone is bombarded with information, nearly every moment, which carries with it both possibilities through the easing access to information but also many pitfalls. In this environment, change is seemingly constant, information is often

DOI: 10.4018/978-1-6684-5585-2.ch011

inconsistent, and confidence in scientific knowledge is declining (Aksoy et al., 2020). This has notable implications for educators – especially for those in the sciences, which, ever since the enlightenment, Western society, in particular, has come to put great trust in what appeared to be an endless success. But now as questions arise and confidence wanes in scientists themselves there is an increasingly important role for science educators. Namely, the role in both re-establishing confidence in scientists but also to not naively grant any undeserved reverence to scientists, who are, humans, engaging in the scientific process that is marked with errors along the never-ending path to improved knowledge.

In this context, some have argued that these trends, and the influences they have on generations raised within them, has led to entirely new theories of learning, such as *connectivism* (Siemens, 2005). Whether or not these theories hold true across time is not as important as a serious reflection on the implications these trends have for instructional practice. One could argue that no subject area is more under the microscope with regards to these changes in the pace and availability of information than the domain of science. Broadly, we can recognize that the environment in which science educators find themselves is like that of the larger world – busy.

1. How can science educators best teach in such an environment?
2. How can decisions be made that maximize the effectiveness of classroom time?
3. How can technology be integrated purposefully to extend learning?

Implications

These questions will be explored through the rest of this chapter. It is the intention of this chapter to present a set of principles – or axioms – about learning that have direct bearing on instructional utilization of technology in the classroom. This will, in turn, provide a guide to an effective integration of technology that promotes learning that lasts within the sciences. First, we will survey literature related to the topic, then we will present a list of essential principles that science educators should consider. Next, several examples will be drawn from real-world K-12 science classrooms that reflect the application of these principles. Following the specific examples, the chapter will conclude with discussion and consideration of implications that these principles have for practice. This will conclude with a heuristic to be applied for the purpose of informing principled integration of technology for science learning that lasts.

LITERATURE REVIEW

To address the topic of how to integrate technology in a principled manner into the science classroom it is first important to consider the broadest and most foundational element underlying the purpose of taking a science class: learning. Learning requires making information your own and that entails that learning is active (Ambrose et al., 2010) as well as encompassing a real change in which something new is retained by the learner (Kirschner et al., 2006). In what follows we will explore learning in the context of science and in consideration of technology.

Learning in Science

One of the particularly intriguing developments over the past few decades in cognitive science is what is referred to as the domain-specificity of learning, which is distinct from what is sometimes called varied ways of learning (Pritchard, 2018). There has been debate over whether breadth or depth was central and while there is no doubt that both play a significant role in maximizing our ability to be successful (Epstein, 2019) much research has pointed to the not-so-obvious fact that some skills are not really transferrable but are rather, domain-specific. Tricot and Sweller (2014) explain that domain-specific knowledge is "memorized information that can lead to action permitting specified task completion over indefinite periods of time" (p. 266). This is to be contrasted with so-called domain-general skills, which are imagined to be able to be used to solve any problem in any area. Researchers examining experts in particular domains have consistently found that the central feature that defines these experts is not some generic skill but rather a huge amount of domain specific knowledge that they hold in long-term memory (Ericsson & Charness, 1994).

In the context of the domain of science, this has also been recognized and seen an increase in emphasis around what the central features of quality learning look like within the field of science. Resulting from these conversations came the development of subject-specific standards that aim to identify underlying conceptual ideas and thinking skills has been ongoing. Within the domain of science this has resulted in the creation of the *Next Generation Science Standards*, which were developed as a research-based approach to transform science education in a way that aided with cultivation of informed citizens who were prepared for college and/or careers and who understood and appreciated the scientific enterprise (Lee et al., 2015). These standards include three dimensions identified in this process to clarify those features that are most central to guide people in effective science learning. As such, the current understanding for directing effective learning within the domain of science means the explicit targeting of science and engineering practices, the emphasis on disciplinary core ideas, and a consistent consideration of crosscutting concepts is central to driving learning in science.

Technology in the Classroom

In the recent past, the United States has increased its emphasis on the use of technology in the nation's schools. In 2008, Congress jointly authorized the nonprofit *Digital Promise* to support comprehensive research in order to provide American citizens with knowledge and skills needed to compete in a global economy. It has been noted that application of technology in educational contexts can serve positively in helping students seek, build, and share knowledge as well as fostering engagement in self-directed learning (Gordon et al., 2013). Additionally, McKnight et al. (2016) found that technology transformed teachers' but the degree to which these are effective rests largely on how they are implemented within the classroom. Teacher use of technology in the classroom is not the only important factor for learning in this environment. It actually takes a good bit of academic learning in order to use technology effective for learning, which means that just because students have exposure to technology all the time and can be called "digital natives" is not sufficient evidence that they know how to use technology to aid in learning (Bennett & Maton, 2010). Independent of whether or not technology is used in the classroom is the extent to which that use is effective and supports student learning. Clark and Mayer (2008) note that:

From the plethora of media comparison research conducted over the past 60 years, we have learned that it's... the instructional methods that cause learning. When instructional methods remain essentially the same, so does the learning, no matter which medium is used to deliver instruction. (p. 14)

In what circumstances and for what purpose do effective teachers integrate technology as part of their pedagogical practice? Building on research around Pedagogical Content Knowledge (Shulman, 1986), researchers adapted the PCK model to answer this question through Technological Pedagogical Content Knowledge (TPACK). TPACK is recognized as an emergent form of knowledge that goes beyond content, pedagogy, and technology components and is the basis of effective teaching with technology (Koehler & Mishra, 2009).

Having set this concise contextual and empirical foundation, we will turn to a theoretical framework for addressing the question: How can teachers effectively integrate technology into the science classroom in a way that leads to learning that lasts?

ESSENTIAL PRINCIPLES SCIENCE EDUCATORS SHOULD USE

The theoretical framework for this study will be grounded upon two bases, each of which will be explored next to provide a baseline for application. First, we will begin with principles of learning derived from cognitive science (Krahenbuhl, 2021; Willingham, 2021). These principles will include basic precepts from cognitive psychology with regards to what learners have in common today and in the past with regards to how our mind makes meaning of information (Weinstein & Sumeracki, 2019). Secondarily, we will consider the unique realities of learning in technologically enhanced environments. This will begin with a specific emphasis on TPACK (Koehler & Mishra, 2009) but also consider principles of learning when that learning occurs within multiple forms of media (Mayer, 2020) as well as various concerns for technology and the mind (Turkle, 2017; Wolf, 2018). Together these construct a theoretical framework that guides the process towards answering the central question for this chapter through the presentation of four guiding principles to inform effective integration of technology in science settings.

When everything is considered, if what we are doing does not directly impact the student, the content, and the lesson then it won't make a change to learning. Furthermore, a change to one of those will have implications for the others. As Illeris (2016) points out, "learning implies the integration of two very different processes, namely an external interaction process between the learner and his or her social, cultural, or material environment, and an internal psychological process of elaboration and acquisition" (p. 10). With these concepts in mind and in consideration of both internal and external processes that bear upon our learning, let us clarify four principles that science educators should adopt and apply when deciding how to integrate technology into their science classrooms to promote learning that lasts.

In the post-pandemic world, it is becoming even more common to see widespread integration of technology. In our efforts to leverage technology to enhance student learning we need to follow some guided principles to help think about when and how to integrate that technology, not simply because it is added in, but because it meaningfully improves student learning. The principles articulated in this chapter outline a simple set of axioms to help guide educators to think about this in a consistent and focused manner.

As surveyed earlier, there are potential negative side-effects of student use of technology in the classroom (and elsewhere) that educators must be aware of (Anthony et al., 2021). Consequently, the first principle operates in a manner that moderates these negative effects by always pausing to ask: does this help learning? If adding technology to a particular topic does not promote learning in some way then we ought not do so. This guides us to the first axiom for principled integration of technology in the science classroom.

Principle 1: Moderation: Integrate Technology When the Task Calls for Technology

The first principle is a moderating principle. Because we have ample scientific data that consistently point to the fact that excessive use of technology can be detrimental to students' holistic development, we open up with a principle to moderate our integration of technology. As the Next Generation Science Standards position statement argues, "K-12 science education should reflect the interconnected nature of science as it is practiced and experienced in the real world." Technology integration in the science classroom should likewise reflect this interconnected nature so that it provides coherent application of technology. Beyond this basis of moderation, however, this also helps to serve with ensuring there is coherence in what we are doing, which is supported by Mayer (2020) as well as reducing the impact on working memory, which helps learning generally (Willingham, 2021).

In effect, this principle tells us that if the task does not call for technology, then we do not add in technology. Technology is to be added in only when it is an essential component for completing the task at hand. The educator who is rich in TPACK recognizes this and as described by Koehler and Mishra (2009) is open-minded seeking of technology use, not for its own sake but for the sake of advancing student learning and understanding.

Although we start with an attitude of moderation, we also recognize that there are incredible possibilities and documented positives that can be gained through wise integration of technology into the classroom (Sancho-Gil, 2020). So, we do not look at technology with a jaded view but rather a skeptical one that is always asking to what extent adding technology into the lesson will help learning. Our second principle guides that thinking to be even more precise as we also consider will this specific integration of technology aid in thinking about the scientific process? One of the widely agreed upon elements of science curricula in the contemporary world is that it should explicitly integrate scientific and engineering design processes (Lee et al., 2015). The second principle applies directly to our consideration of using technology – wise educators will integrate technology not in an ad hoc manner but in a way that targets these central ideas guiding effective science curricula.

Principle 2: Precision: Integrate Technology to Support Scientific Inquiry

If the first principle is aimed at moderation, the second is aimed at precision. With increased emphasis that is being placed on disciplinary core ideas in science, one of the essential aspects that comes in any science classroom is increased priority to engaging in scientific inquiry. Educators should capitalize upon specific opportunities to use technology as a tool to help support students while engaging in scientific inquiry. So, our second principle is to be targeted and precise in our application of technology. Learning a new tool can become an add-on to the content and skills that are required so keep it focused: integrate technology consistently to support scientific inquiry so that the same technological tools can be used across time, in varied contexts, but all aimed towards supporting scientific inquiry.

The first two principles aid us in developing a clear basis for when and in what matter technology should be specifically brought into the science classroom to promote scientific thinking and learning it is also necessary to think about how we can integrate it to promote deep and sustained learning. Cognitive science has demonstrated that challenging students to retrieve information from their long-term memory instead of simply googling it is much better for their learning (Roediger & Butler, 2011). Additionally, evidence has consistently confirmed that spreading the practice students complete across time leads to learning that endures (Carpenter et al., 2012). Furthermore, the use of technology is often associated with higher levels of motivation (Gordon et al., 2013). This can be particularly important to leverage because practice, while beneficial for learning, is not necessarily something that is enjoyed. This is part of why the third principle places emphasis on how science educators can leverage technology to draw upon the benefits of both technology and practice. In doing so, learning will be aided by design.

Principle 3: Extension: Integrate Technology to Practice Scientific Skills and Knowledge

Having provided a moderated and precise means to decide when to apply technology into the science classroom, the next principles aim to help think about how we ought to do so. In particular, the preponderance of evidence with regards to learning points to the fairly obvious but sometimes ignored fact that practice is required to make learning stick (Dunlosky et al., 2013). The third principle could then be conceived of as an extension principle. When we integrate technology into our science classrooms it should include opportunities to practice specific scientific skills or engage in inquiry that leads to the building of essential scientific knowledge. And when it comes to the science classroom it is nearly a perfect setting in which students can develop technological knowledge and skills that will be applied in future settings. In the context of scientific inquiry, the process is so universally relevant that it provides high leverage for learners to become college and career ready, and through including technology into the practice of this it will build technological literacy, which has been noted as important for this end (Hobbs, 2017; Ohler, 2013).

As learners' progress in a particular sub-field of science and their knowledge grows it is also important to challenge them to extend their thinking and apply ideas in more advanced ways. So, the fourth and final principle charges science educators to think carefully about the timely application of technology where students can apply technological tools to push their thinking and enrich their learning.

Principle 4: Timely Application: Integrate Technology to Apply Scientific Concepts When Ready

The final principle encourages educators to think deeply about both the when and how. Because students are novices, we need to ensure that they have foundational knowledge in place before pushing too far ahead to application or they will become frustrated (Willingham, 2021). But once foundational knowledge has been established in our students the time is right to think about application (Krahenbuhl, 2021). The use of technology to apply scientific inquiry offers a great means through which to do this. For instance, students might be challenged to utilize digital information that is transmitted over long distances and assessing its impact on the communication and dissemination of knowledge. One local teacher had her students participate in Zoom-based discussions with other elementary schools around the world focused on what the scientific process looked like. Part of the focus of this activity was not just to place emphasis on the scientific process but also to reflect on the role that technology plays in that regard. This allowed her students to discuss, after the fact, distinctions in quality and speed of communication in different settings to consider the impact that technology offers in promoting scientific inquiry across global contexts. This sort of approach fits very neatly with the Next Generation Science standards but also cultivates metacognitive thinking which is recognized as an important factor in learning and academic achievement (Fleur et al., 2021). The fourth principle places emphasis on the important instructional consideration of timely application. Science educators should carefully identify those application elements that are the priority within their domain and consider where these can be applied through technological medium. After having spent time ensuring students are equipped with the foundational knowledge, the wise educator will then integrate technology to apply these scientific concepts.

Through application of these, science educators will be equipped to integrate technology more wisely for the specific purpose of extending and enriching student learning in the field of science. Next, let us turn to four specific examples derived from real-world classrooms for how educators might allow these principles to guide their thinking on the integration of technology into the science classroom.

Applying the Principles in Science Teaching

In order to explore what application of these principles in the science classroom looks like we surveyed the area and identified several highly effective science educators. Furthermore, these educators had interacted with our faculty and adhered to these principles within their classrooms. The following section will outline just a few specific examples of applying these principles to guide science instruction that leads to learning that lasts. Each of the four principles will be provided with an example to guide this discussion. Now, we will revisit the first principle – that of moderation.

Example 1: Pushing Pause – Moderation Principle

Susan A. Educator teaches fourth grade specials in her rural school district. Her district is one that is pushing for increasing integration of technology and she is one of the leading adopters. Susan, however, has also chosen some areas where integration of technology is just an add-on and specifically removed them from her classroom. Specifically, she used to have the students use the Internet and basic searches to find information about a unit before beginning, which she was told would build excitement in the students and give them increased voice. Although she had known that sometimes these google searches led to off-task distraction and that they often found misrepresented information, she had also noticed instances in which students would encounter outright false information. When she encountered these principles, this was one of the first areas she pushed pause on. She recognized that the tool through which students explored information for the first time had significant ramifications on their subsequent learning. And because these students were not yet equipped with sufficient knowledge about the subject matter to discern among good quality, lower quality, and poor-quality information, she decided that a more curated opening, led by her as the expert, would be important.

This choice reflected an excellent application of this first principle. Within the field of cognitive science, a widely recognized cognitive bias that is at work in all of us is what is referred to as *anchoring bias* (Wilson et al., 1996). This bias means that we tend to use the first information we encounter about a topic as an anchor from which other information is affixed. This means that if the first information is unhelpful, or not central, it can have harmful effects on learning. So, Susan pushed pause on the use of technology to aid in pre-unit searches and instead ensured that the first content they encountered was curated by her and would consistently be on target and accurate. Susan is the same teacher we will look at for application of the second principle, in which she specifically integrated technology into a unit so as to support scientific inquiry.

Example 2: Mixing In – Precision Principle

Susan has always enjoyed teaching students to think deeply about the interactions between humans and the environment. One of the units that Susan taught was completely altered in order to apply the precision principle. Specifically, in a class activity in which Susan typically challenged her students to create a tactile model to represent the relationship between basic needs in an environment. This required students to think about relationships between basic needs, such as shelter, water, and food and considering different plants and animals, including humans, with the larger environment in which they lived. Typically, this activity had been completed simply using hands-on materials such as cotton balls to underscore "fur" and Crisco to illustrate "blubber" and students would see why these features of polar animals helped them survive in their environment. But by including simple technology such as thermometers, students now operate in teams to try and equip their "animal" with the best protection for a polar – or another environment. The "animal" was made of two pieces of paper stapled together in the shape of an animal surrounded by plastic wrap with the thermometer placed in the middle. A small hole in the paper on one side of the animal was cut open so the thermometer could be read. Students cover their animal with varied layers and make predictions about what it will do for the temperature that is read inside the "animal". After a few attempts, students see how these features make a difference for life in these types of environments.

In a secondary context this could easily be adapted to leverage specific online technological tools to allow students to get real-time responses using computer simulations of real-world challenges, such as global climate change. This could be followed by comparing predictions of those models with what they would expect and a discussion around the assumptions of such modeling. Given science's focus on the explicit use of modeling the world around us educators can draw widely from those resources to get their students thinking, precisely, about the ways in which technology can be brought in to support the practice of scientific inquiry while considering the implications of doing so, as well.

Example 3: Practicing the Essentials – Extension Principle

In our next example we will turn from Susan to Samantha, who is a middle school science teacher. Her district has moved to one-to-one devices and, as a result, use of technology is a consistent – and required – part of her work. One illustration of the extension principle in action for integration of technology is through her use of gaming tools on the technological devices to build in low-stakes retrieval practice for her students, which is one of the most widely documented ways to help learning stick (Weinstein & Sumeracki, 2019). A common tool for this is Kahoot although her school uses several applications that function in slightly differing ways but for the same purpose of retrieval. But each of the tools, when applied, serves the same function: to provide extended retrieval practice of essential knowledge.

For instance, in a unit aimed to determine the impact of competitive, symbiotic, and predatory interactions in an ecosystem, Samantha chose several places throughout the unit to provide spaced practice (Carpenter, 2012) using these technological tools for the purpose of retrieval. During one of the retrieval games students explore the different types of symbiosis and then, the meaning and importance of symbiosis is matched among hosts and symbiotic relationships based upon their respective needs. She builds the games in the applications ahead of time, including use of Nearpod to make direct instruction more engaging and through use of specific drawing slides, she can share out good examples with the class to discuss what aspects make each strong. In this way, her students encountered strong benefits of application of the extension principle in the science classroom.

One other example for application of this principle could come from challenging students to distill key scientific concepts down to the essentials. It has been noted that being able to explain concepts that may be complex in simplified forms is an important illustration of one's understanding of it (Krahenbuhl, 2021). And one way in which science educators could do this is through challenging students near the end of a unit, or as a challenge task to extend the learning for those who grasp something more quickly, to construct a digital story to teach about the concept to a novice. One teacher, working in a unit school context (K-12), had his students at the end of each quarter divide up essential topics and prepare digital stories that they would then take and share with students in the elementary grades. The use of technology allowed for easy sharing of these and the focus on essential elements of the areas gave extended practice and forced students to retrieve, to elaborate, and to present simple summaries of these ideas leading to improved learning for the students (Weinstein & Sumeracki, 2019).

Example 4: Allowing Choice Once Foundations Are in Place – Timely Application

Samantha teaches another unit in which the students have to demonstrate an understanding of the concept of location as well as how maps and globes represent geographical locations and contexts. In the past, this has been completed largely through the use of atlases and maps derived from the district-provided textbooks. However, recently she has moved to establishing the foundational knowledge students need to know for reading and interpreting maps and then using technology to allow for a guided discovery of what they find using Google Earth and/or Google Maps.

In this lesson, specifically, the students learn about the seven continents of the world. The first day of the unit is spent establishing some key foundations for understanding and interpreting the maps that they will encounter. The second day is aimed at clarifying and comprehending essential terminology regarding basic geographical features they will encounter and be on the lookout for. The third day provides guided, in-class concept development and practice of these concepts. On the fourth day, the students work in small groups to explore the different continents and consider a series of questions by studying Google Earth for the continent. Each team develops their own representation of the continent placing emphasis on differing features – physical features, political features, and/or economic features – and these are shared electronically. After this is completed, Samantha often chooses a few good examples to underscore relationships among these differing functional maps to help discuss how these interact with human-environment interactions. For example, one of the more common illustrations is to see patterns of settlements near bodies of water and then noticing trade routes following those same means. By using the online platform and sharing these maps, students build their skills through application of global positioning satellite (GPS) to inform their thinking about geographic issues.

In a secondary context, educators could apply this principle by letting students consider wave properties and their role in communication of information in digital settings. For instance, students could first spend time in class exploring the properties of waves and their uses for digital communication. Then, students might be challenged to solve a problem for how to transfer a picture, stored as the values of an array of pixels, could be sent and shared. Students could use various technological forms of communica-

Table 1. Principles for effective technology integration in science classrooms

Principle	Description	Application Questions
Moderation	Ensuring coherence between activity and technology to moderate its integration	• When I add technology into a lesson does it align with the tasks directly? • Will the technology being integrated divert or converge thinking on the content?
Precision	Focusing specific technology integration to cultivate scientific thinking and inquiry	• What technology tools do I have access to that can help learners in scientific inquiry? • How frequently do I complete lab-type activities? Could these be enhanced with technology?
Extension	Leverage technology to engage students in practice to help learn essential science knowledge and skills	• How can I use technology to promote student practice of essential scientific knowledge? • How can I use technology to promote student practice of essential scientific skills?
Timely Application	Careful consideration of when and how to integrate technology	• Does technology distract time I cannot lose in this lesson? • Can I frontload training in technology tools I will use again and again early in the semester? • How can I use technology as an opportunity to go deeper into learning?

tion to examine which transferred the product at higher and lower images and consider why these might have occurred bringing the application back to the central topic of wave properties.

These brief examples provide a specific look at how these principles can be applied in the science classroom. By applying the first principle – the principle of moderation – the science educator thinks carefully about *when* to integrate technology. The second principle – the principle of precision – extends this by focusing on targeted times and contexts to integrate technology. The third principle – the principle of extension – encourages educators to leverage technological tools for the much-needed practice that students need in order to learn. And the fourth principle – the principle of timely application – considers opportunities when students, through use of specific technology, can access content they would not be able to otherwise, but does so *after* ensuring the students are equipped to use the technology and the access it affords wisely.

DISCUSSION AND RECOMMENDATIONS

Through this chapter we have articulated four guiding principles that can help science educators wisely integrate technology into their classrooms. These ideas are supported by cognitive science and have also been tested and found helpful by real-world K-12 educators as illustrated by the examples provided. To sum up the ideas in this chapter a table outlines these principles, describes their basic benefit, and lists a few questions to consider in applying them.

This table provides a helpful outline of guiding questions to consider so one can self-assess their integration of these principles into their practice. Through considering these, educators will improve their effective utilization of technology in science classrooms in ways that will promote learning that lasts.

To close out this chapter we will discuss and share recommendations to consider with regards to the application of these axioms to guide principled integration of technology in the science classroom. By adopting these principles derived from the body of literature related to cognition, to specific learning in science, and in technology integration educators can improve future practice. With regards to educators there are two recommendations to consider. First, it would be wise to integrate technology for scientific endeavors that mirror real-world application. To avoid excessive application of technology just for its own sake, a wise educator chooses carefully when to use technology to enrich a science lesson. In this case, the single most important first decision should be to ask: does adding technology to this particular lesson mirror what professionals would actually do? It is important to note that we are not asking students to do the same type of task – they are novices and not equipped to learn much by doing what experts do (Willingham, 2021). Rather, the question being asked is: does focusing on technology make sense in this context? If it does not for professionals, adding it on to those learning in science seems an unwise, misdirected choice. As noted by the *Digital Promise*, educators ought to focus on networks connecting us with people and with ideas and that in doing so, these networks can emphasize stories that inspire ideas and action. Through committing to building technology-integrated lessons only in cases where there are real-world applications evident we will be helping our students move towards achieving the aims of this promise as well as being efficient with our resources. And second, technology should be integrated into instruction at the right time and in the right context and not otherwise. It is critical to be cautious with the extent to which we integrate technology into learning in any context for there are documented, serious, and negative byproducts of digital literacy and digital immersion. Children, in particular, can have adverse consequences for developing social and academic skills (Turkle, 2017; Hurwitz & Schmitt, 2020). And beyond that, excessive learning in a technologically infused context can alter how our minds operate, which may have negative ramifications on brain development, especially before the prefrontal cortex is completely finished with its development (Wolf, 2018). In particular, children who do not yet have a fully developed prefrontal cortex, have less ability to self-regulate than adults (McCarter, 2019). This issue is particularly important to consider for elementary and middle school educators in that it will increase the likelihood that such students will become completely swept up in the tool and lose attention to the task at hand as they will struggle at regulating their thoughts away from the concreteness of the tool itself (Ionnidis et al., 2018).

Future research ought to consider the applications of these principles and their impact and efficiency for practice. It is recognized that change to instructional practice that does not bear directly on the teacher, the student, or the content to be learned will fail to make meaningful impact (City et al., 2009). There is a need to examine best practices in a field for their coherence with cognition to emphasize evidence-informed practices. Sometimes, best practice is more a collection of the prevailing preferences of educators but that may not be in accord with evidence (Kirschner et al., 2006). In this way, there are two specific actions research ought to explore. First, future research should seek to empirically examine the veracity of claims of best practice through the lens of randomized control trials and second, future research should seek to construct harmonized, systematized models that provide specific actions science educators ought to take at different stages in student development. These two areas provide real opportunity for logical next steps to improve science educators' integration of technology into the future.

CONCLUSION

The intention of this chapter was to present a set of principles – or axioms – about learning that have direct bearing on instructional utilization of technology in the classroom. These were articulated and shared to provide a guide to an effective integration of technology that promotes learning that lasts within the sciences. It was the intent of the authors to not only discuss these concepts from a theoretical perspective, but to also transform *thought into action* where practitioners can apply these concepts into building effective technology into their daily lesson plans. The specific examples provided were informed by application of these principles in real-world K-12 science classrooms. Finally, closing discussion considered application of these principles broadly as well as identified recommendations for practice and for research. It is hoped that this contribution to the body of work on integrating technology into the science classroom will equip practitioners and researchers alike to reconsider best practice and to take steps to improve practice for the sake of student learning.

REFERENCES

Aksoy, C. G., Eichengreen, B., & Saka, O. (2020). *Revenge of the experts: Will COVID-19 renew or diminish public trust in science?* SRC Discussion Paper No. 96. London School of Economics and Political Science and Economic and Social Research Council.

Ambrose, S. A., Bridges, M., Lovett, M., DiPietro, M., & Norman, M. (2010). *How learning works: Seven Research-Based Principles for Smart Teaching*. Jossey-Bass.

Anthony, W. L., Zhu, Y., & Nower, L. (2021). The relationship of interactive technology use for entertainment and school performance and engagement: Evidence from a longitudinal study in a nationally representative sample of middle school students in China. *Computers in Human Behavior*, *122*, 106846. doi:10.1016/j.chb.2021.106846

Bennett, S., & Maton, K. (2010). Beyond the 'digital natives' debate: Towards a more nuanced understanding of students' technology experiences. *Journal of Computer Assisted Learning*, *34*(5), 321–331. doi:10.1111/j.1365-2729.2010.00360.x

Carpenter, S. K., Cepeda, N. J., Rohrer, D., Kang, S. H. K., & Pashler, H. (2012). Using spacing to enhance diverse forms of learning: Review of recent research and implications for instruction. *Educational Psychology Review*, *24*(3), 369–378. doi:10.100710648-012-9205-z

City, E. A., Elmore, R. F., Fiarman, S. E., Teital, L., & Lachman, A. (2009). *Instructional rounds in education: A network approach to improving teaching and learning*. Harvard Education Press.

Clark, R. C., & Mayer, R. E. (2008). Learning by viewing versus learning by doing: Evidence-based guidelines for principled learning environments. *Performance Improvement*, *47*(9), 5–13. doi:10.1002/pfi.20028

Dunlosky, J., Rawson, K. A., Marsh, E. J., Nathan, M. J., & Willingham, D. T. (2013). Improving students' learning with effective learning techniques: Promising directions from cognitive and educational psychology. *Psychological Science in the Public Interest, 14*(1), 4–58. doi:10.1177/1529100612453266 PMID:26173288

Epstein, D. (2019). *Range: Why Generalists Triumph in a Specialized World.* Penguin.

Ericsson, K. A., & Charness, N. (1994). Expert performance: Its structure and acquisition. *The American Psychologist, 49*(8), 725–747. doi:10.1037/0003-066X.49.8.725

Fleur, D. S., Bredeweg, B., & van den Bos, W. (2021). Metacognition: Ideas and insights from neuro- and educational sciences. *Nature: Science of Learning, 6*(13), 13. doi:10.1038/s41539-021-00089-5

Gordon, N., Brayshaw, M., & Grey, S. (2013). Maximizing gain for minimal pain: Utilizing natural game mechanics. *Innovation in Teaching and Learning in Information and Computer Sciences, 12*(1), 27–38.

Hobbs, R. (2017). *Create to learn: Introduction to digital literacy.* John Wiley & Sons.

Hurwitz, L. B., & Schmitt, K. L. (2020). Can children benefit from early internet exposure? Short- and long-term links between internet use, digital skill, and academic performance. *Computers & Education, 146*, 1–11.

Illeris, K. (2016). *Learning, Development, and Education: From Learning Theory to Education and Practice, the Selected Works of Knud Illeris.* Routledge.

Ionnidis, K., Treder, M. S., Chamberlain, S. R., Kraly, F., Redden, S. A., Stein, D. J., & Grant, J. E. (2018). Problematic internet use as an age-related multifaceted problem: Evidence from a two-site survey. *Addictive Behaviors, 81*, 157–166.

Kirscher, P. A., Sweller, J., & Clark, R. E. (2006). Why minimal guidance during instruction does not work: An analysis of the failure of constructivist, discovery, problem-based, experiential, and inquiry-based teaching. *Educational Psychologist, 41*(2), 75–86.

Koehler, M. J., & Mishra, P. (2009). What is technological pedagogical content knowledge? *Contemporary Issues in Technology & Teacher Education, 9*(1), 60–70.

Krahenbuhl, K. S. (2021). *Building Effective Learning Environments: A Framework for Merging the Best of Old and New Practices.* Routledge.

Lee, C., Miller, E., & Januszyk, R. (2015). *Next generation science standards (NGSS) for all students.* NSTA Press.

Mayer, R. E. (2020). *Multimedia Learning* (3rd ed.). Cambridge University Press.

McCarter, S. A. (2019). Adolescence. In E. D. Hutchison (Ed.), *Dimensions of Human Behavior: The Changing Life Course* (pp. 189–229). Sage Publications.

McKnight, K., O'Malley, K., Ruzic, R., Horsley, M. K., Franey, J. J., & Bassett, K. (2016). Teaching in a digital age: How educators use technology to improve student learning. *Journal of Research on Technology in Education, 48*(3), 194–211.

Ohler, J. (2013). *Digital storytelling in the classroom: New media pathways to literacy, learning, and creativity*. Corwin Press. doi:10.4135/9781452277479

Pritchard, A. (2018). *Ways of Learning: Learning Theories for the Classroom* (4th ed.). Routledge.

Roediger, H. L. III, & Butler, A. C. (2011). The critical role of retrieval practice in long-term retention. *Trends in Cognitive Sciences*, *15*(1), 20–27.

Sancho-Gil, J. J. (2020). Digital technology as a trigger for learning promises and realities. *Digital Education Review*, *37*, 195–207.

Shulman, L. (1986). Those who understand: Knowledge growth in teaching. *Educational Researcher*, *15*(2), 4–14.

Siemens, G. (2005). Connectivism: A learning theory for the digital age. *International Journal of Instructional Technology & Distance Learning*, *2*(1), 3–10.

Tricot, A., & Sweller, J. (2014). Domain-specific knowledge and why teaching generic skills does not work. *Educational Psychology Review*, *26*, 265–283.

Turkle, S. (2017). *Alone together: Why We Expect More from Technology and Less from Each Other*. Basic Books.

Weinsten, Y., & Sumeracki, M. (2019). *Understanding How We Learn: A Visual Guide*. Routledge.

Willingham, D. T. (2021). *Why don't Students Like School? A Cognitive Scientist Answers Questions About How the Mind Works and What it Means for the Classroom*. Jossey-Bass.

Wilson, T. D., Houston, C. E., Etling, K. M., & Brekke, N. (1996). A new look at anchoring effects: Basic anchoring and its antecedents. *Journal of Experimental Psychology*, *125*(4), 387.

Wolf, M. (2018). *Reader, Come Home: The Reading Brain in the Digital World*. Harper-Collins.

Chapter 12
Roles of Technology in the Science Classroom:
Meta-Analysis, Example Vignettes, and Guidance for Teacher Practitioners and Educators

Andria C. Schwortz
Quinsigamond Community College, USA

Barbara Hickman
University of Wyoming, USA

Andrea C. Burrows Borowczak
University of Central Florida, USA

Daniel A. Dale
University of Wyoming, USA

Adam D. Myers
University of Wyoming, USA

ABSTRACT

The ability to use technology within the context of science is important for science students; however, the integration of technology within science is not implemented consistently across the curriculum, nor across grade and age levels. More insight is needed as to how science and technology are integrated in the classroom. Technology usage takes many forms in the classroom, including educational and instructional technology, programming and computational thinking, mathematics modeling, science-specific tools, and dataset usage. This chapter consists of three main sections: a meta study of forms of technology used in science education, a series of vignettes of actual use of technology in the classroom, and guidance on thoughtful adoption of technology.

DOI: 10.4018/978-1-6684-5585-2.ch012

INTRODUCTION

The combination of science, technology, engineering, and mathematics (STEM) into a common integrated STEM (or iSTEM) curriculum is an educational goal within many countries, such as Canada (Achieve, 2010a), China (Hong Kong: Achieve, 2010a; Inner Mongolia: Gao et al., 2018), England (Achieve, 2010a), Korea (Song, 2017), Malaysia (Bunyamin, 2020; Khairani, 2017), the United States or USA, (National Research Council [NRC], 2010, 2013), Vietnam (Thuy et al., 2020), and more (Gao et al., 2020; Moore et al., 2020). Research has shown that iSTEM increases student engagement (Struyf et al., 2019). Students who are members of underrepresented groups can also be helped by iSTEM (Guzey et al., 2020; Moore et al., 2020). In addition, iSTEM has the potential to improve students' "interest, engagement, attitude, and motivation" (Gao et al., 2020, p. 3). Integrated STEM may, therefore, be the frontier in STEM education (see also Jackson et al. 2020). While iSTEM may help students, instructors often receive insufficient training in how to integrate STEM content in the classroom (Thuy et al., 2020), and research is limited into how instructor professional development (PD) programs prepare teachers to integrate STEM (Burrows et al., 2021). Many instructors don't even consider integrating STEM to be an important aspect of being a STEM teacher (Holincheck & Galanti, 2022).

When examining the integration of technology specifically within STEM, this is most often addressed by using the technology as a tool within science education, rather than making the technology the content, much the way that mathematics is used as a tool within science education (Wofford, 2009).

Another motivator for integrating STEM is the role of authentic STEM. Since technology usage is part of the authentic activities undertaken by STEM researchers, the role of authentic STEM within iSTEM education needs to be understood (Schwortz, 2021; Resnick et al., 2018). Providing students with classroom activities that teach them the skills used in actual scientific practice is seen as a route to engage them in their learning (Achieve, 2010a, 2010b; National Governors Association Center for Best Practices & Council of Chief State School Officers, 2010; Microsoft, 2017; NRC, 2013). Authentic STEM is a term used when students are scaffolded through using the same skills used by actual scientists in their careers (Burrows et al., 2021).

When students perform authentic STEM activities, they work collaboratively to solve a problem or find meaning, ask and answer questions, and reflect on their learning (Spuck, 2014). Technology as authentic iSTEM is used in several content areas of instruction, including environmental science (Resnick et al., 2018) and astronomy (Schwortz, 2021). Professional development in astronomy especially holds promise for providing authentic STEM experiences (Burrows et al., 2021). The USA science guidelines call for teachers of primary and secondary students to utilize authentic classroom experiences in both science (*Next Generation Science Standards* [NGSS]; NRC, 2013) and mathematics classes (*Common Core State Standards for Mathematics,* CCSS; National Governors Association Center for Best Practices & Council of Chief State School Officers, 2010).

STEM education needs to be based on real-world situations and problems (Dale et al., 2019; Langen, et al., 2014) – and real-world STEM is integrated (Moore, et al., 2020). When students engage in authentic STEM activities, they may make better decisions about their own lives and studies (Spuck, 2014). In addition, they experience improved motivation in the sciences (Hellgren & Lindberg, 2017), as students from underrepresented groups report more appreciation for active classrooms than do students from more represented groups (Buck et al., 2020; Rainey et al., 2019). Postsecondary students are shown to benefit from authentic science activities in many fields, such as psychology (Brothen, 1984), environmental science (Carey & Gougis, 2017), astronomy (Kobulnicky & Dale, 2016), and physics (Wilcox &

Lewandowski, 2016). Using technology within an authentic STEM setting can improve students' sense of self-efficacy (Carey & Gougis, 2017), though some risk has been indicated that the use of technology, specifically, could engage stereotype threat for students with a low sense of self-efficacy (Ball et al., 2019). Technology can also be used to bring resources to students from outside the classroom, whether by working information into a typical classroom setting or by making informal education settings (such as museums or zoos) more accessible within the classroom (Ponners & Asim, 2016).

The two terms "educational technology" and "instructional technology" are often used when approaching technology in the classroom. Instructional technology is sometimes used to describe technology that only the instructor controls, such as slideshows (e.g., MS Powerpoint or Google Slides) used to convey information to students (Kilty, 2019; Reiser, 2012). Educational technology, on the other hand, commonly refers to technology that the students themselves control (e.g., Ng et al., 2020). However, these usages are not consistent, as other sources use instructional technology to refer to any use of technology to improve learning and performance (whether used by the instructor or student), like the concept of using media within a classroom setting (e.g., Reiser, 2012, discusses these two conflicting usages). Whichever definitions of educational and instructional technology are used, this use of technology or media is intended as an aide to the learning of a particular content area, rather than it being the content area itself (Kilty, 2019). These definitions are not only inconsistent, but also are limited and do not describe any details of how the technology is used. More accurate terminology is needed.

In the service of defining terminology, the authors of this chapter (or we) focus primarily on K-16 education – that is ages 12 to 22 years, including primary/elementary school; middle/intermediate/junior high school; secondary/high school; and postsecondary school/college/university. Other educational activities associated with school and teaching include homework assignments, flipped classrooms, online classes, and emergency remote instruction due to the recent pandemic.

This chapter also focuses on the natural sciences, such as the life sciences (biology, ecology, neurology, physiology, etc.), and physical sciences (chemistry, physics, astronomy, geology, etc.). These science disciplines use scientific methods and can be referred to as "hard science" instead of the social "soft science" (Jalilifar et al., 2017). Medicine is often referred to as either an applied science, or as a combination of art and science (Francis, 2020; Herman, 2001; Panda, 2006). Although medicine is not explicitly the target of this chapter, when it overlaps with science (such as biology) it has been included. Mathematics is often interwoven in science disciplines and is highlighted in the chapter as well.

Technology is harder to define, however. Are electronic circuits technology, even without using computers or smartphones? Is taking digital photos inherently technology, even if students are just looking at the photos and not mathematically analyzing them? Is graphing on a digital calculator technology, and if not, why is graphing on a computer? Is creating a concept map or flowchart on a computer more technological than doing so on paper with a pencil? We chose an expansive definition, so if in doubt, it is considered technology. This approach has resulted in the inclusion of some forms of technology that purists might have preferred not to count, such as using cellphone GPS data for digital mapping, physical dummies for learning medical procedures (known as medical simulations) or using websites to host photos and other static material.

Even before the COVID-19 pandemic, it was well-accepted that technology in the classroom is of interest not only to education researchers, but to instructors as well (Dexter & Richardson, 2020; Hechter & Vermette, 2013). In 2017, the Department of Education's Office of Educational Technology (2017) published an updated USA national technology plan that highlighted changes in the tone of the discussion since the 2010 plan had been written. The updated plan no longer included a debate about whether technology had a place in classrooms, but rather how using technology could improve student learning and outcomes. Given the increasing acceptance of school-based educational technology, it is perhaps no surprise that the USA spent $13.2 billion on educational technology in 2015, investing $4.9 billion on hardware and $8.38 billion on software (Baker, 2017). LearnPlatform (Kuykendall, 2022) reported that between August and December 2021, 74 different digital learning tools were used by K-12 students in U.S. school districts, while educators used 86. Districts accessed "1,449 tech solutions each month during the 2020-21 school year" (para 4).

Framework of Technology and Science

Unfortunately, published research that showcases mechanisms to integrate technology specifically with the rest of iSTEM is limited. Two recent studies investigated how technology was used in student teachers' classrooms (Kilty & Burrows, 2021; Schwortz & Burrows, 2021). However, a clear definition of iSTEM that includes technology is still elusive (Moore et al. 2020), and more insight is needed in the field. This is especially true considering the potential teacher changes or preferences in technology use over time. The COVID-19 pandemic also gave researchers a unique opportunity to investigate the mass adoption of remote instruction at all levels of instruction. Thus, the time is ripe for research into how technology is integrated into the rest of STEM education, and this chapter focuses specifically on how it is integrated into the science classroom.

Working within the above definitions, and to begin to make use of this opportunity, the research questions explored include:

1. What research exists on integrating technology within science classrooms?
2. Do science teachers describe their own teaching with technology in the same way as reflected in the research?
3. How can teachers thoughtfully prepare to bring technology into their classroom?

The remainder of this chapter consists of a meta-analysis of recent research performed on the topic of technology usage in science education. Within this framework, we share examples in instructora' own words, including reflection on how these vignettes compare to the results of the meta-analysis. We also provide guidance on how instructors can thoughtfully plan for beginning to use technology in their own classes, including guidance on how they can make a case for funding from their administration. A separate appendix lists many current forms of technology used by science instructors in the second section, both including examples and links where appropriate.

METHODOLOGY

"What research exists on integrating technology within science classrooms?"

This section of the chapter identifies the ways in which technology is used in science classrooms according to research. We performed a meta-analysis of articles in science education journals, guided by the more detailed question, "During 2021, what forms of technology usage were published in the context of the K-12 science classroom or teacher education?"

Research Design

Decker and McGill (2019) provided a list of 32 STEM education journals in their Table 2 (p. 6). We used their list as a starting point, removed journals in the non-natural-science fields, and expanded the list to include journals Decker and McGill excluded based on focusing on different education levels or based on additional science fields. We targeted papers published in the year 2021 as the most recent full year before performing this study (during 2022), and thus eliminated journals no longer in publication. See Figure 1 for a flowchart of this process. A total of 86 articles from 12 journals were included, as shown in Table 1.

Information about each selected article was recorded in a spreadsheet, starting with bibliographic information, link, abstract, science subject area, and short descriptions of the age/grade level and of the usage of technology. The short descriptions were then input into a qualitative data analysis system (NVivo) and coded for themes. The themes were organized into types or categories of technology use, starting with five a priori types identified in Schwortz (2021): (a) instructional and educational technology, (b) computational thinking (CT) and programming, (c) mathematical modeling, (d) science-specific technology (SST), and (e) datasets. The themes were iteratively analyzed and compared to these five categories.

Findings

Table 2 shows the grade levels and content areas of the included articles. These were more heavily weighted toward biology and physics and toward the undergraduate level. We found a comparatively high number of physics papers with an unknown grade level, primarily due to choosing the journal *The Physics Teacher* to represent physics education research. It included a section dedicated to proposing teaching strategies, without the expectation of specifying the grade level. Technology was utilized in a wide variety of ways. Examples of some of the descriptions of the articles' subject area, student age or grade level, and technology usage are presented in Table 3.

CATEGORIES OF TECHNOLOGY USAGE

The themes were then arranged into types of technology usage in science classrooms. We eliminated one category and discovered additional ones, resulting in eight main categories of technology usage:

1. information curation,
2. communication tools,
3. everyday technology,
4. science-specific technology (SST),
5. visualization tools,
6. computational thinking (CT) and programming,
7. mathematical modeling, and
8. datasets and data mining.

Note. Different types of technology were then organized into these categories, and in some cases, specific brands and products are given as examples. Some of these types of technology usage fall into

Figure 1. PRIMSA 2020 Flow Diagram of the Selection of Articles (Page et al., 2021).

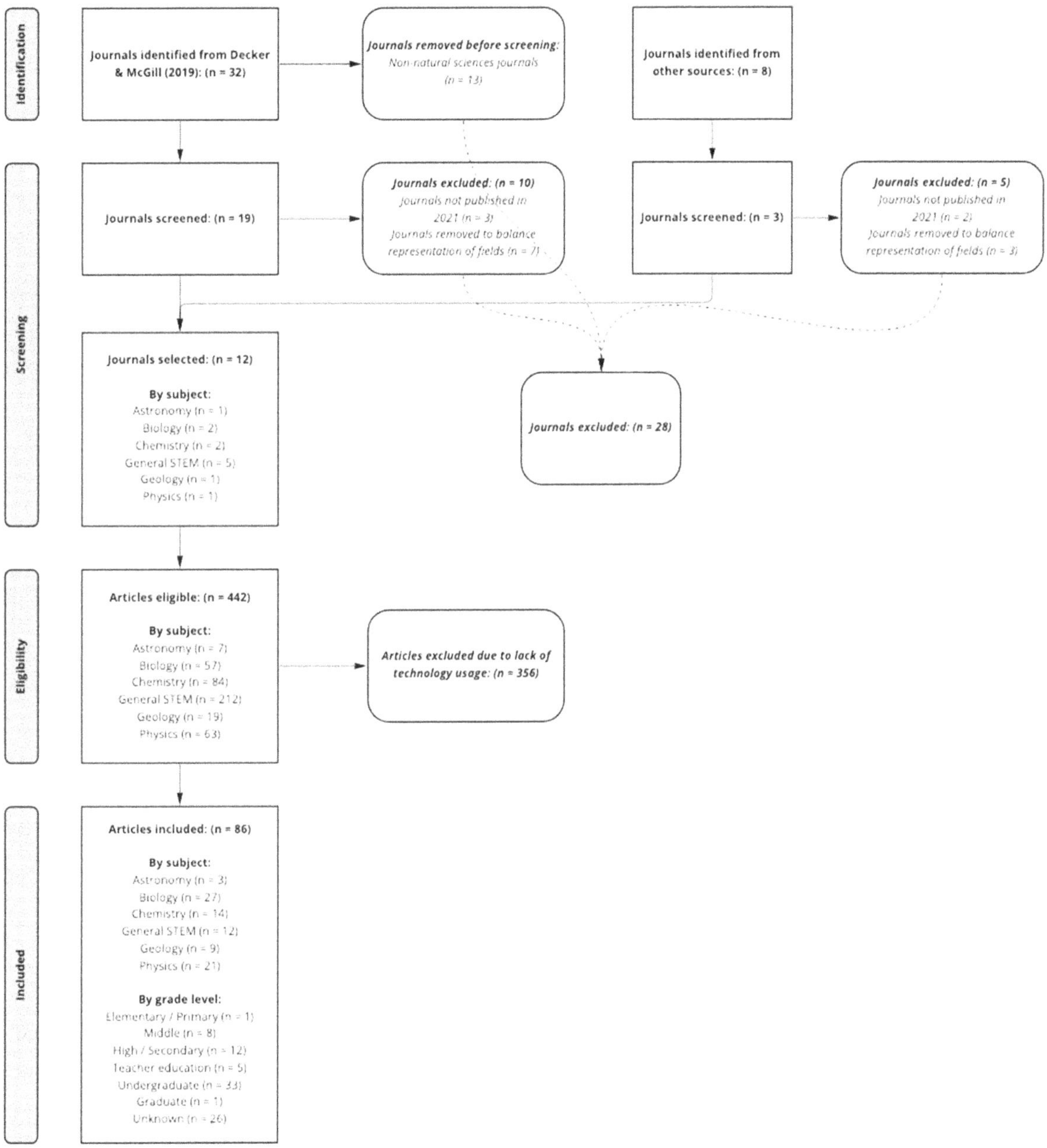

more than one category. For example, medical simulations are both a visualization tool (so that medical students can learn more about the organs and processes within the body and how to interact with them) and a science-specific technology (as the use of medical simulations is specific to biology and the health sciences). An additional catch-all technology category of "other" was used for examples of technology that did not directly fit into one of our eight categories. These were organized using a Euler diagram, as shown in Figure 2, and are further described below.

Table 1. List of journals included in this study and number of articles within each journal fitting the criteria for being about technology use in science education

Subject	Journal Name	No. of Articles in Journal	No. of Articles Included	% Articles Included
Biology	*CBE--Life Sciences Education*	42	6	14.3%
Biology	*Journal of Microbiology and Biology Education*	15	14	93.3%
Chemistry	*Chemistry Education: Research and Practice (UnivChemEdu)*	41	6	14.6%
Chemistry	*Journal of Chemical Education*	43	7	16.3%
General STEM	*International Journal of Science and Mathematics Education*	10	1	10.0%
General STEM	*International Journal of Science Education*	104	7	6.7%
General STEM	*Journal of Research in Science Teaching*	54	6	11.1%
General STEM	*Journal of STEM Education*	20	3	15.0%
General STEM	*Research in Science & Technological Education*	24	6	25.0%
Geology	*Journal of Geoscience Education*	19	9	47.4%
Physics	*The Physics Teacher*	63	19	30.2%
Astronomy	*Astronomy Education Journal*	7	2	28.6%
Total		**442**	**86**	**19.5%**

Note. We read each article title and abstract, searching for studies that were designed for K-16, that involved a science educational activity or process and that provided information about how technology was utilized by students or educators. For articles where this information was not included in the title or abstract, the article itself was read, searching for these items. A total of 442 articles were examined, of which 86 fit our criteria. We did not intend this study to be fully exhaustive, as per Decker and McGill (2019), but a search for representative examples from major fields of natural sciences.

We counted the number of articles that included each type of technology (see Table 4). Because most studies involved more than one form of technology or focused on a type of technology that bridged two

Table 2. Crosstab listing number of articles included at each grade level and science subject area

Grade Level	Astronomy	Biology	Chemistry	General STEM	Geology	Physics	Total Articles Included
Elementary School				1			**1**
Middle School		3		2	2	1	**8**
High School		2	2	5		3	**12**
Teachers			1	2	1	1	**5**
Undergraduate	3	21	2	1	3	3	**33**
Graduate					1		**1**
Unknown		1	9	1	2	13	**26**
Total Articles Included	**3**	**27**	**14**	**12**	**9**	**21**	**86**
# Articles Eligible	7	57	84	212	19	63	442
% Articles Included	42.9%	47.4%	16.7%	5.7%	47.4%	33.3%	19.5%

Table 3. Example descriptions of select articles

Subject	Age/Grade	Technology Usage
Astronomy	Undergraduate	Visualization tool, authentic dataset
Biology	High school	Graphing including via computer
Biology	Undergraduate	Online mentoring, virtual research experiences
Biology	Undergraduate	Remote instruction, social media
Chemistry	Teacher	PowerPoint animations of molecular motions
General STEM	Elementary School	Online inquiry science lessons
Geology	Grad	Digital field notebooks, sensors, photos, mapping
Geology	Teacher	Authentic datasets
Physics	High School	Digital camera on telescope
Physics	Middle school	Simulations for virtual labs
Physics	Teacher	CAD for science concepts

Note. Articles selected to show diverse examples of the descriptions.

categories, the total count of instances of the technology (142) is greater than the number of articles examined (86). Dividing the number of instances of technology by the number of articles, each examined article included 1.7 forms of technology on average.

Next, we describe the eight categories of technology use in science classrooms discovered through the meta-analysis. Many take the place of nontechnological teaching pedagogies, but some are completely

Figure 2. Euler Diagram Showing Eight Main Categories of Technology Usage Within Science Education.
Note. Items which overlap two or more categories fall into all overlapped categories.

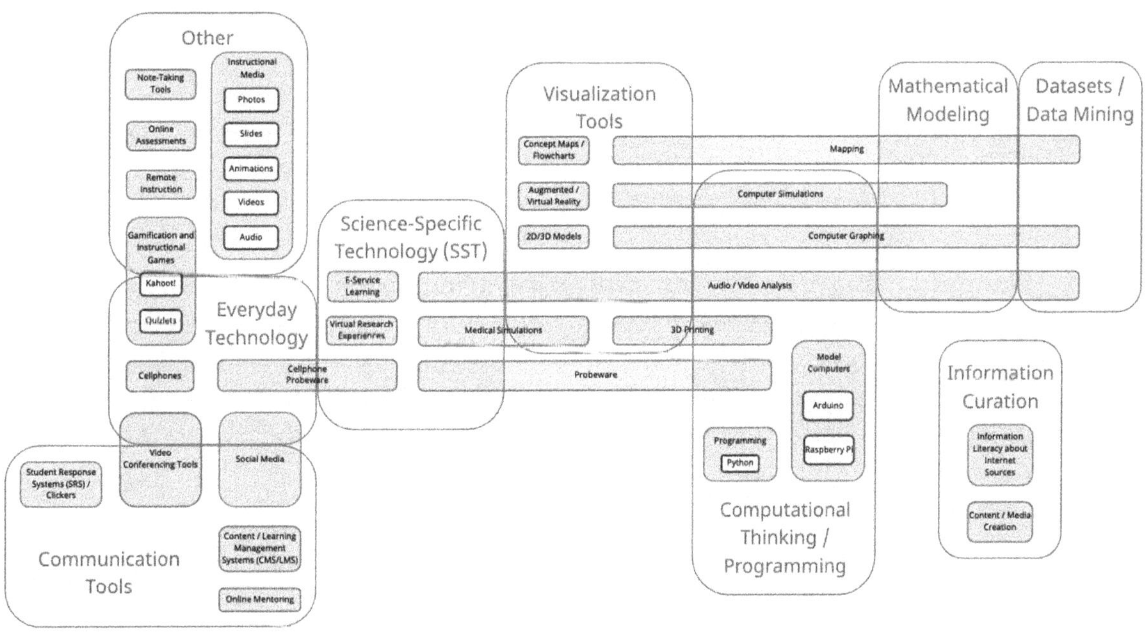

Table 4. Number and percent of articles containing each type of technology by subject area

No./% of Articles	Astronomy	Biology	Chemistry	General STEM	Geology	Physics	Total
Information Curation	0 / 0%	2 / 7.4%	5 / 36%	7 / 58%	1 / 11%	0 / 0%	15 / 17.4%
Communication Tools	0 / 0%	21 / 78%	9 / 64%	4 / 33%	2 / 22%	2 / 9.5%	38 / 44.2%
Everyday Technology	0 / 0%	3 / 11%	1 / 7.1%	1 / 8.3%	1 / 11%	2 / 9.5%	8 / 9.3%
Science-Specific Technology	1 / 33%	3 / 11%	3 / 21%	2 / 17%	2 / 22%	14 / 67%	25 / 29.1%
Visualization Tools	1 / 33%	6 / 22%	7 / 50%	3 / 25%	3 / 33%	7 / 33%	27 / 31.4%
Computational Thinking & Programming	0 / 0%	0 / 0%	0 / 0%	1 / 8.3%	0 / 0%	5 / 24%	6 / 7.0%
Mathematical Modeling	0 / 0%	2 / 7.4%	0 / 0%	0 / 0%	0 / 0%	2 / 9.5%	4 / 4.7%
Datasets & Data Mining	3 / 100%	2 / 7.4%	0 / 0%	0 / 0%	3 / 33%	0 / 0%	8 / 9.3%
Other	0 / 0.0%	2 / 7.4%	2 / 14%	3 / 25%	3 / 33%	1 / 4.8%	11 / 12.8%

Note. Each cell's count (e.g., Everyday Technology x Chemistry) is divided by the number of included articles in that subject (see Table 2), so that the percent given is the percent of that subject's articles which have that use of technology (e.g., 1 Chemistry article used Everyday Technology, which represents 7.1% of the 14 Chemistry articles). The percentage in the Total column indicates the percent of all included articles that had that form of technology.

new. Each category starts with a description and some general examples, followed by select quotations from the studies surveyed. See Figure 2 for a visual representation of these categories and examples.

Information Curation

Activities or processes where learners engage in evaluation and collection of internet sources, and/or instructors provide external sources of information. General examples include projects where students exercise information literacy to evaluate the reliability of webpages, or where collaboratively come up with a list of useful sources on a topic.

[Students'] Internet-specific justification beliefs suggested that they did not consider it highly important to justify knowledge claims encountered on the Internet by means of internal and external sources. (Cheng, et al., 2021, pg. 993.)

This study provided a rare opportunity for students to confront science information as potentially false claims—a context that students encounter increasingly often in our contemporary information landscape. Furthermore, this study is one of few that experimentally tests the effects of an intervention to support critique and evaluation in a science classroom setting, building upon the extant body of literature on the pivotal role of critique in the assessment of scientific claims (Tseng, et al., 2021, pg. 1170-1171).

Communication Tools

Enhance or enable communication between two or more students, between an instructor and students, between students and content area experts, etc. General examples follow.

1. Content/learning management systems (CMS/LMS): Canvas, Blackboard, Moodle, or Google Classroom.
2. Student response systems (SRS) or clickers: i>Clicker or QClick standalone devices, phone- or web-based systems such as Poll Everywhere.
3. Video conferencing tools with real-time audio and video conversations: Skype, Zoom, or Blackboard Collaborate.
4. Social media and microblogging: Twitter, Facebook, Instagram, Snapchat, Microsoft Teams, Discord, Blogger, LiveJournal, Medium, Substack.
5. Online mentoring: a use of or motivation for technology rather than a specific form of technology.

How did upper-elementary students respond to an embedded assessment prompt in an online technology enhanced Grade 5 Genetics unit in order to construct a scientific explanation around variation of traits based on their class trait survey? (Ibourk & Kendrick, 2021, pg. 1174.)

In our sample population, neither undergraduate mentees nor graduate mentors said that their experiences substantially altered their future plans. This may speak to the general success of virtual mentorship but may also have resulted from the respondents having accomplished their goals before the pandemic started or may also reflect how many felt the situation was a temporary one. (Speer, et al., 2021, pg. 7.)

Everyday Technology

Technology that students would be using outside the classroom already. General examples include cellphones, games (also listed under "Other"), social media (also listed under "Communication Tools").

The application of constructivist mobile learning contributed to the fact that the students in Group E were, in comparison to students in Group C, able to show: the faster determination of plants, the easier adoption of more complex botanical content, and the ability to perform more complex and deliberate activities in the determination of plants. (Anđić, et al., 2020, pg. 412.)

TikTok duets allow users to make commentary on an already existing video. Both videos appear side by side allowing the user to respond to and pause the original video while still offering their specific commentary. In an academic setting, this can function in the same way as a debate or discussion would in a classroom. This is one option to foster more discussion among students in an innovative manner. (Lungu & Lungu, 2021, pg. 4.)

Science-Specific Technology

A tool consisting of "data collection" and analysis, "simulations and modeling tools," or "online collaborative tools" (Hsu & Sharma, 2008, p. 214), only relevant within a science context. Examples: probeware (devices used to measure physical characteristics, such as position, temperature, or acidity), video or audio analysis, medical simulators (which are also a visualization tool), cellphone-based probeware (e.g., external devices attached to a cellphone, or inherent measurement tools within the phone such as a phone's accelerometer, GPS, or pressure sensor, which bridge the gap between everyday technology of cellphones, and the SST of probeware). See Kersaint et al. (2013) for more examples of software designed for science education and online science resources. "We investigated the characteristics of a medical simulation-based informal learning environment (MSBE) of a science center." (Tal & Dallashe, 2020, pg. 1.)

The absence of vegetation facilitates the use of modern photogrammetric techniques for the cost-effective generation of high-resolution digital outcrop models (DOMs). These georeferenced models can be used in pre-fieldwork activities to help prepare for traditional geological fieldwork, during fieldwork to record observations, and post-fieldwork to conduct quantitative geological analyses... This integration is strengthened if key geoscientific data, like geological and topographical maps, subsurface profiles, borehole data, remote sensing data, geophysical data and DOMs can be integrated through a common database, such as the Svalbox database that we present in this commentary. (Senger, et al., 2020, p. 123.)

Visualization Tools

Tools that allow a learner to build a mental model of a process or something invisible. General examples include 2D and 3D models, augmented or virtual reality, concept maps or flowcharts, medical simulations (which are also SST), geographical mapping (also listed under "Mathematical Modeling" and "Datasets and Data Mining"), designing one's own computer simulations (also "Computational Thinking and Programming"). "Science on a Sphere® utilizes four projectors to display dynamic global data on a large sphere." (Semmens, et al., 2020, p. 73.)

In this study, 25 chemistry student teachers, who had studied the concept of molecular motions and had taken the course of Chemistry Instructional Media and Technology, were assigned to make expressed models of water molecules' motions in the form of animations with PowerPoint, the most familiar program and installed on students' computers. (Guspatni, 2021, p. 312.)

Computational Thinking and Programming

For the purposes of this chapter, we have combined CT and programming into a single category. This category includes actual programming (such as using Python) and building and using model computers (such as Arduinos and Raspberry Pi, which can also be used with probeware). The category also includes "practices such as abstraction, modeling, and decomposition" (The Association for Computing Machinery, 2016, p. 3), as well as data visualization and pattern recognition (Krauss & Prottsman, 2018).

"For this purpose, an Arduino Uno Microcontroller and BMP180 pressure sensor are used. The BMP180 can also measure temperature, so another temperature sensor is not needed." (Salar, 2021, p. 202.)

In general, the peak of a wave packet travels at a speed equal to the magnitude of the group velocity that can be vastly different than the phase speeds of the component harmonic waves, even if the entire set of harmonic waves only differ among themselves slightly in frequency and wavelength. It is difficult for students to understand the origin of this difference from verbal descriptions and static graphs of wave functions at successive instants in times, such as are typically presented in textbooks. However, considerable insight is possible using the free online software Desmos. (Mungan, 2021, p. 30.)

Mathematical Modeling

Instances where "mathematics is used to describe or represent phenomena"; representations using symbols or numbers, manipulatives (physical objects), images (Hjalmarson et al., 2020; see also Kilty, 2019), and visualization of the data (Resnick et al., 2018); applying statistical analysis to data (National Research Council, 2010).

This paper presents a strategy for studying Hubble's law and the accelerating expansion of the universe using actual data. Along with understanding the physical concepts, an explicit goal is to develop skills for analyzing data in terms of a model. (Hyde, 2021, p. 159.)

Teachers that teach mathematics or another science subject in addition to biology would have more learning opportunities, which in turn could explain their feelings of motivation and confidence. This finding might be clarified in future research by documenting learning opportunities in the area of quantitative methods or data visualization in the subject-specific teacher education programs. (Enzingmüller & Prechtl, 2021, p. 11.)

Datasets and Data Mining

Working with sets of data that are too large to be analyzed one by one and datasets are "any collection of data, often arranged in a spreadsheet" (Schwortz, 2021, p. 3). This category includes spreadsheets and other representations of data (American Association of Physics Teachers [AAPT], 2016; Campbell & Abd-Hamid, 2013), "process data; represent data visually; prepare documents and presentations that are authentic to the discipline" (AAPT, 2016, p. 4)

Under our retooled approach, instructors navigate the Macaulay Library archives to identify the appropriate sound files needed for lessons and request the sound files from the library well in advance of class (6 weeks)... To avoid quantitative measurement errors, instructors should use uncompressed (e.g., .wav or .aiff) sound formats. Along with the sound files, instructors may obtain metadata files containing information for each sound file (i.e., date, location, time, sex, age, recordist, and video, if available). (Pinou, et al., 2021, p. 1.)

DISCUSSION

Overall, the authors identified twelve journals of science education research published in 2021, spanning the topics of astronomy, biology, chemistry, general STEM, geology, and physics. From these journals, 86 articles were found to discuss the use of technology in a science classroom. A process of iterative coding for themes identified eight categories of technology usage. The commonly used terms "educational technology" and "instructional technology" were not used due to conflicting usages of them in the literature (Schwortz, 2021), and more importantly, they are often too vague to be useful in this context.

Generally, in the articles there was an overemphasis (more than 12.5%) on research about information curation, communication tools, science-specific technology, and visualization tools, and an underemphasis of research into uses of everyday technology, computational thinking and programming, mathematical modeling, and datasets and data mining. There were also a number of technology uses that might not seem to fit into the other categories, such as notetaking tools and online assessments, as well as forms which may have overlapped with other categories, such as gamification and instructional games overlapping with Everyday Technology (III). Similarly, remote instruction and instructional media may overlap with Communication Tools (II).

The theme of remote instruction was extensively represented in the articles due to the time frame selected (the second year of the COVID-19 pandemic, with many schools recently switching to emergency remote instruction), and in biology, specifically, due a special issue of the *Journal of Microbiology and Biology Education* (2021, Volume 22, Issue 1, March-April) focused on emergency remote instruction. *The Physics Teacher* included a section of teaching strategies, and the issues selected had numerous examples of probe ware (digital devices that take data from physical experiments for computer-based analysis) and computational thinking. The chemistry and geology articles seemed to have a plethora of visualization tools. Based on this information, the authors wondered how educators would describe their technology use via the eight identified technology themes.

Educators' Own Words: Vignettes

"Do science teachers describe their own teaching with technology in the same way as reflected in the research?"

For comparison to the research, we solicited educators at all levels (early childhood through advanced postsecondary) for examples of how they use technology in the context of science instruction. The solicitation took place through social media including Reddit and Twitter, and it was a convenience sample. Due to this convenience sample focusing on known contacts of the authors and social media used by a plurality of Americans (Reddit: Clement, 2022; Twitter: Dean, 2022), we expected that the respondents would skew toward teaching older age groups (specifically, secondary and postsecondary), toward physics, and toward educators in the US. Many of these volunteers included specific brands and websites for their technology, which are listed together in Appendix: Resources. Although respondents taught multiple grade levels and various disciplines, they were focused from middle school instruction through postsecondary work (students aged 12 and beyond). These vignettes were chosen to showcase the technology themes identified in this chapter.

Necole: 4 Technology Themes - Science-Specific Technology, Datasets/ Data Mining, Mathematical Modeling, and Communication Tools

For example, Necole is a sixth-grade science teacher at Powell Middle School (USA). She describes her own use of technology through four identified themes from this chapter including Science-Specific Tools, Datasets and Data Mining (i.e., analysis of data), Mathematical Modeling, and Communication Tools (using a learning management system). She states,

I have used technology such as Vernier probeware to collect data. After data collection, the students use spreadsheet programs to make graphs for data analysis. One of the most significant changes in using technology in my classroom came when the school adopted the use of the learning management system (LMS) Canvas. Canvas provides 24-hour, 7-days-a week access to instructional materials. Canvas was a huge game changer for my students, parents, and myself. Canvas allows me to design blended learning opportunities for my students.

Michael: 3 Technology Themes –Science-Specific Technology, Mathematical Modeling, and Communication Tools

Another teacher, Michael, who teaches both middle and high school at Acera School (USA) found one main theme identified in this chapter which was a focus on Science-Specific Technology in his teaching of molecular biology. Michael related,

I have been conducting molecular biology assays with my students, like a CRISPR-Cas9 lab, using the technology that keeps science afloat, through authentic experiences. The goal is for students to make a very specific edit in the genome of bacteria to change their color. This lab requires following a 40-step process using micropipettes, water baths, and incubators at specific temperatures, and computers to record and analyze the results. While the outcome is predetermined, the skills of following procedures and learning to use the appropriate technology are crucial and are included in the NGSS standards. In one investigation, students culture the microbes in their microbiome and then purify its DNA. This requires centrifuges and pipettes, and when the DNA has been isolated, a PCR machine is needed to amplify it for sequencing. Indeed, learning how to use a traditional thermocycler PCR machine is absolutely essential for today's science students and it's even better if it is a RT-PCR machine, which can be used to deliver relative abundances of sequences in real time.

Although Michael's teaching focused on one theme, his students present their work with two other themes of Mathematical Modeling (e.g., digital format, including data tables, graphs, charts, and statistical measurements) and Communication Tools.

Casey: 4 Technology Themes - Communication Tools, Visualization Tools, Science-Specific Technology, and Mathematical Modeling

Another teacher, Casey, brings us an international perspective as an 11th grade high school chemistry teacher at University Preparatory (Turkey). She mainly uses the theme of Communication Tools (i.e., Google Slides) to guide discussion to make predictions about the pressure of gasses. Casey explained,

As students file into their classroom after break, they see questions projected on the board using Google Slides: "What is kinetic molecular theory? How do you think gas would behave if I pumped gas particles into this closed chamber?" and an example visual from PhET. I pass out a student guide to facilitate student collection of data while using the computer simulation. The guide includes instructions on how to use the simulator and questions such as, "What is the initial pressure (in atm) in the chamber?" and, "When you slide the handle on the left of the chamber and slide it to the right as far as it will go, does the volume go up or down?

Casey also identified PhET as an example of two other themes including Visualization Tools and Science-Specific Technology. Readers can access PhET animated, interactive simulations for science and mathematics topics on a website (phet.colorado.edu). Her students use the three main identified themes of Visualization Tools, Math Modeling, and Communication Tools with PhET. The students provide data in the simulations which they then graph to understand the relationships between properties and finally share with their peers via Google Slides.

Evan: 1 Theme - Everyday Technology

Evan, a close-to-retiring K-12 teacher in the USA, also uses PhET for its interactive simulations. He also emphasizes using students' cellphones as a tool, a use of the technology theme: Everyday Technology. Evan answers,

Many of my peers still ban cellphones in their classes. But I begin the first day of school by collectively saying to my physics students, "Do you have a phone? Get it out, put it on the table in front of you. It is a tool. I expect you to have it and to be able to use it. It is bad form to let the battery run down. I never want to hear that you do not have enough space for videos or pictures because you have too many saved from your families' vacation to Disney World. It is your calculator, it is your level, it is your stopwatch and timer, and that is just for starters." My entire focus has changed in my many years of teaching from delivering content and being the source of information to teaching my students how to work with information that they have at their fingertips, either from labs they have done or from the internet.

Chip: 1 Technology Theme – Computational Thinking/Programming

At the collegiate level, Chip teaches introductory undergraduate physics at the University of Wyoming (USA). He guides his students to perform coding via Matlab, a use of the technology theme of Computational Thinking/Programming. He states,

As part of each assignment, I include a required coding and plotting exercise related to the present problem set. For example, students compute and graph the position, velocity, and acceleration of two racecars. The graph can be interpreted to see which one is ahead at any point in the race. When teaching projectile motion, students graph the distance traveled by a cannon ball as a function of the angle of launch. By inspection from the graph, students can see that a 45-degree angle results in the greatest distance. And I show students how to write a loop so that they can plot the shape of a traveling wave and see it move in a short animation.

Chip views this early introduction of programming as essential to STEM majors, as they can later transfer their learning to other languages such as Python. By learning programming early, it sets a foundation for their later classes, research projects, and internships.

As University of Wyoming undergraduates advance in their learning, they progress to Adam's advanced undergraduate physics classes.

Adam: 4 Technology Themes –Computational Thinking/Programming, Communication Tools, Mathematical Modeling, and Visualization Tools

Adam's classes also have a focus on Python as a Computational Thinking/Programming tool in physics and astronomy. He describes,

The first third of each semester focuses on learning the basics of writing and debugging software. The final two thirds progress to solving a broad range of problems in physics and astronomy, such as Fourier transforms and spectral analysis, planetary habitability, magnetostatics, particle motion, and statistics appropriate to analyzing scientific data. At the end of each third of the course, one week is spent on writing code to display increasingly sophisticated plots. The first week of plotting instruction focuses on graphing simple equations. The second and third weeks focus on plotting results that inform realistic problems in physics and astronomy.

Adam supplements his teaching of programming with additional Communication Tools such as video conferencing, and clickers. His students present their results through graphs, again a form of Mathematical Modeling and Visualization Tools.

Here we showcase that the educators (ranging from middle school through postsecondary teaching) volunteered descriptions of all the technology themes except for Information Curation. We find this curious as science and mathematics instructors often teach the value of questioning sources of information, yet this was not highlighted in any vignette. The K-12 educators taught science (i.e., general science, life science, and physical science), with two teachers going into detail on their biology lab and chemistry class. The postsecondary educators taught physics and astronomy. Only one of the vignettes was from outside the USA, and as expected due to the social media use, the USA was represented more than any other nationality. None of the volunteers were elementary school educators, which fits the trend from the meta-analysis and the expectation from this convenience sample.

Many of these vignettes show usages of technology that match what was seen in the meta-analysis, from geographical mapping to concept mapping, from programming to probe ware, from websites to word processing. In addition, the instructors reported that students enjoy these uses of technology. These vignettes also demonstrate more practical examples that readers of this chapter may wish to attempt in their own classrooms. For further specific ideas of what technology to use, see the Appendix.

As shown in this chapter so far, at the large scale, many countries' national standards focus on integrating STEM education, and at the small scale, technology usage in the science classroom is useful, common, and enjoyable. However, teachers sometimes feel a need for more technology preparation (Burrows et al., 2021; Thuy et al., 2020). So, what can be done to encourage more instructors to implement thoughtful technology use in their classrooms? And if you are a instructor and ready to take the plunge, how can you do so? The next section provides some guidance for instructors and teacher educators who wish to begin to use technology in the classroom. Peer leaders are especially encouraged to look at the advice for implementation across a school.

RECOMMENDATIONS

"How can instructors thoughtfully prepare to bring technology into their classroom?"

This section provides guidance for the thoughtful implementation of technology in the science classroom. This section outlines the process for introducing technology in a systematic way, which may be useful when making a request for funding of new technology. This section is organized by starting with motivation for following a systemic process in implementation, followed by how implementation science addresses this need, and a practical guide to doing so.

Motivation

Despite the yearly expenditure of tens of billions of dollars on classroom technology, a 2019 study by Glimpse reported that on average 67% of software licenses went unused. The Glimpse study posited that one of the reasons for the underutilization of technology was a lack of specific goals around both expectations for use by teachers and the impact of technology on student learning outcomes. Sheffield et al. (2018) made the same observation and detailed a highly structured technology integration implementation plan that included specific behavioral and learning outcomes.

In addition to a lack of clear goals, there are possible environmental (e.g., equipment, broadband access) barriers to successful use and integration of classroom technology. The attitudes and beliefs of teachers toward technology and their "insufficient level of technological, pedagogical, and content knowledge (TPACK) also continue to impact classroom educational technology integration" (Paulus et al., 2020, p. 193; see also Mueller et al., 2008). Teachers must address their technology knowledge and skills to their specific pedagogy and content, instead of using some "generic TPACK" applicable to any content domain (Bull et al., 2019). The field of implementation science can help address these barriers.

Implementation Science in Theory

Implementation science can be defined as the "scientific study of methods to promote the systematic uptake of research findings and other evidence-based practices into routine practice, and, hence, to improve the quality and effectiveness" (Bauer et al., 2015, p. 3). Implementation science suggests that instructor peer leaders should plan for initiatives, innovations, and interventions and that stages of planning should begin long before the funding request is signed, or the new computer program downloaded.

The first stage of a specific implementation is the exploration stage, and it may well be the most critical for integrating technology into a classroom. The goal of this stage is to assess the need for change, to identify the overall fit of the initiative into other existing programs or practices, and to identify with specificity the problem or challenge the initiative may solve or improve. The exploration stage also assesses potential barriers to implementation related to "funding, staffing, referrals, and system changes" (National Implementation Research Network, n.d., para 3). The exploration stage thus recommends that six factors be considered before any new initiative is started, consisting of (a) need, (b) evidence, (c) fit, (d) usability, (e) capacity, and (f) supports (Metz & Louison, 2018).

The other stages in an implementation cycle include installation, initial implementation, and full implementation (Fixsen et al., 2015). These stages are not linear, and an organization might "move among the stages due to changes in staff, funding, leadership or unsuccessful attempts at employing the intervention with high fidelity" (Duda & Wilson, 2015, p. 12). The next section provides examples of a school applying these principles to begin adopting new technology into the classroom.

Applied Implementation Science

In April 2019, *Education Week* published a special report on educational technology and innovation that included a story about Guerrero, a principal in Houston who was attempting to implement a laptop program at her school with one laptop per student (Davis, 2019). Guerrero described how this principal had modeled her approach to her new educational technology practice on a 1995 article, "Leading Change: Why Transformation Efforts Fail," by retired Harvard Business School professor named Kotter.

In a list of strategies like the stages of implementation science, Kotter (1995) detailed eight factors that influence the success of a new idea. They included communicating a vision, empowering others to act on that vision, and creating a process that encourages and supports systemic adoption and consistent practice of the initiative. While this specific implementation initially came from the top down, it was successful due to collaboration from all levels within the school.

In the same article (Davis, 2019), Kohn, then the director of a leadership/management program at Rice University, reflected on failed innovations at the K-12 level and was quoted as saying that,

districts roll out big ed-tech-device programs using tactics in direct opposition to those recommended in business settings: They do it all at once with little thought to the goals behind the technology, they don't get buy-in, they don't build on small successes. (para 7)

A focused approach to the six factors and critical questions that are part of the exploration stage might help the adoption of technology find a more positive reception at schools. That process might take place within the context of an established professional learning structure, or a different structure might be needed. In this discussion of failed technology initiatives, it found that buy-in at all levels of the school were necessary for success.

Tam (2015) described a typical PD model where instructors have a role as passive learners, and Paulus et al. (2020) explained how that model would likely not allow instructors to develop the self-efficacy necessary for them to consistently and enthusiastically integrate new technology into their classrooms. While Paulus et al. (2020) recommended using a professional learning community (PLC) structure to address implementation planning, some schools may find that the exploration stage questions do not fit into their PLC format and they might need a different venue in which to address those critical issues. The role of district and site leadership is crucial for success in implementation. Leaders must create a collaborative process to make decisions about the purchase and use of technology and to support efficacy in instructor learning (Dexter & Richardson, 2020). Clearly instructors must be involved participants, or even initiators, in the adoption of technology at a school for the change to be effective.

Guidance for Instructors

To address TPACK challenges and improve the rate of use and efficacy of integrated technology before instructors begin to use technology or place it in the hands of students, they can take distinct actions to assist in successful implementation and integration of the technology. Pearson (Glimpse, 2019) recommended "two important foundational questions" be addressed about purchasing and implementing technology in the classroom:

1. Are these digital tools and technology things that we teachers actually want?
2. Are we as a school district ready to implement this technology?

If the answer to both questions is yes, then peer leaders among teachers must next ask themselves, "Are we prepared to set clear expectations for usage and implementation fidelity?" (Glimpse, 2019, para 6). Pearson's questions get to the core of how to create and sustain a K-12 environment that encourages and supports effective and meaningful technology integration to engage, challenge, and accelerate student learning.

Regardless of where and how the implementation work takes place, the same themes appear in research and anecdotal examples of classroom success with technology implementation and integration:

1. Establishing a clear vision for the implementation: need, problem, and setting goals and expectations.
2. Finding and using evidence to support the implementation.
3. Determining how the new technology will integrate into other systems and goals that are already in place.
4. Creating and using a process of selection and adoption that supports classrooms and teachers in learning and using the new technology.
5. Planning for the long-term alignment of resources that need to be in place to support the use of the new technology.

When going through this list, the eight forms of technology usage discussed earlier in this chapter may be utilized. Specifically, instructors can discuss how the need or problem matches the common forms of technology usage. For example, is it critical to teach students specific skills they will need to pursue a career in science? Then, the answer might be Science-Specific Technology. Is the problem that students cannot distinguish reliable websites from conspiracy theories? Then Information Curation (and the separate field of information literacy) might be relevant to use. Do you already have curriculum goals for students to apply mathematics skills to scientific data? To meet this goal, perhaps using a Mathematical Modeling or Visualization Tools with computers and graphing can tie to Datasets and Data mining (or real-world scientific data) from an online repository.

The purchase of technology is just one part of a much longer planning process that should include deliberate and thoughtful efforts involving teachers. Using specific steps in considering the stages of implementation may help ensure that educators have the resources and supports to use and integrate new technology effectively within the context of their classrooms. And framing a purchase request within these terms shows that the initiative has a higher chance of success, and, thus, may give the request a higher chance of success as well.

This chapter shows the many uses of technology in the science classroom that have been revealed by systematic research. If you already use technology in your science classroom, you may wish to reflect upon how your usage fits into these categories and consider where your use of technology is stronger or whether you could potentially expand into new areas. Mathematical modeling could be a particularly fruitful form of technology to incorporate in a science classroom, as it also brings in mathematics, allowing the further integration of STEM. If you do not use technology, the eight categories of technology usage discovered herein may serve as ideas to begin brainstorming ways you could use it in the future.

CONCLUSION

This chapter's meta-study showed that many distinct forms of technology usage are happening in classrooms within formally observed settings, and many of these are also self-reported as being put into practice by instructors. Further work is needed to confirm the usefulness of the eight identified categories of technology usage in science education. Researchers may wish to guide future work toward equal proportions of technology forms studied. Further exploration of an "other" category may reveal more useful themes, and the large number of papers discussing instructional media and remote instruction may lead to an additional category or turn out to just be part of Communication Tools (II). In addition, there were tantalizing signs that different fields of science may lend themselves to different forms of technology supporting the findings of Kilty & Burrows (2021). This needs to be investigated in more depth with larger sample sizes, including safeguards to make sure that any such distinction is not caused by bias in the publication of education articles in each field (as opposed to being due to actual technology uses by the instructors in each field), and changes to education made due to responses to the COVID-19 pandemic need to be explicitly acknowledged.

This chapter addressed three broad questions through a variety of explorations: (a) a meta-study to answer what research currently exists on integrating technology within science classrooms, (b) instructors' own words to show examples of how they describe their own use of technology, and (c) guidance based on implementation science regarding how instructors can thoughtfully prepare to use technology in their classrooms.

Addressing the first question, 86 articles published in 12 science education research journals in the year 2021 were selected. These studies spanned the life and physical sciences, including astronomy, biology, chemistry, ecology, geology, neurology, physics, and physiology, and were inclusive of medicine when overlapping biology or similar subjects. The technology in these articles was found to primarily fall into one of eight categories, representing a significant body of research into how technology is integrated with science. This specific form of integrating STEM had been poorly understood previously (Moore et al. 2020), yet this work shows such a body of research exists, even though more general conclusions were lacking. These eight forms of technology usage in science classrooms represent a detailed theory or taxonomy of technology use within the science classroom.

However, some limitations do exist to this section of the chapter. The meta-study was not intended to be exhaustive but to explore examples of such usage, and further work might expand the categories. Additional work can strive to find a more representative sample of different science fields or to compare different science fields to each other. The choice of forms of technologies used and studied were clearly affected by the COVID-19 pandemic, and future meta-studies may need to account for this bias. Further examination of the populations addressed by these studies would be another route of future inquiry.

Several vignettes from K-16 educators using their own words to describe how they use technology in their science teaching were provided within the discussion section. No elementary school teachers were among this group of volunteers, so we do not include any self-descriptions from this population. Few of the respondent instructors mentioned use of datasets and datamining, similarly to the result of 9.3% of papers in the meta-study (see Table 4) included in this category. Future research could focus on researching elementary teachers' own perceptions of their integration of science and technology.

Lastly, to address the third question, we provided guidance on how instructors can thoughtfully begin to use technology in their classrooms, using lessons taken from implementation science. Instructors buy-in and participation in the exploration and decision process increase the likelihood for a particular project to be successful. A list of questions and considerations were provided in this section for instructor peer leaders to consider when requesting funding for technology in the classroom, or for instructors that are new to technology to consider at the outset of technology usage.

This chapter provides guidance to current instructors and teacher educators about forms of technology that are commonly used in the classroom and ways to implement new technology in a systematic way that is more likely to lead to long-term usage. Instructors are encouraged to explore the eight forms of technology identified here and, if already using technology in the classroom, to investigate where they could strengthen their technology usage. Teacher educators may wish to see if additional instances of teaching with datasets and data mining can be added to the curriculum, as this category was minimally addressed by instructors in their vignettes and was one of the less represented categories among those researched. Technology is an important component of science education, according to not only many countries' education standards – Canada (Achieve, 2010a), China (Hong Kong: Achieve, 2010a; Inner Mongolia: Gao et al., 2018), England (Achieve, 2010a), Korea (Song, 2017), Malaysia (Bunyamin, 2020; Khairani, 2017), the USA (National Research Council [NRC], 2010, 2013), Vietnam (Thuy et al., 2020), and more (Gao et al., 2020; Moore et al., 2020) – but also according to instructors (Dexter & Richardson, 2020; Hechter & Vermette, 2013; and the chapter vignettes). This chapter highlights examples and provides guidance for instructors to improve use or begin their use of technology.

There continues to be a need for additional clarity surrounding technology (Kilty, 2019; Reiser, 2012). Although there is some research into this subject (e.g., Kilty & Burrows, 2021; Schwortz & Burrows, 2021), there is a shortage of research on how students learn these uses of technology and less on how instructors learn to implement technology education in their classrooms. Further work is needed to prepare students for the current usage of technology in STEM education, and to prepare educators to teach these topics to students. We can assume that technology adoptions are not going to decrease, and technology use will increase in all levels of teaching and in all disciplines, and understanding not just what technology is used, but how and why it is used and integrated by instructors, is crucial in moving the science education conversation forward.

REFERENCES

Achieve. (2010a). *Connecting science standards with assessment: A snapshot of three countries' approaches – England, Hong Kong, and Canada*. Author.

Achieve. (2010b). *International science benchmarking report: Taking the lead in science education: Forging Next-Generation Science Standards*. Author.

American Association of Physics Teachers Undergraduate Curriculum Task Force. (2016). *AAPT recommendations for computational physics in the undergraduate physics curriculum*. https://www.aapt.org/Resources/upload/AAPT_UCTF_CompPhysReport_final_B.pdf

Anđić, B., Cvjetićanin, S., Lavicza, Z., Maričić, M., Novović, T., & Stešević, D. (2020). Mobile and printed dichotomous keys in constructivist learning of biology in primary school. *Research in Science & Technological Education*, *39*(4), 393–420. doi:10.1080/02635143.2020.1763290

Baker, E. (2017). How school districts can save (billions) on edtech. *Technology for Education Consortium*. https://marketbrief.edweek.org/wp-content/uploads/2017/03/How_School_Districts_Can_Save_Billions_on_Edtech.pdf

Ball, C., Huang, K.-T., Rikard, R. V., & Cotten, S. R. (2019). The emotional costs of computers: An expectancy-value theory analysis of predominantly low-socioeconomic status minority students' STEM attitudes. *Information Communication and Society*, *22*(1), 105–128. doi:10.1080/1369118X.2017.1355403

Bauer, M. S., Damschroder, L., Hagedorn, H., Smith, J., & Kilbourne, A. M. (2015). An introduction to implementation science for the non-specialist. *BMC Psychology*, *3*(1), 32. doi:10.118640359-015-0089-9 PMID:26376626

Bonilla, S., & MacPherson, A. (2021). Fighting "bad science" in the information age: The effects of an intervention to stimulate evaluation and critique of false scientific claims. *Journal of Research in Science Teaching*, *58*(8), 1152–1178. doi:10.1002/tea.21696

Brothen, T. (1984). Three computer-assisted laboratory exercises for introductory psychology. [SAGE Publications.]. *Teaching of Psychology*, *2*(11), 105–107. doi:10.120715328023top1102_14

Buck, G. A., Francis, D. C., & Wilkins-Yel, K. G. (2020). Research on gender equity in STEM education. In C. C. Johnson, M. J. Mohr-Schroeder, T. J. Moore, & L. D. English (Eds.), *Handbook of research on STEM education* (pp. 289–299). Taylor & Francis. doi:10.4324/9780429021381-27

Bull, G., Hodges, C., Mouza, C., Grant, M., Archambault, L., Borup, J., Ferdig, R.E., & Schmidt-Crawford, D. A. (2019). Conceptual dilution. *Contemporary Issues in Technology and Teacher Education, 19*(2). https://citejournal.org/volume-19/issue-2-19/editorial/editorial-conceptual-dilution

Bunyamin, M. A. H., Talib, C. A., Ahmad, N. J., & Surif, J. (2020). Current teaching practice of physics teachers and implications for integrated STEM education. *Universal Journal of Educational Research.*, *8*(5A), 18–28. doi:10.13189/ujer.2020.081903

Burrows, A. C., Borowczak, M., Myers, A., Schwortz, A. C., & McKim, C. (2021). Integrated STEM for teacher professional learning and development: "I need time for practice." *Education Sciences*, *11*(21), 1–23. doi:10.3390/educsci11010021

Burrows, A. C., Swarts, G. P., Hutchison, L., Katzmann, J. M., Thompson, R., Freeman, L., Schanke, A., Kilty, T., & Reynolds, T. (2021). Finding spaces: Teacher education technology competencies (TETCs). *Education in Science*, *11*(11), 733. doi:10.3390/educsci11110733

Campbell, T., & Abd-Hamid, N. H. (2013). Technology use in science instruction (TUSI): Aligning the integration of technology in science instruction in ways supportive of science education reform. *Journal of Science Education and Technology*, *22*(4), 572–588. doi:10.100710956-012-9415-7

Carey, C. C., & Gougis, R. D. (2017). Simulation modeling of lakes in undergraduate and graduate classrooms increases comprehension of climate change concepts and experience with computational tools. *Journal of Science Education and Technology*, *1*(26), 1–11. doi:10.100710956-016-9644-2

Cheng, C. H., Bråten, I., Yang, F. Y., & Brandmo, C. (2021). Investigating structural relationships among upper-secondary school students' beliefs about knowledge, justification for knowing, and Internet-specific justification in the domain of science. *Journal of Research in Science Teaching*, *58*(7), 980–1009. doi:10.1002/tea.21689

Clement, J. (2022, July 27). *Regional distribution of desktop traffic to Reddit.com as of May 2022 by country*. Statista. https://www.statista.com/statistics/325144/reddit-global-active-user-distribution/

Dale, D. A., Sutter, J., & Kloster, D. (2019). Asking real-world questions with inquiry-based labs. *The Physics Teacher, 57*(8), 547–550. doi:10.1119/1.5131122

Davis, M. (2019, April 23). Harvard Business Review, MBA lessons guide principals' ed-tech leadership. *Education Week*. https://www.edweek.org/technology/harvard-business-review-mba-lessons-guide-principals-ed-tech-leadership/2019/04

Dean, B. (2022, January 5). *How many people use Twitter in 2022?* [New Twitter stats]. Backlinko. https://backlinko.com/twitter-users

Decker, A., & McGill, M. (2019). A systematic review exploring the differences in reported data for pre-college educational activities for computer science, engineering, and other STEM disciplines. *Education Sciences, 9*(2), 69. doi:10.3390/educsci9020069

Dexter, S., & Richardson, J. W. (2020). What does technology integration research tell us about the leadership of technology? *Journal of Research on Technology in Education, 52*(1), 17–36. doi:10.1080/15391523.2019.1668316

Duda, M., & Wilson, B. (2015). *Using implementation science to close the policy to practice gap* (A Literate Nation White Paper). https://www.wilsonlanguage.com/wp-content/uploads/2016/06/Implementation_Science_White_Paper_for_Literate_Nation.pdf

Enzingmüller, C., & Prechtl, H. (2021). Constructing graphs in biology class: Secondary biology teachers' beliefs, motivation, and self-reported practices. *International Journal of Science and Mathematics Education, 19*(1), 1–19. doi:10.100710763-019-09975-2

Fixsen, D., Blase, K., Metz, A., & Van Dyke, M. (2015). Implementation science. International Encyclopedia of the Social and Behavioral Sciences, 11, 695-702. doi:10.1016/B978-0-08-097086-8.10548-3

Francis, G. (2020). Medicine: Art or science? *Lancet, 395*(10217), 24–25. doi:10.1016/S0140-6736(19)33145-9 PMID:31982047

Gao, S., Wang, J., & Zhong, Z. (2018). Influence of science instruction reform on academic performance of eighth grade students in Chinese Inner-Mongolia Autonomous Region. *Compare: A Journal of Comparative Education, 48*(6), 879–895. doi:10.1080/03057925.2017.1365285

Gao, X., Li, P., Shen, J., & Sun, H. (2020). Reviewing assessment of student learning in interdisciplinary STEM education. *International Journal of STEM Education, 7*(1), 24. doi:10.118640594-020-00225-4

Glimpse. (2019, May 14). *Glimpse K12 analysis of school spending shows that two-thirds of software license purchases go unused* [Press release]. https://www.glimpsek12.com/blog-posts/glimpse-k12-analysis-of-school-spending-shows-that-two-thirds-of-software-license-purchases-go-unused

Guspatni, G. (2021). Student-generated PowerPoint animations: A study of student teachers' conceptions of molecular motions through their expressed models. *Chemistry Education Research and Practice, 22*(2), 312–327. doi:10.1039/D0RP00229A

Guzey, S. S., Caskurlu, S., & Kozan, K. (2020). Integrated STEM pedagogies and student learning. In C. C. Johnson, M. J. Mohr-Schroeder, T. J. Moore, & L. D. English (Eds.), *Handbook of research on STEM education* (pp. 65–75). Taylor & Francis. doi:10.4324/9780429021381-8

Hechter, R. P., & Vermette, L. (2013). Technology integration in K-12 science classrooms: An analysis of barriers and implications. *Themes in Science and Technology Education*, *6*(2), 73–90.

Hellgren, S., & Lindberg, S. (2017). Motivating students with authentic science experiences: Changes in motivation for school science. *Research in Science & Technological Education*, *35*(4), 409–426. doi:10.1080/02635143.2017.1322572

Herman, J. (2001). Medicine: The science and the art. *Medical Humanities*, *27*(1), 42–46. doi:10.1136/mh.27.1.42 PMID:23670550

Hjalmarson, M. A., Holincheck, N. B., Courtney, K., & Galanti, T. M. (2020). Learning models and modeling across the STEM disciplines. In C. C. Johnson, M. J. Mohr-Schroeder, T. J. Moore, & L. D. English (Eds.), Handbook of research on STEM education (pp. 223–233). Taylor & Francis.

Holincheck, N., & Galanti, T. M. (2022). Are you a STEM teacher?: Exploring PK-12 teachers' conceptions of STEM education. *Journal of STEM Education: Innovations and Research*, *23*(2), 23–29.

Hsu, P.-S., & Sharma, P. (2008). A case study of enabling factors in the technology integration change process. *Journal of Educational Technology & Society*, *11*, 213–228.

Hyde, J. M. (2021). Exploring Hubble Constant Data in an Introductory Course. *The Physics Teacher*, *59*(3), 159–161. doi:10.1119/10.0003654

Ibourk, A., & Kendrick, M. (2021). *Elementary students' explanation of variation of traits and teacher's feedback using an online embedded assessment tool.* doi:10.1080/09500693.2021.1903618

Jackson, C., Tank, K. M., Appelgate, M. H., Jurgenson, K., Delaney, A., & Erden, C. (2020). History of integrated STEM curriculum. In C. C. Johnson, M. J. Mohr-Schroeder, T. J. Moore, & L. D. English (Eds.), *Handbook of research on STEM education* (pp. 169–183). Taylor & Francis. doi:10.4324/9780429021381-17

Jalilifar, A., White, P., & Malekizadeh, N. (2017). Exploring nominalization in scientific textbooks: A cross-disciplinary study of hard and soft sciences. *International Journal of English Studies*, *17*(2), 1–20. doi:10.6018/ijes/2017/2/272781

Kersaint, G., Ritzhaupt, A. D., & Liu, F. (2013). Technology to enhance mathematics and science instruction: Changes in teacher perceptions after participating in a yearlong professional development program. *Journal of Computers in Mathematics and Science Teaching*, *33*, 73–101.

Khairani, A. Z. (2017). Assessing urban and rural teachers' competencies in STEM integrated education in Malaysia. *MATEC Web of Conferences, 87*, 1-5.

Kilty, T. J. (2019). *Technology integration and science education: An instructional design and technology perspective* (Publication No. 13882883) [Doctoral dissertation, University of Wyoming]. ProQuest Dissertations Publishing.

Kilty, T. J., & Burrows, A. C. (2021). Secondary science preservice teachers: Technology integration in methods and residency. *Journal of Science Teacher Education*, *32*(5), 578–600. doi:10.1080/1046560X.2021.1907514

Kobulnicky, H., & Dale, D. (2016). A community mentoring model for STEM undergraduate research experiences. *Journal of College Science Teaching*, *45*(6), 17–23. doi:10.2505/4/jcst16_045_06_17

Kotter, J. (1995, May-June). Leading change: Why transformation efforts fail. *Harvard Business Review*. https://hbr.org/1995/05/leading-change-why-transformation-efforts-fail-2

Krauss, J., & Prottsman, K. (2018). *Computational thinking and coding for every student: The teacher's getting started guide*. Corwin Press.

Kuykendall, K. (2022, March 31). LearnPlatform Launches Certification for Ed Tech Providers Meeting ESSA Evidence and Student Data, Privacy Standards. *THE Journal*. https://thejournal.com/articles/2022/03/31/learnplatform-launches-certification-for-ed-tech-providers-meeting-essa-evidence-and-data-standards.aspx

Langen, T. A., Mourad, T., Grant, B. W., Gram, W. K., Abraham, B. J., Fernandez, D. S., Carroll, M., Nuding, A., Balch, J. K., Rodriguez, J., & Hampton, S. E. (2014). Using large public datasets in the undergraduate ecology classroom. *Frontiers in Ecology and the Environment*, *12*(6), 362–363. doi:10.1890/1540-9295-12.6.362

Lungu, B., & Lungu, M. (2021). Exploring the effects on student learning and engagement of COVID-19: An innovative and interdisciplinary approach. *Journal of Microbiology & Biology Education*, *22*(1), ev22i1.2429. Advance online publication. doi:10.1128/jmbe.v22i1.2429 PMID:33884068

Metz, A., & Louison, L. (2018). *The Hexagon tool: Exploring context*. National Implementation Research Network.

Microsoft. (2017). *Why Europe's girls aren't studying STEM*. https://ncmedia.azureedge.net/ncmedia/2017/02/Microsoft_girls_in_STEM_final-Whitepaper.pdf

Moore, T. J., Johnston, A. C., & Glancy, A. W. (2020). STEM integration: A synthesis of conceptual frameworks and definitions. In C. C. Johnson, M. J. Mohr-Schroeder, T. J. Moore, & L. D. English (Eds.), *Handbook of research on STEM education* (pp. 3–16). Taylor & Francis. doi:10.4324/9780429021381-2

Mueller, J., Wood, E., Willoughby, T., Ross, C., & Specht, J. (2008). Identifying discriminating variables between teachers who fully integrate computers and teachers with limited integration. *Computers & Education*, *51*(4), 1523–1537. doi:10.1016/j.compedu.2008.02.003

Mungan, C. E. (2021). Using desmos to understand the difference between phase and group velocity. *The Physics Teacher*, *59*(1), 30–33. doi:10.1119/10.0003012

National Governors Association Center for Best Practices & Council of Chief State School Officers. (2010). *Common core state standards for mathematics*. Authors.

National Implementation Research Network. (n.d.). *Topic 3: Exploration stage*. https://nirn.fpg.unc.edu/module-4/topic-3-exploration

National Research Council. (2010). *New worlds, new horizons in astronomy and astrophysics*. National Research Council.

National Research Council. (2013). *Next generation science standards: For states, by states*. National Academies Press.

Ng, O.-L., Shi, L., & Ting, F. (2020). Exploring differences in primary students' geometry learning outcomes in two technology-enhanced environments: Dynamic geometry and 3D printing. *International Journal of STEM Education*, *7*(1), 50. doi:10.118640594-020-00244-1

Page, M. J., McKenzie, J. E., Bossuyt, P. M., Boutron, I., Hoffmann, T. C., Mulrow, C. D., Shamseer, L., Tetzlaff, J. M., Akl, E. A., Brennan, S. E., Chou, R., Glanville, J., Grimshaw, J. M., Hróbjartsson, A., Lalu, M. M., Li, T., Loder, E. W., Mayo-Wilson, E., McDonald, S., ... Moher, D. (2021). The PRISMA 2020 statement: An updated guideline for reporting systematic reviews. *Systematic Reviews*, *10*(1), 89. Advance online publication. doi:10.118613643-021-01626-4 PMID:33781348

Panda, S. C. (2006). Medicine: Science or art? *Mens Sana Monographs*, *4*(1), 127–138. doi:10.4103/0973-1229.27610 PMID:22013337

Paulus, M., Villegas, S. G., & Howze-Owens, H. (2020). Professional learning communities: Bridging the technology integration gap through effective professional development. *Peabody Journal of Education*, *95*(2), 193–202. doi:10.1080/0161956X.2020.1745610

Pinou, T., Pitzrick, M. S., Rahaman, A., & Young, M. A. (2021). Letter to the editor. *Journal of Microbiology & Biology Education*, *22*(1), e00096-21. Advance online publication. doi:10.1128/jmbe.v22i1.2199 PMID:33953820

Ponners, P., & Asim, S. (2016). Beyond the classroom walls: Technology infusion advancing science education. *The Delta Kappa Gamma Bulletin*, *83*(1), 61–67. https://www.proquest.com/docview/1822382441

Rainey, K., Dancy, M., Mickelson, R., Stearns, E., & Moller, S. (2019). A descriptive study of race and gender differences in how instructional style and perceived professor care influence decisions to major in STEM. *International Journal of STEM Education*, *1*(6), 6. doi:10.118640594-019-0159-2

Reiser, R. A. (2012). What field did you say you were in? Defining and naming our field. In R. A. Reiser & J. V. Dempsey (Eds.), *Trends and Issues in Instructional Design and Technology* (3rd ed., pp. 1–7). Pearson.

Resnick, I., Kastens, K. A., & Shipley, T. F. (2018). How students reason about visualizations from large professionally collected data sets: A study of students approaching the threshold of data proficiency. *Journal of Geoscience Education*, *66*(1), 55–76. doi:10.1080/10899995.2018.1411724

Salar, R. (2021). An arduino-based experiment designed to investigate gas pressure. *The Physics Teacher*, *59*(3), 202–204. doi:10.1119/10.0003665

Schwortz, A. C. (2021). *Physics and Astronomy Mentoring and Professional Development: Equity and Empowerment Across K-20 STEM Spaces* (Publication No. 2621664057) [Doctoral dissertation, University of Wyoming]. ProQuest Dissertations Publishing. https://www.proquest.com/docview/2621664057

Schwortz, A. C., & Burrows, A. C. (2021). Authentic science experiences with STEM datasets: Post-secondary results and potential gender influences. *Research in Science & Technological Education*, *39*(3), 347–367. doi:10.1080/02635143.2020.1761783

Semmens, K., Fosbenner, L., Maxfield, K., Carr, R. H., Ramage, J., O'Leary, J., Larouche, C., Hayward, J., & Henderson, M. (2020). Teaching Earth systems connections to middle school students in an informal learning environment with Science on a Sphere®, personal relevance and arts-based activities. *Journal of Geoscience Education*, *69*(1), 71–84. doi:10.1080/10899995.2020.1829396

Senger, K., Betlem, P., Birchall, T., Buckley, S. J., Coakley, B., Eide, C. H., Flaig, P. P., Forien, M., Galland, O., Gonzaga, L. Jr, Jensen, M., Kurz, T., Lecomte, I., Mair, K., Malm, R. H., Mulrooney, M., Naumann, N., Nordmo, I., Nolde, N., ... Smyrak-Sikora, A. (2020). Using digital outcrops to make the high Arctic more accessible through the Svalbox database. *Journal of Geoscience Education*, *69*(2), 123–137. doi:10.1080/10899995.2020.1813865

Sheffield, R., Blackley, S., & Moro, P. (2018). A professional learning model supporting teachers to integrate digital technologies. *Issues in Educational Research*, *28*, 487–510. http://www.iier.org.au/iier28/sheffield.pdf

Song, M. (2017). Teaching integrated STEM in Korea: Structure of teacher competence. *LUMAT-B: International Journal on Math. Science and Technology Education*, *2*(4), 61–72.

Speer, J. E., Lyon, M., & Johnson, J. (2021). Gains and losses in virtual mentorship: A descriptive case study of undergraduate mentees and graduate mentors in stem research during the covid-19 pandemic. *CBE Life Sciences Education*, *20*(2), ar14. doi:10.1187/cbe.20-06-0128 PMID:33734867

Spuck, T. (2014). Putting the "authenticity" into science learning. In T. Spuck, L. Jenkins, T. Rust, & R. Dou (Eds.), Best practices in STEM education: Innovative approaches from Einstein Fellow alumni (Vol. 27, pp. 107-139). Peter Lang.

Struyf, A., De Loof, H., Boeve-de Pauw, J., & Van Petegem, P. (2019). Students' engagement in different STEM learning environments: Integrated STEM education as promising practice? *International Journal of Science Education*, *41*(10), 1387–1407. doi:10.1080/09500693.2019.1607983

Tal, T., & Dallashe, S. (2021). Engaging the hands, heads, and hearts in a medical simulation informal learning environment. *Journal of Research in Science Teaching*, *58*(6), 759–789. doi:10.1002/tea.21677

Tam, A. C. F. (2015). Exploring teachers' beliefs about teacher learning in professional learning communities and their influence on collegial activities in two departments. *Compare: A Journal of Comparative Education*, *45*(3), 422–444. doi:10.1080/03057925.2013.872025

The Association for Computing Machinery, Code.org, Computer Science Teachers Association, Cyber Innovation Center, & National Math and Science Initiative. (2016). *K-12 computer science framework*. http://www.k12cs.org

Thuy, N. T. T., Bien, N. V., & Quy, D. X. (2020). Fostering teachers' competence of the integrated STEM education. *Jurnal Penelitian dan Pembelajaran IPA, 2*(6), 116.

Tseng, A. S. (2017). *Reimagining the role of technology in education: 2017 national education technology plan update*. Office of Educational Technology.

Wilcox, B. R., & Lewandowski, H. J. (2016). Open-ended versus guided laboratory activities: Impact on students' beliefs about experimental physics. *Physical Review. Physics Education Research*, *12*(2), 020132. doi:10.1103/PhysRevPhysEducRes.12.020132

Wofford, J. (2009). K-16 computationally rich science education: A ten-year review of the Journal of Science Education and Technology (1998-2008). *Journal of Science Education and Technology*, *18*(1), 29–36. doi:10.100710956-008-9127-1

APPENDIX

Table 5. Web-based technology tools that can be used in the science classroom, with a description of their purposes. See also Burrows et al. (2021) for more discussion of some of these examples.

Resource	Link	Purpose	Category/ies
Anaconda	https://www.anaconda.com/	Python programming tool, including Jupyter notebooks	Computational Thinking / Programming
Arduino	https://www.arduino.cc/	Microcontroller (model computers)	Computational Thinking / Programming
Atom Visualizer	Android phone app	Augmented reality and 3D simulations of atoms	Visualization Tools
BigBangAR	phone app	Augmented reality about the evolution of the universe	Visualization Tools, Other
BioInteractive	https://www.biointeractive.org/	Biology data	Datasets / Data Mining
Blooket	https://www.blooket.com/	Quizzes, unit review, pre-assessment	Other
BrainPop	https://www.brainpop.com/	Videos, formatives, animation software, mind mapping software, gamification	Information Curation, Visualization Tools, Other
BreakoutEDU	https://breakoutedu.com/	Educational games, gamification, unit review	Other
BuzzIn.live	https://buzzin.live/	Online buzzer system for games, surveys	Other
Canvas	https://www.instructure.com/canvas	LMS system free to individual teachers and their students	Communication Tools
Choice Boards/Menus with Google Slides and Google Classroom	combination usage	Digital "choose your own adventure" / decision tree navigation	Other
CK12	https://www.ck12.org/	Free online textbooks	Information Curation, Other
Classcraft	https://www.classcraft.com/	Classroom management	Communication Tools
CS First	https://csfirst.withgoogle.com/	Programming instruction	Computational Thinking / Programming
DataCorral	https://datacorral.uwyo.edu/	Source of publicly available environmental data, via University of Wyoming	Datasets / Data Mining
Desmos	https://www.desmos.com/	Math tool	Math Modeling
Discord	https://discord.com/	Communication tool including text-based chat, and audio/video conferencing	
Fanschool (formerly Kidblog)	https://go.fan.school/kidblog	Student content curation	Information Curation
Flipgrid	https://info.flipgrid.com/	Discussion around videos	Information Curation, Visualization Tools
Formative	https://www.formative.com/	Formative and summative assessment, lots of tools for learning accommodations	Communication Tools, Other
GeoGebra	https://www.geogebra.org/?lang=en	Math, whiteboard, and more	Math Modeling, Visualization Tools
Geometer's Sketchpad	https://www.keycurriculum.com/	Math visualization	Math Modeling, Visualization Tool
Google Classroom	https://classroom.google.com/	Course/learning management system (CMS/LMS)	Communication Tools
Google Drive	https://drive.google.com/	File storage and office programs	Information Curation
Google Earth	https://earth.google.com/	Mapping tool (more complex and more powerful than Google Earth)	Visualization Tools
Google Jamboard	https://jamboard.google.com/	Digital notice board	Communication Tools, Visualization Tools
Google Maps	http://maps.google.com/	Mapping tool (simpler and easier to use than Google Earth)	Visualization Tools
Google Sites	https://sites.google.com/	Webpage hosting	Information Curation
Google Slides escape rooms	combination usage	Problem solving and exploration	Visualization Tools, Other
Google Surveys	https://surveys.google.com/	Surveys, quizzes, polling	Communication Tools, Other
Happy Atoms	https://happyatoms.com/	Augmented reality about molecules	Visualization Tools
iMovie	https://www.apple.com/imovie/	Movie creation, stop-motion, animation	Visualization Tools
Kami	https://www.kamiapp.com/	Write on top of digital documents	Other
Kessler Science	https://www.keslerscience.com/	Lesson plans, labs, etc. for elementary and middle school science	Other
Khan Academy	https://www.khanacademy.org/	Videos, online lessons, quizzes, etc. on many STEM subjects	Other
Lego Mindstorms	https://www.lego.com/en-us/themes/mindstorms	Build programmable LEGO robots	Computational Thinking / Programming
Lightspeed	https://www.lightspeedsystems.com/	Online student safety tool	Other
Loom	https://www.loom.com/	Screencasting	Information Curation
Maple	https://www.maplesoft.com/products/Maple/	Math tool	Math Modeling, Visualization Tool, Computational Thinking / Programming
Mathematica	https://www.wolfram.com/mathematica/	Math tool	Math Modeling, Visualization Tool, Computational Thinking / Programming
Mathigon Polypad	https://mathigon.org/polypad	Math visualization, graphing, and more	Math Modeling, Visualization Tools

continued on following page

Table 5. Continued

Resource	Link	Purpose	Category/ies
Matlab	https://www.mathworks.com/products/matlab-online.html	Math tool	Math Modeling, Visualization Tools, Computational Thinking / Programming
Merge Cube	https://mergeedu.com/cube	Augmented reality and 3D simulations	Visualization Tools
Micro:bit	https://microbit.org/	Microcontroller (model computers)	Computational Thinking / Programming
Miro	https://miro.com/	Concept mapping, flow charts, Venn diagrams, and Euler diagrams — Figures 1 and 2 in this chapter were created using Miro.	Visualization Tools
MS Excel	-	Spreadsheet	Math Modeling, Datasets / Data Mining
Murison	https://www.mursion.com/	Practice giving presentations and other skills with an AI audience	Communication Tools, Other
Nature Cams	assorted	Webcams for nature, can be directly studied, or used to "keep a calm vibe" while students independently working	Other
Nearpod	https://nearpod.com/	Interactive classroom tool, slideshows, formative assessment, & learning tool	Communication Tools
NetLogo	https://ccl.northwestern.edu/netlogo/	Programming simulations about natural and social phenomena	Math Modeling, Visualization Tool, Computational Thinking / Programming
Newslea	https://newsela.com/	Current events readings	Information Curation, Other
Padlet	https://padlet.com/	Digital notice board	Communication Tools
Pasco Interfaces	https://www.pasco.com/products/interfaces-and-dataloggers	Devices to connect Pasco probeware to computers for data analysis	Science-Specific Technology
Pasco Probeware	https://www.pasco.com/products/sensors	Digital sensors for taking measurements in lab	Science-Specific Technology
PhET	https://phet.colorado.edu/	Physics and other science simulations, good for labs or active demonstrations	Visualization Tools, Other
Poll Everywhere	https://www.polleverywhere.com/	Surveys and student response system (SRS)	Communication Tools, Other
Quizizz	https://quizizz.com	Online quizzes and quiz-based games, unit review or as pre-assessment	Other
Quizlet and Quizlet Live	https://quizlet.com/	Online quizzes and quiz-based games, unit review or as pre-assessment	Other
Raspberry Pi	https://www.raspberrypi.org/	Microcontroller (model computers)	Computational Thinking / Programming
Savvas Learning Elevate Science	https://www.savvas.com/index.cfm?locator=PS3eR5&utm_content=elevate%20Science%20K-8	Science content for elementary and middle school	Other
Screencast-O-Matic	https://screencast-o-matic.com/	Screencasting	Information Curation
Skype	https://www.skype.com	Video conferencing	Communication Tools
Swivl	https://www.swivl.com/	Audio / video tools	Visualization Tools
TinkerCAD	https://www.tinkercad.com/	Free version of Autocad to 3D design project builds or 3D print models	Visualization Tools, Computational Thinking / Programming
Trinket	https://trinket.io/	Programming instruction	Computational Thinking / Programming
Twitter	https://twitter.com	Social media, microblogging	Communication Tools
Tynker	https://www.tynker.com/	Hour of Code, introductory coding	Computational Thinking / Programming
Vernier Interfaces	https://www.vernier.com/product-category/?category=interfaces	Devices to connect Vernier probeware to computers for data analysis	Science-Specific Technology
Vernier Probeware	https://www.vernier.com/product-category/?category=sensors	Digital sensors for taking measurements in lab	Science-Specific Technology
VR VEXcode Project	https://vr.vex.com/	Programming, such as for robotics, Scratch, Python, C++	Computational Thinking / Programming
Vyond (formerly GoAnimate)	https://www.vyond.com/	Stop animation / movies	Information Curation, Visualization Tools
W3Schools	https://www.w3schools.com/	Programming instruction	Computational Thinking / Programming
Wolfram Alpha	https://www.wolframalpha.com/	Math tool and general encyclopedia	Visualization Tools, Other
YouTube	https://www.youtube.com/	Science videos, hosting of videos, enrichment, more	Information Curation, Other
Zoom	https://zoom.us/	Video conferencing	Communication Tools
Zooniverse	https://www.zooniverse.org/	Citizen science projects including many topics in life sciences, physical sciences, and social sciences	Datasets / Data Mining, Science-Specific Technology, Other

Chapter 13
Using Vicarious Experiences to Artificially Create the "Sweet Spot":
Modeling Pedagogy and Technology Integration

Jeremy Unruh
University of Houston-Clear Lake, USA

Michelle Giles
https://orcid.org/0000-0001-5473-9175
University of Houston-Clear Lake, USA

Jana Willis
https://orcid.org/0000-0003-4339-9675
University of Houston-Clear Lake, USA

ABSTRACT

Teacher education programs continue to struggle to address challenges that arise from course sequencing, isolated content courses, and untethered field experiences. Applying strategies that mimic classroom practices through vicarious experiences that include repeated exposure to standards-based, interdisciplinary curriculum could bridge many of these gaps. This chapter explores the use of an environmental education curriculum as a foundation for modeling best practices in technology integration based on technological pedagogical content knowledge (TPACK). Building effective STEM integration models could better prepare teacher candidates to be inclusive of STEM in their future classrooms.

INTRODUCTION

In a perfect instructional setting there exists a "sweet spot" where technology, pedagogy, and content

DOI: 10.4018/978-1-6684-5585-2.ch013

intersect, often referred to as TPACK (Koehler, et al., 2013; Mishra & Koehler, 2006; Shulman, 1986). However, too often teacher candidates engage in coursework where the "sweet spot" is not realized. For many education programs, content courses in STEM areas often exist outside of the educational colleges. Additionally, considerations of pedagogy, especially field-based classroom experiences, are often limited in content scope and occur late in the teacher candidate's program of study. These challenges are further complicated with the reality that technology may or may not be included in any of these courses or experiences. In the case of this study, the technology course is a stand-alone course and comes early in the teacher education program sequence, prior to any content or pedagogy courses, often as one of the initial courses taken before acceptance into the teacher education program. This means teacher candidates in the technology course have not engaged in any previous coursework that provided them with foundational knowledge in curriculum, lesson planning, lesson development, or effective technology use. The debate surrounding stand-alone technology courses and when in the program sequence the teaching of technology should occur is one that has existed for many years (Foulger et al., 2019; Schmidt-Crawford et al., 2018; Wang, 2006). At the researchers' institution the debate was a "a chicken and egg problem"; teacher candidates need to be proficient in technology, while content and methods course instructors prefer to have the teacher candidates enter their courses with technology skills and experiences, reducing the need to include the teaching of technology in their courses. In response to the separation of technology skill development and pedagogy at their institution, the researchers adopted environmental education (EE) curriculum materials that were standards-based, interdisciplinary, and cross-grade level. The lessons in the curriculum were inclusive of authentic tasks, modeled inquiry and project-based learning (PBL), followed the 5E lesson plan, and were developed and field tested by certified classroom teachers. The EE materials were used to provide repeated exposure to experience-based EE lessons, which mimicked vicarious mastery experiences. Vicarious learning is as an instructional method that allows individuals to learn from the experiences of others (Bandura, 1986, 1997; Thompson & Sorbet, 2020; Wang, Ertmer, & Newby, 2004), especially when individuals lack prior experience or lack self-efficacy about their capabilities (Bandura, 1997). Studies have shown that vicarious experiences can be used to teach new skills and enhance self-efficacy (Maddox, 2011). The question under investigation then became: could vicarious experiences with sound educational content, pedagogical models, and effective technology integration impact teacher candidates' perceptions of curriculum, lesson planning, and uses of technology?

In a previous study, a purposeful sample of 424 teacher candidates in a course that employed the same environmental education (EE) foundational curriculum were administered the Technology and Teaching Efficacy Scale to evaluate efficacy levels in teaching and technology use (Tanguma et al., 2004). Findings from the earlier study suggested teacher candidates who engaged in a course that repeatedly exposed them to sound experience-based EE lessons that mimicked vicarious mastery experiences gained understanding/awareness of content knowledge and practices in teaching, leading to increased sense of preparedness, and teaching self-efficacy (Giles & Willis, 2019). Outcomes from that study support the hypothesis for this current study, that EE curriculum could provide vicarious experiences that could support artificial creation of the "sweet spot"; positively impacting the teacher candidate's cross-curricular pedagogy, lesson planning, and technology use.

BACKGROUND

Self-Efficacy and Technology

Bandura (1994) defined self-efficacy "as people's beliefs about their capabilities to produce designated levels of performance that exercise influence over events that affect their lives" (p. 10). His research indicated that individuals who possessed high levels of self-confidence in their abilities would approach difficult tasks as challenges rather than obstacles (Bandura, 1994). Therefore, higher levels of self-efficacy produce higher levels of performance. Teacher candidates often enter technology training with low self-efficacy. For technology novices, new technologies seem threatening and overwhelming, resulting in intensification of existing beliefs about their inability to master technology tasks. Yan and Piper (2003) in a study of elementary and secondary teachers found self-efficacy to have a significant positive impact on technology use. Barton and Dexter, 2020, "suggest a holistic system of learning supports teachers' access to three sources of self-efficacy information (verbal persuasion, vicarious experiences, and mastery experiences) and prompts teachers' reflection on source information such that it can impact their self-efficacy" (p. 89). In their study, they interviewed middle school math and science teachers from two schools who participated in a leadership program designed to support teachers' technology integration. Kent and Giles (2017) found positive teacher-efficacy in technology proficiency to be essential for effective educational technology integration in early-childhood classrooms. Results of a study of 525 primary school teachers by Hermans et al. (2008) supported the "hypothesis that teacher beliefs are significant determinants in explaining why teachers adopt computers in the classroom" (p. 1).

According to Bandura's self-efficacy theory, beliefs "are constructed from four principle sources of information" (Bandura, 1997, p. 79): (a) enactive mastery experience, (b) vicarious experience, (c) verbal persuasion, and (d) physiological and affective states. Modeling and observations can influence the self-efficacy of participants during vicarious experiences (Bandura et al., 1977; Jacobs et al., 2020; Tschannen-Moran & Hoy, 2007). Self-efficacy is especially sensitive to vicarious experience where individuals have little prior experience or are uncertain about their capabilities (Bandura, 1997). Vicarious experiences can be used to teach new skills and enhance self-efficacy (Maddux, 2002; Maddux & Kleiman, 2020).

Self-Efficacy and Science

Research has shown that many preservice elementary teachers have low self-efficacy beliefs when teaching science because of a lack of understanding of science concepts (Bautista, 2011; Bleicher & Lindgren 2005; Norris, et al., 2018; Menon & Sadler, 2016; Menon & Sadler, 2018; Schibeci & Hickey 2000; Schriver & Czerniak 1999; Trundle et al. 2002). Teacher candidates need exposure to good science teaching and learning if they are going to gain understanding and improve their self-efficacy (Bautista, 2011). Within the area of Environmental Science, education teachers' lack of understanding and experience is further complicated by limited exposure and narrow curricula that are not inclusive of Project Based Learning (PBL) strategies (Genc, 2015; Li, 2013). PBL moves the learner beyond content knowledge acquisition to creating a space where learners can work their way to a solution in their own way (Krajcik & Blumenfeld, 2006); where prior knowledge and experience can be engaged through new tasks (Roessingh & Chambers, 2011). Many teachers' understandings of PBL as an inquiry-based strategy does not align with their science education curriculum. This often hinders their ability to implement

PBL in their classrooms (Al-Balushi & Al-Aamri, 2014; Maeng et al., 2020; Nariman & Chrispeels, 2016). Barriers to PBL implementation include teachers' discomfort with their new role as facilitators and the resulting loss of class time; students' inexperience with cooperative or collaborative learning; and designing appropriate assessment and evaluation instruments for such projects (Henderson, et al., 2021). Teachers fear PBL hinders their ability to fully cover the curriculum and to prepare students for standardized assessments (Al-Balushi & Al-Aamri, 2014; Maeng et al., 2013; Whitaker, 2019). Providing models of PBL through curriculum and classroom assignments could positively increase teacher candidates use of PBL in their future classrooms.

Creating a Positive Experience

PBL integration into the classroom can be made easier with technology. There is no standard template on how to integrate technology and efforts often face resistance from educators that believe technology does more harm than good; creating its own barriers (Calik et al., 2015; Dunaetz, 2022; Maeng et al., 2013; Koehler & Mishra, 2009). Promoting development of deep scientific understanding through inquiry-based activities will not happen through technological advances that focus on the technology rather than how to approach teaching subject matter with the technology (Maeng et al., 2013; Koehler & Mishra, 2009; Seneviratne, et al., 2019; Tanak, 2020). Technology changes rapidly and pressure exists to ensure teachers learn, practice, reflect, and share new ways to integrate and implement technology in the classroom (Calik et al., 2015; Fernández-Batanero et al., 2021; Koehler & Mishra, 2009). Many preservice and in-service teachers lack technology implementation skills or theoretical understanding, and if they do not build these competencies during their preservice programs, they will not use it in their future classroom (Calik et al., 2015; Fernández-Batanero et al., 2021; Kohler & Mishra, 2009). Positive learning experiences during preservice and professional development could build high self-efficacy to use both PBL and technology in the classroom (Büssing et al., 2019; Calik et al., 2015; Korucu-Kis & Ozmen, 2019; Willis et al., 2014).

EXPLORATION

The Technology Course

The course used in the study is a required undergraduate course focused on the exploration of a variety of technologies that can be used to enhance teaching and learning. Since the course comes early in the teacher candidates' educational journey, most if not all the teacher candidates have never written a lesson plan or taught a classroom lesson. Add to that the fact that in this classroom there are teacher candidates seeking certification across all grade levels and a variety of content levels. So, finding a common curriculum for modelling lesson planning and technology integration is further complicated. To bridge these gaps in knowledge and experience as well as trying to find a common curriculum, the teacher candidates used lessons from an EE curriculum to serve as their foundation to guide their technology integration efforts and understanding of the lesson plan process. Each course unit followed a PBL model where a new technology was introduced (e.g., digital collage, video, multimedia, comic strips, apps/games, presentations, data collection, learning management systems); followed by an assigned project designed to mimic a section of a 5E lesson plan that used the new technology. Teacher candidates selected a lesson from the EE curriculum, then mapped the lesson using a visual mapping tool that allowed them to identify the components of the lesson as it related to the 5E components of the lesson. The teacher candidates then applied the new technology for that unit 5E component of the lesson to support teaching, learning, or assessment. For example, in the Engage unit, teacher candidates were asked to identify an app or game that could be used to engage students in the lesson content. They had to describe how the app or game would be used and their rationale for their decision. This cycle was repeated with each unit in the course. The course concluded with a final PBL assignment that required teacher candidates, in groups of 5 members, to develop their own 5E lesson plan inclusive of integrated technology for each 5E component. Instructionally, the course provided scaffolding of technology skills as well as pedagogical practices modeled in each of the EE lessons. The course used in this study was developed following a scaffolded design model that offered teacher candidates learning opportunities that supported incremental technology skill development, where new skills then could be built upon existing skills (Willis, 2015). The intent was to simultaneously build competence and confidence, through accomplishments that helped teacher candidates become more confident in their use of technology and awareness of lesson planning. Equally, the repeated exposure to standards-based curricula provided experiences that could positively influence candidates' sense of preparedness and teaching self-efficacy. Bandura noted that enactive mastery positively impacts efficacy beliefs (Bandura, 1994); "All of the vicarious modes of influence – whether conveyed through effective actual modeling, symbolic modeling, videotaped self-modeling, or cognitive self-modeling – enhance efficacy beliefs and improve performance" (p. 95).

The Teacher Candidates

A purposeful sample of preservice teacher candidates enrolled in a teacher education program at a 4-year university was solicited for participation in this study. The participants consisted of undergraduate students (n=513) enrolled in multiple sections of an introductory technology course across multiple years who completed the end of course survey. The survey was a required assignment in the course, however there was no mechanism in the course that prevented teacher candidates from opting to not complete that assignment. Only completed surveys were used in the data analysis. The teacher candidate population was representative of the university population with approximately 83.0% female and 17.0% male. Ages ranged from 16-60 years. Reported race/ethnicity was 63.0% White, 19.0% Hispanic, 3.0% Black/African American, and 2.0% Asian.

LOOKING FOR SOLUTIONS

Instrument

A researcher developed a retrospective pre-post survey that was used for assessment to evaluate teacher candidates' responses to questions related to lesson plan components and technology for individual parts of a lesson. In this type of survey design, the respondents take the survey once at the end of the intervention but are asked to respond to the same item twice, once as their past self (before the intervention) and once as their current self. The decision to use a retrospective pre-post was the result of earlier studies conducted by these researchers with a similar population. Results of these studies found that when traditional pre-post surveys were used, teacher candidates often inflated their abilities due to lack of knowledge, low self-efficacy, or fear of reporting low performance (Giles & Willis, 2019; Lang & Savageau, 2017). The survey questions were drawn from an in-course assessment of the EE materials used in the technology course (Appendix A). To address the validity of the instrument, the questions were developed by the researchers several years ago and were vetted by a panel of experts in technology, environmental education, and pedagogy. The questions selected for this study were focused on lesson planning and technology use, *I could find cross curricular ways to approach a lesson topic, I viewed lessons as a sequence of activities, and I knew multiple ways to use technology to support subject area content.* The survey questions were rated on a 5-point Likert type scale (1 = *strongly disagree*, 2 = *disagree*, 3 = *neutral*, 4 = *agree*, 5 = *strongly agree*). The overall instrument showed strong reliability with a Cronbach's alpha = .91, and the three sets of pre/post items included in the survey had an alpha coefficient of .77.

Data Collection Procedures

Prior to data collection, the researchers obtained permission from the university level Institutional Review Board. Participants enrolled in all sections of the course were participated in the survey as it was a required assignment in the course. Participants were administered the survey on the last day of class across multiple sections for two semesters. The survey was distributed using SurveyMonkey and accessed only by the researchers. All documents were stored on a protected desktop and data stick that will be destroyed after 3-years.

Results

Due to the ordinal nature of the data, they were not normally distributed (Shapiro-Wilk $p < .05$). Because the items were paired, they were analyzed using the Wilcoxon Signed Rank Test for paired data in order to assess if statistically significant mean differences existed from pre- to post. All comparisons were significantly different, with higher agreement for each item in the post than in the pretest see Table 1. This indicates that after participating in the vicarious experiences, teachers' self-reported awareness of cross-curricular pedagogy, lesson planning, and technology use were higher. As an instructional approach, the inclusion of opportunities for repeated exposure to sound experience-based practices that mimic vicarious mastery experiences could promote gains in understanding/awareness of content knowledge and practices in teaching, leading to increased sense of preparedness, and teaching self-efficacy (Giles & Willis, 2019).

As noted earlier in the chapter, the teacher candidates enrolled in this course arrive with little to no knowledge related to curriculum, lesson planning, or the effective use of technology in the classroom to enhance teaching and learning. The lack of foundational knowledge coupled with missing commonalities among the course population, where teacher candidates were seeking certifications that spanned all grades levels (EC-12), and all content areas (e.g., science, math, history, social studies, elementary education) made finding a common curriculum for modelling lesson planning and technology integration a bit challenging. The researchers included the EE curriculum, technology tools, and integration strategies to create a vicarious "sweet spot" where content (EE), pedagogy, (interdisciplinary curriculum with model lesson plans), and technology (tools and integration strategies) came together. In question one *I could find cross-curricular ways to approach a lesson topic*, the significant growth from pre to post informed the researchers that the use of the EE materials helped teacher candidates better understand

Table 1. Means, standard deviations and results of Wilcoxon signed rank test

	Mean	SD	Z Score	P Value
Before taking this course, I could find cross-curricular ways to approach a lesson topic.	3.15	1.238	15.19	< .001
After taking this course I could find cross-curricular ways to approach a lesson topic.	4.39	.838		
Before taking this course, I viewed lessons as a sequence of activities.	3.57	1.182	13.00	< .001
After taking this course I viewed lessons as a sequence of activities.	4.43	.774		
Before taking this course, I knew multiple ways to use technology to support subject area content.	3.28	1.302	14.73	< .001
After taking this course I knew multiple ways to use technology to support subject area content.	4.54	.772		

interdisciplinary curriculum that supported integrated instruction in the classroom, It is of interest to note that this question had the lowest mean on the pre assessment, an indication that more models of cross-curricular instruction would benefit teacher candidates, possibly helping them move away from siloed instruction to more interdisciplinary lessons (Trevathan & Willis, 2019). The results of question two, *I viewed lessons as a sequence of activities,* also showed significant growth from pre to post assessment, this question had the lowest mean difference of the three questions for the pre assessment, indicating a significant gap in foundational understanding that should be addressed. Helping teacher candidates more fully understand the interconnectivity of lesson plans is an essential skill set for all teachers so that they can prepare lessons that will move their own students along a continuum of learning, scaffolding their knowledge. Not unlike the deliberate design of the technology course, where students' basic technology skills were slowly built over the course of the semester to improve competence and confidence in technology skills. The positive results of the analysis of question three, *I knew multiple ways to use technology to support subject area content.,* was a win for the researchers. Moving teacher candidates closer to the understanding that technology can be integrated throughout the lesson cycle, to enhance content knowledge development, is foundational within TPACK for effective seamless technology integration to occur (Atabek, 2020; Estes, & Dailey-Hebert, 2018; Kenward, Hipp, & Baker, 2011); Zhao & Bryant, 2006). The growth from pre to post for all three questions support the researchers' intent to mimic a "sweet spot" within the technology to course that could provide teacher candidates with vicarious experiences that could positively impact their knowledge and skills.

IMPLICATIONS FOR PRACTICE

The intent of this book chapter was to provide teacher educators with strategies that would support development of effective models for using interdisciplinary curriculum to provide vicarious experiences that could support artificial creation of the "sweet spot", where technology, pedagogy, and content intersect; positively impacting the teacher candidate's self-efficacy levels in teaching and technology use. This could impact development of effective models for preparing teacher candidates to seamlessly integrate technology into the curriculum when technology and content are separated in coursework and program sequencing. Additionally, this chapter introduced a possible model for preparing teacher candidates integrate technology into the content instruction. This model where content (EE), pedagogy, (interdisciplinary curriculum with model lesson plans), and technology (tools and integration strategies) came together to provide vicarious experiences could be replicated and included in technology content, or methods courses to support content and skill development.

FUTURE RESEARCH DIRECTIONS

The rapidly changing nature of technology development and the increasing expectations of technology use in the classroom dictates that the exploration of adaptive instructional models for preparing teacher candidates to be tomorrow's innovators be an ongoing research agenda. Future studies are necessary to identify barriers and solutions to successful preparation of future teachers, ensuring they possess high levels of self-efficacy and competence in their ability to integrate technology into their classrooms to support teaching and learning. Examining practices that prepare candidates to adapt to new technologies with confidence, while seamlessly building competency should be the focus of all teacher preparation programs. Future studies should explore application of these practices in professional development models for inservice teachers in an effort to avoid stagnation in current technology use and to promote development of future thinkers who possess an inquisitive mindset and who have the resilience to experiment with technologies that are newly available and that will be implemented into the classrooms of tomorrow.

CONCLUSION

The course explored in this chapter continues to evolve both in response to rapidly changing technologies but also to address the increasing emphasis on interdisciplinary STEM content across all grade levels. Preparing teacher candidates to be knowledgeable and confident in STEM content has the potential to strengthen the weak STEM pipeline both in terms of course selection and career choices of the students in their classrooms. The use of vicarious experiences that artificially create the "sweet spot", where technology, pedagogy, and content intersect, could be the basis for development of replicable teaching models for both technology integration efforts as well as interdisciplinary STEM efforts.

REFERENCES

Al-Balushi, S. M., & Al-Aamri, S. S. (2014). The effect of environmental science projects on students' environmental knowledge and science attitudes. *International Research in Geographical and Environmental Education*, *23*(3), 213–227. doi:10.1080/10382046.2014.927167

Atabek, O. (2020). Experienced educators' suggestions for solutions to the challenges to technology integration. *Education and Information Technologies*, *25*(6), 5669–5685. doi:10.100710639-020-10243-y

Bandura, A. (1986). *Social foundations of thought and action: A social cognitive theory*. Prentice Hall.

Bandura, A. (1994). Self-efficacy. In V. S. Ramachaudran (Ed.), *Encyclopedia of human behavior* (Vol. 4, pp. 71–81). Academic Press.

Bandura, A. (1997). *Self-efficacy: The exercise of control.* Freeman.

Bandura, A., Adams, N. E., & Beyer, J. (1977). Cognitive processes mediating behavioral change. *Journal of Personality and Social Psychology, 35*(3), 125–139. doi:10.1037/0022-3514.35.3.125 PMID:15093

Barton, E. A., & Dexter, S. (2020). Sources of teachers' self-efficacy for technology integration from formal, informal, and independent professional learning. *Educational Technology Research and Development, 68*(1), 89–108. doi:10.100711423-019-09671-6

Bautista, N. U. (2011). Investigating the use of vicarious and mastery experiences in influencing early childhood education majors' self-efficacy beliefs. *Journal of Science Teacher Education, 22*(4), 333–349. doi:10.100710972-011-9232-5

Bleicher, R. E., & Lindgren, J. (2005). Success in science learning and preservice science teaching self-efficacy. *Journal of Science Teacher Education, 16*(3), 205–225. doi:10.100710972-005-4861-1

Büssing, A. G., Schleper, M., & Menzel, S. (2019). Emotions and pre-service teachers' motivation to teach the context of returning wolves. *Environmental Education Research, 25*(8), 1174–1189. doi:10.1080/13504622.2018.1487034

Çalik, M., Ebenezer, J., Özsevgeç, T., Küçük, Z., & Artun, H. (2015). Improving science student teachers' self-perceptions of fluency with innovative technologies and scientific inquiry abilities. *Journal of Science Education and Technology, 24*(4), 448–460. doi:10.100710956-014-9529-1

Dunaetz, D. R. (2022). When technology does more harm than good: Technostress in missionary contexts. *Journal of the Evangelical Missiological Society, 2*(1), 112–128. doi:10.31219/osf.io/w7b32

Estes, J. S., & Dailey-Hebert, A. (2018). Modeling technology integration in teacher preparation programs. In *Handbook of Research on Pedagogical Models for Next-Generation Teaching and Learning* (pp. 82–97). IGI Global. doi:10.4018/978-1-5225-3873-8.ch005

Fernández-Batanero, J. M., Román-Graván, P., Reyes-Rebollo, M. M., & Montenegro-Rueda, M. (2021). Impact of educational technology on teacher stress and anxiety: A literature review. *International Journal of Environmental Research and Public Health, 18*(2), 548. doi:10.3390/ijerph18020548 PMID:33440738

Foulger, T. S., Wetzel, K., & Buss, R. R. (2019). Moving toward a technology infusion approach: Considerations for teacher preparation programs. *Journal of Digital Learning in Teacher Education, 35*(2), 79–91. doi:10.1080/21532974.2019.1568325

Genc, M. (2015). The project-based learning approach in environmental education. *International Research in Geographical and Environmental Education, 24*(2), 105–117. doi:10.1080/10382046.2014.993169

Giles, M., & Willis, J. (2019). Technology proficiency self-evaluation: What are they reporting? In *Society for Information Technology & Teacher Education International Conference* (pp. 1831-1834). Association for the Advancement of Computing in Education (AACE).

Henderson, K. J., Coppens, E. R., & Burns, S. (2021). Addressing barriers to implementing problem-based learning. *AANA Journal*, *89*(2). PMID:33832571

Hermans, R., Tondeur, J., van Braak, J., & Valcke, M. (2008). The impact of primary school teachers' educational beliefs on the classroom use of computers. *Computers & Education*, *51*(4), 1499–1509. doi:10.1016/j.compedu.2008.02.001

Jacobs, K. G., Kugler, J., Chi, J., Stuart, E., Merrell, S. B., & Rassbach, C. (2020). A mixed methods approach to understanding curricular impact of a capstone course on the self-efficacy of fourth-year medical students. *Cureus*, *12*(8). Advance online publication. doi:10.7759/cureus.9537 PMID:32905172

Kent, A. M., & Giles, R. M. (2017). Preservice teachers' technology self-efficacy. *SRATE Journal*, *26*(1), 9–20.

Kenward, K., Hipp, S., & Baker, M. (2011, March). Seamless technology integration for pre-service teachers. In *Society for Information Technology & Teacher Education International Conference* (pp. 2542-2544). Association for the Advancement of Computing in Education (AACE).

Koehler, M., & Mishra, P. (2009). What is technological pedagogical content knowledge (TPACK)? *Contemporary Issues in Technology & Teacher Education*, *9*(1), 60–70.

Koehler, M. J., Mishra, P., & Cain, W. (2013). What is technological pedagogical content knowledge (TPACK)? *Journal of Education*, *193*(3), 13–19. doi:10.1177/002205741319300303

Korucu-Kis, S., & Ozmen, K. S. (2019). Exherent and inherent value beliefs about technology: Missing pieces in the puzzle of technology integration? *International Journal of Educational Technology*, *6*(1), 1–11.

Krajcik, J. S., & Blumenfeld, P. C. (2006). Project-based learning. In R. K. Sawyer (Ed.), *The Cambridge handbook of the learning sciences* (pp. 317–333). Cambridge University Press.

Lang, D., & Savageau, J. A. (2017). *Starting at the end: Measuring learning using retrospective pre-post evaluations. Center for Health Policy and Research (CHPR).* Publications.

Li, H. (2013). *Educational Change towards PBL (Problem Based Learning): An Organizational Perspective*. River Publishers.

Maddux, J. E. (2002). Self-efficacy: The power of believing you can. In S. Lopez & C. Snyder (Eds.), *The Oxford handbook of positive psychology* (2nd ed., pp. 335–344). Oxford University Press.

Maddux, J. E., & Kleiman, E. M. (2020). 31 Self-Efficacy. The Oxford Handbook of Positive Psychology, 443.

Maeng, J. L., Whitworth, B. A., Bell, R. L., & Sterling, D. R. (2020). The effect of professional development on elementary science teachers' understanding, confidence, and classroom implementation of reform-based science instruction. *Science Education*, *104*(2), 326–353. doi:10.1002ce.21562

Menon, D., & Sadler, T. D. (2016). Preservice elementary teachers' science self-efficacy beliefs and science content knowledge. *Journal of Science Teacher Education*, *27*(6), 649–673. doi:10.100710972-016-9479-y

Menon, D., & Sadler, T. D. (2018). Sources of science teaching self-efficacy for preservice elementary teachers in science content courses. *International Journal of Science and Mathematics Education*, *16*(5), 835–855. doi:10.100710763-017-9813-7

Mishra, P., & Koehler, M. J. (2006). Technological pedagogical content knowledge: A framework for teacher knowledge. *Teachers College Record*, *108*(6), 1017–1054. doi:10.1111/j.1467-9620.2006.00684.x

Nariman, N., & Chrispeels, J. (2016). PBL in the era of reform standards: Challenges and benefits perceived by teachers in one elementary school. *The Interdisciplinary Journal of Problem-Based Learning*, *10*(1). Advance online publication. doi:10.7771/1541-5015.1521

Norris, C. M., Morris, J. E., & Lummis, G. W. (2018). Preservice teachers' self-efficacy to teach primary science based on 'science learner' typology. *International Journal of Science Education*, *40*(18), 2292–2308. doi:10.1080/09500693.2018.1528645

Roessingh, H., & Chambers, W. (2011). Project-based learning and pedagogy in teacher preparation: Staking out the theoretical mid-ground. *International Journal on Teaching and Learning in Higher Education*, *23*(1), 60–71.

Schibeci, R. A., & Hickey, R. (2000). Is it natural or processed? Elementary school teachers and conceptions about materials. *Journal of Research in Science Teaching: The Official Journal of the National Association for Research in Science Teaching*, *37*(10), 1154–1170. doi:10.1002/1098-2736(200012)37:10<1154::AID-TEA7>3.0.CO;2-6

Schmidt-Crawford, D. A., Lindstrom, D., & Thompson, A. D. (2018). Addressing the "why" for integrating technology in teacher preparation. *Journal of Digital Learning in Teacher Education*, *34*(3), 132–133. doi:10.1080/21532974.2018.1465783

Schriver, M., & Czerniak, C. M. (1999). A comparison of middle and junior high science teachers' levels of efficacy and knowledge of developmentally appropriate curriculum and instruction. *Journal of Science Teacher Education*, *10*(1), 21–42. doi:10.1023/A:1009472629345

Seneviratne, K., Hamid, J. A., Khatibi, A., Azam, F., & Sudasinghe, S. (2019). Multi-faceted professional development designs for science teachers' self-efficacy for inquiry-based teaching: A critical review. *Universal Journal of Educational Research*, *7*(7), 1595–1611. doi:10.13189/ujer.2019.070714

Shulman, L. S. (1986). Those who understand: Knowledge growth in teaching. *Educational Researcher*, *15*(2), 4–14. doi:10.3102/0013189X015002004

Tanak, A. (2020). Designing TPACK-based course for preparing student teachers to teach science with technological pedagogical content knowledge. *Kasetsart Journal of Social Sciences*, *41*(1), 53–59.

Tanguma, J., Underwood, D., & Mayo, N. (2004). A validation study of the technology and teaching efficacy scale. *Southeast Educational Research Association Annual Conference*, Dallas, TX.

Thompson, A. E. & Sorbet, S. (2020). The impact of self-efficacy on leveraging technology in the classroom. *Leveraging Technology to Improve School Safety and Student Wellbeing*, 244-261. doi:10.4018/978-1-7998-1766-6

Trevathan, M., & Willis, J. (2019, March). Teacher education, experience, training, and course factors in STEM implementation: A look at engineering. In *Society for Information Technology & Teacher Education International Conference* (pp. 2070-2075). Association for the Advancement of Computing in Education (AACE).

Trundle, K. C., Atwood, R. K., & Christopher, J. E. (2002). Preservice elementary teachers' conceptions of moon phases before and after instruction. *Journal of Research in Science Teaching: The Official Journal of the National Association for Research in Science Teaching*, *39*(7), 633–658. doi:10.1002/tea.10039

Tschannen-Moran, M., & Hoy, A. W. (2007). The differential antecedents of self-efficacy beliefs of novice and experienced teachers. *Teaching and Teacher Education*, *23*(6), 944–956. doi:10.1016/j.tate.2006.05.003

Wang, L., Ertmer, P. A., & Newby, T. J. (2004). Increasing preservice teachers' self-efficacy beliefs for technology integration. *Journal of Research on Technology in Education*, *36*(3), 231–250. doi:10.1080/15391523.2004.10782414

Wang, Y. M. (2006). Stand-alone computer courses in teachers' IT training. *EDUCAUSE Quarterly*, *29*(3), 8.

Whitaker, S. C. (2019). *Middle School Teachers' Perceptions of Project-based Learning as It Impacts First Year Implementation* [Doctoral dissertation]. University of South Carolina.

Willis, J., Weiser, B., & Kirkwood, D. (2014). Bridging the gap: Meeting the needs of early childhood students by integrating technology and environmental education. *International Journal of Early Childhood Environmental Education*, *2*(1), 140–155.

Willis, J. M. (2015). Examining technology and teaching efficacy of preservice teacher candidates: A deliberate course design model. *Current Issues in Education (Tempe, Ariz.)*, *18*(3).

Zhao, Y., & Bryant, F. L. (2006). Can teacher technology integration training alone lead to high levels of technology integration? A qualitative look at teachers' technology integration after state mandated technology training. *Electronic Journal for the Integration of Technology in Education*, *5*(1), 53–62.

Compilation of References

Abrahamson, D., Nathan, M. J., Williams-Pierce, C., Walkington, C., Ottmar, E., Soto, H., & Alibali, M. W. (2020). The future of embodied design for mathematics teaching and learning. *Frontiers in Education*, *5*(147), 147. doi:10.3389/feduc.2020.00147

Acedo, C., & Hughes, C. (2014). Principles for learning and competencies in the 21st-century curriculum. *Prospects*, *44*(4), 503–525. doi:10.100711125-014-9330-1

Achieve. (2010a). *Connecting science standards with assessment: A snapshot of three countries' approaches – England, Hong Kong, and Canada*. Author.

Achieve. (2010b). *International science benchmarking report: Taking the lead in science education: Forging Next-Generation Science Standards*. Author.

Adam, A. (2017). A framework for seeking the connections between technology, pedagogy and culture: A study in the Maldives. *Journal of Open, Flexible, and Distance Learning*, *21*(1), 35–51.

Aikenhead, G. S. (2005). Science-based occupations and the science curriculum: Concepts of evidence. *Science Education*, *89*(2), 242–275. doi:10.1002ce.20046

Akarsu, M., Okur Akçay, N., & Elmas, R. (2020). *STEM eğitimi yaklaşımının özellikleri ve değerlendirilmesi* [Characteristics and evaluation of STEM education approach]. Boğaziçi Eğitim Fakültesi Dergisi, Özel Sayı.

Aksoy, C. G., Eichengreen, B., & Saka, O. (2020). *Revenge of the experts: Will COVID-19 renew or diminish public trust in science?* SRC Discussion Paper No. 96. London School of Economics and Political Science and Economic and Social Research Council.

Al-Balushi, S. M., & Al-Aamri, S. S. (2014). The effect of environmental science projects on students' environmental knowledge and science attitudes. *International Research in Geographical and Environmental Education*, *23*(3), 213–227. doi:10.1080/10382046.2014.927167

Al-Dheleai, Y., & Tasir, Z. (2019). Web 2.0 for fostering students' social presence in online learning-based interaction. *Journal of Technology and Science Education*, 13-19.

Alemdar, M., Lingle, J. A., Moore, R., & Wind, S. A. (2017a). Developing an engineering design process assessment using think-aloud interviews. *International Journal of Engineering Education*, *33*(1(B)), 441–452.

Alexander, K. (2015). Instruments, sensors, and apps are turning smartphones into engineering measurement tools. *Mechanical Engineering (New York, N.Y.)*, *137*(9), 42–45. doi:10.1115/1.2015-Sep-3

Alibali, M. W., & Nathan, M. J. (2012). Embodiment in mathematics teaching and learning: Evidence from learners' and teachers' gestures. *Journal of the Learning Sciences*, *21*(2), 247–286. doi:10.1080/10508406.2011.611446

Almazán, J., Bergasa, L. M., Yebes, J. J., Barea, R., & Arroyo, R. (2013). Full auto-calibration of a smartphone on board a vehicle using IMU and GPS embedded sensors. *IEEE Intelligent Vehicles Symposium (IV)*, 1374-1380. 10.1109/IVS.2013.6629658

Ambrose, S. A., Bridges, M., Lovett, M., DiPietro, M., & Norman, M. (2010). *How learning works: Seven Research-Based Principles for Smart Teaching*. Jossey-Bass.

American Association of Physics Teachers Undergraduate Curriculum Task Force. (2016). *AAPT recommendations for computational physics in the undergraduate physics curriculum.* https://www.aapt.org/Resources/upload/AAPT_UCTF_CompPhysReport_final_B.pdf

Anastopoulou, S., Sharples, M., & Baber, C. (2011). An evaluation of multimodal interactions with technology while learning science concepts. *British Journal of Educational Technology*, *42*(2), 266–290. doi:10.1111/j.1467-8535.2009.01017.x

Anderson, J. L., & Wall, S. D. (2016). Kinecting physics: Conceptualization of motion through visualization and embodiment. *Journal of Science Education and Technology*, *25*(2), 161–173. doi:10.100710956-015-9582-4

Anđić, B., Cvjetićanin, S., Lavicza, Z., Maričić, M., Novović, T., & Stešević, D. (2020). Mobile and printed dichotomous keys in constructivist learning of biology in primary school. *Research in Science & Technological Education*, *39*(4), 393–420. doi:10.1080/02635143.2020.1763290

Anthony, W. L., Zhu, Y., & Nower, L. (2021). The relationship of interactive technology use for entertainment and school performance and engagement: Evidence from a longitudinal study in a nationally representive sample of middle school students in China. *Computers in Human Behavior*, *122*, 106846. doi:10.1016/j.chb.2021.106846

Antonio, A., Martin, N., & Stagg, A. (2012). Engaging higher education students via digital curation. In *Proceedings of the 29th annual conference of the Australasian society for computers in learning in tertiary education (ASCILITE 2012)* (pp. 1-5). Australasian Society for Computers in Learning in Tertiary Education (ASCILITE).

Apedoe, X. S., Reynolds, B., Ellefson, M. R., & Schunn, C. D. (2008). Bringing engineering design into high school science classrooms: The heating/cooling unit. *Journal of Science Education and Technology*, *17*(5), 454–465. doi:10.100710956-008-9114-6

Araka, E., Maina, E., Gitonga, R., & Oboko, R. (2020). Research trends in measurement and intervention tools for self-regulated learning for e-learning environments—Systematic review (2008–2018). *Research and Practice in Technology Enhanced Learning*, *15*(6), 1–21. doi:10.118641039-020-00129-5

Ashcraft, M. H. (2002). Math anxiety: Personal, educational, and cognitive consequences. *Current Directions in Psychological Science*, *11*(5), 181–185. doi:10.1111/1467-8721.00196

Ashe, D., & Bibi, S. (2011, December). Unpacking TPACK and students' approaches to learning: Applying knowledge in pieces to higher education teaching and learning. In Ascilite (pp. 128-132). Academic Press.

Atabek, O. (2020). Experienced educators' suggestions for solutions to the challenges to technology integration. *Education and Information Technologies*, *25*(6), 5669–5685. doi:10.100710639-020-10243-y

Atman, C. J., Adams, R. S., Cardella, M. E., Turns, J., Mosborg, S., & Saleem, J. (2007). Engineering design processes: A comparison of students and expert practitioners. *Journal of Engineering Education*, *96*(4), 359–379. doi:10.1002/j.2168-9830.2007.tb00945.x

Avsar, H., Fischer, J., & Rodden, T. (2015). Target size guidelines for interactive displays on the flight deck. *34th Digital Avionics Systems Conference*. 10.1109/DASC.2015.7311400

Azevedo, R., Witherspoon, A., Graesser, A., McNamara, D., Chauncey, A., Siler, E., . . . Lintean, M. (2009). MetaTutor: Analyzing self-regulated learning in a tutoring system for biology. In Artificial intelligence in education (pp. 635-637). IOS Press.

Azizi, S. M., & Khatony, A. (2019). Investigating factors affecting on medical sciences students' intention to adopt mobile learning. *BMC Medical Education*, *19*(1), 1–10. doi:10.118612909-019-1831-4 PMID:31638977

Aznar-Díaz, I., Hinojo-Lucena, F. J., Caceres-Reche, M. P., & Romero-Rodríguez, J. M. (2020). Analysis of the determining factors of good teaching practices of mobile learning at the Spanish University. An explanatory model. *Computers & Education*, *159*, 104007. doi:10.1016/j.compedu.2020.104007

Baker, E. (2017). How school districts can save (billions) on edtech. *Technology for Education Consortium*. https://marketbrief.edweek.org/wp-content/uploads/2017/03/How_School_Districts_Can_Save_Billions_on_Edtech.pdf

Ball, C., Huang, K.-T., Rikard, R. V., & Cotten, S. R. (2019). The emotional costs of computers: An expectancy-value theory analysis of predominantly low-socioeconomic status minority students' STEM attitudes. *Information Communication and Society*, *22*(1), 105–128. doi:10.1080/1369118X.2017.1355403

Bandura, A. (1986). *Social foundations of thought and action: A social cognitive theory*. Prentice Hall.

Bandura, A. (1994). Self-efficacy. In V. S. Ramachaudran (Ed.), *Encyclopedia of human behavior* (Vol. 4, pp. 71–81). Academic Press.

Bandura, A. (1997). *Self-efficacy: The exercise of control*. Macmillan.

Bandura, A., Adams, N. E., & Beyer, J. (1977). Cognitive processes mediating behavioral change. *Journal of Personality and Social Psychology*, *35*(3), 125–139. doi:10.1037/0022-3514.35.3.125 PMID:15093

Baran, E., Canbazoglu Bilici, S., & Uygun, E. (2016). Investigating the impact of a Technological Pedagogical Content Knowledge (TPACK)-based professional development program on science teachers' TPACK. In M. Herring, M. J. Koehler, & P. Mishra (Eds.), *Handbook of Technological Pedagogical Content Knowledge* (2nd ed., pp. 271–283). Routledge.

Baran, E., Canbazoglu-Bilici, S., Mesutoglu, C., & Ocak, C. (2019). Investigating the impact of an out-of-school STEM education program on students' attitudes towards STEM and STEM careers. *School Science and Mathematics*, *119*(4), 223–235. doi:10.1111sm.12330

Barsalou, L. W. (2008). Grounded Cognition. *Annual Review of Psychology*, *59*(1), 617–645. doi:10.1146/annurev.psych.59.103006.093639 PMID:17705682

Barton, E. A., & Dexter, S. (2020). Sources of teachers' self-efficacy for technology integration from formal, informal, and independent professional learning. *Educational Technology Research and Development*, *68*(1), 89–108. doi:10.100711423-019-09671-6

Bates, A. T. (2015). *Teaching in a digital age*. https://opentextbc.ca/teachinginadigitalage/

Bauer, M. S., Damschroder, L., Hagedorn, H., Smith, J., & Kilbourne, A. M. (2015). An introduction to implementation science for the non-specialist. *BMC Psychology*, *3*(1), 32. doi:10.118640359-015-0089-9 PMID:26376626

Bautista, N. U. (2011). Investigating the use of vicarious and mastery experiences in influencing early childhood education majors' self-efficacy beliefs. *Journal of Science Teacher Education*, *22*(4), 333–349. doi:10.100710972-011-9232-5

Beagrie, N. (2006). Digital curation for science, digital libraries, and individuals. *International Journal of Digital Curation*, *1*(1), 3–16.

Bednarz, S. W., & van der Schee, J. (2006). Europe and the United States: The implementation of geographic information systems in secondary education in two contexts. *Technology, Pedagogy and Education, 15*(2), 191–205. doi:10.1080/14759390600769573

Beichner, R. J. (1994). Testing student interpretation of kinematics graphs. *American Journal of Physics, 62*(8), 750–762. doi:10.1119/1.17449

Bender, W. N. (2017). *20 strategies for STEM instruction*. Learning Sciences International.

Benli, K. P., Dillman, B., Louelh, R., Poirier-Quinot, M., & Darrasse, L. (2015). Illustrating the quantum approach with an Earth magnetic field MRI. *European Journal of Physics, 36*, 1–20. doi:10.1088/0143-0807/36/3/035032

Bennett, S., & Maton, K. (2010). Beyond the 'digital natives' debate: Towards a more nuanced understanding of students' technology experiences. *Journal of Computer Assisted Learning, 34*(5), 321–331. doi:10.1111/j.1365-2729.2010.00360.x

Berglas-Shapiro, T., Eylon, B.-S., & Scherz, Z. (2017). A technology-enhanced intervention for self-regulated learning in science. *Teachers College Record, 119*(13), 1–26. doi:10.1177/016146811711901301

Berne, P., Morales, A. L., Langstaff, D., & Invalid, S. (2018). Ten principles of disability justice. *WSQ. Women's Studies Quarterly, 46*(1), 227–230. doi:10.1353/wsq.2018.0003

Betts, B., & Payne, N. (2016). From content to curation. In A. Anderson & B. Betts (Eds.), *Ready, set, curate* (pp. 7–13). Association for Talent Development.

Billinghurst, M., & Duenser, A. (2012). Augmented reality in the classroom. *Computer, 45*(7), 56–63. doi:10.1109/MC.2012.111

Birchfield, D., & Johnson-Glenberg, M. (2010). A next gen interface for embodied learning: SMALLab and the geological layer cake. *International Journal of Gaming and Computer-Mediated Simulations, 2*(1), 1–10. doi:10.4018/jgcms.2010010105

Bittel, A., Elazzazi, A., & Bittel, D. (2016). Accuracy and precision of an accelerometer-based smartphone app designed to monitor and record angular movement over time. *Telemedicine Journal and e-Health, 22*(4), 302–309. doi:10.1089/tmj.2015.0063 PMID:26447774

Bleicher, R. E., & Lindgren, J. (2005). Success in science learning and preservice science teaching self-efficacy. *Journal of Science Teacher Education, 16*(3), 205–225. doi:10.100710972-005-4861-1

Boda, P. A. (2022). On the methodological and epistemological power of epistemic (de) centering as a reflexive praxis of resistance toward disability justice. *Qualitative Research Journal.*

Boda, P. A. (In press). How do we disrupt the labeling of disability and difference as deficit in urban schools? In S. R. Steinberg & E. Adjapong (Eds.), 19 Urban questions: Teaching in the city (3rd ed.). New York, NY: Peter Lang.

Boda, P. A. (2018). Exclusion from participation in science: Confessions from an ally on the other side of the fence. In M. Koomen, S. Kahn, C. Atchison, & T. Wild (Eds.), *Toward inclusion of all learners through science teacher education* (pp. 301–311). Sense/Brill Publishing. doi:10.1163/9789004368422_033

Boda, P. A. (2019). Conceptualizing the margins in science education: The limits of multicultural analyses. *Cultural Studies of Science Education, 14*(2), 493–514. doi:10.100711422-019-09926-x

Boda, P. A., Nusbaum, E., & Kulkarni, S. (2022). From 'what is' toward 'what if' through intersectionality: Problematizing ableist erasures and coloniality in racially-just research. *International Journal of Research & Method in Education, 45*(4), 1–14. doi:10.1080/1743727X.2022.2054981

Boda, P. A., & Riley Miller, A. (2022). Educational technologies for multicultural science learning: Comparing reform and critical design studies. In M. Atwater (Ed.), *International handbook of research in multicultural science education*. Springer. doi:10.1007/978-3-030-83122-6_3

Bodzin, A. (2011). The implementation of a geospatial information technology (GIT)-supported land use change curriculum with urban middle school learners to promote spatial thinking. *Journal of Research in Science Teaching*, *48*(3), 281–300. doi:10.1002/tea.20409

Bodzin, A., & Cirruci, L. (2009). Integrating geospatial technologies to examine urban land use change: A design partnership. *The Journal of Geography*, *108*(4-5), 186–197. doi:10.1080/00221340903344920

Bodzin, A., Fu, Q., Kulo, V., & Peffer, T. (2014). Examining the enactment of a geospatial curriculum design approach on students' geospatial thinking and reasoning. *Journal of Science Education and Technology*, *23*(4), 562–574. doi:10.100710956-014-9488-6

Bodzin, A., Fu, Q., Peffer, T., & Kulo, V. (2013). Developing energy literacy in U.S. middle level students using the geospatial curriculum approach. *International Journal of Science Education*, *35*(9), 1561–1589. doi:10.1080/09500693.2013.769139

Bodzin, A., Peffer, T., & Kulo, V. (2012). The efficacy of educative curriculum materials to support geospatial science pedagogical content knowledge. *Journal of Technology and Teacher Education*, *20*(4), 361–386.

Bonilla, S., & MacPherson, A. (2021). Fighting "bad science" in the information age: The effects of an intervention to stimulate evaluation and critique of false scientific claims. *Journal of Research in Science Teaching*, *58*(8), 1152–1178. doi:10.1002/tea.21696

Boor, I., Gerritsen, D., de Greef, L., & Rodermans, J. (2021). *Meaningful assessment in interdisciplinary education*. Amsterdam University Press. doi:10.5117/9789463729048

Boote, S. (2014). Assessing and understanding line graph interpretations using a scoring rubric of organized cited factors. *Journal of Science Teacher Education*, *25*(3), 333–354. doi:10.100710972-012-9318-8

Börner, K., Bueckle, A., & Ginda, M. (2019). Data visualization literacy: Definitions, conceptual frameworks, exercises, and assessments. *Proceedings of the National Academy of Sciences of the United States of America*, *116*(6), 1857–1864. doi:10.1073/pnas.1807180116 PMID:30718386

Bourke, T., & Lidstone, J. (2015). Mapping geographical knowledge and skills needed for pre-service teachers in teacher education. *SAGE Open*, *5*(1). doi:10.1177/2158244015577668

Broadbent, J., Panadero, E., & Fuller-Tyszkiewicz, M. (2020). Effects of mobile-app learning diaries vs online training on specific self-regulated learning components. *Educational Technology Research and Development*, *68*(5), 2351–2372. doi:10.100711423-020-09781-6

Brophy, S., Klein, S., Portsmore, M., & Rogers, C. (2008). Advancing engineering education in P-12 classroom. *Journal of Engineering Education*, *97*(3), 369–387. doi:10.1002/j.2168-9830.2008.tb00985.x

Brothen, T. (1984). Three computer-assisted laboratory exercises for introductory psychology. [SAGE Publications.]. *Teaching of Psychology*, *2*(11), 105–107. doi:10.120715328023top1102_14

Brown, B. A. (2021). *Science in the city: Culturally relevant STEM education*. Harvard Education Press.

Bryan, L. A., Moore, J. T., Johnson, C. C., & Roehrig, G. H. (2016). Integrated STEM Education. In C. C. Johnson, E. E. Peters-Burton, & T. J. Moore (Eds.), *STEM Road Map: A Framework for Integrated STEM Education* (pp. 23–37). Routledge, Taylor & Francis Group.

Buck, G. A., Francis, D. C., & Wilkins-Yel, K. G. (2020). Research on gender equity in STEM education. In C. C. Johnson, M. J. Mohr-Schroeder, T. J. Moore, & L. D. English (Eds.), *Handbook of research on STEM education* (pp. 289–299). Taylor & Francis. doi:10.4324/9780429021381-27

Buechley, L., Eisenberg, M., Catchen, J., & Crockett, A. (2008, April). The LilyPad Arduino: Using computational textiles to investigate engagement, aesthetics, and diversity in computer science education. In M. Czerwinski & A. Lund (Chairs), *Proceedings of the SIGCHI conference on human factors in computing systems* (pp. 423–432). New York: Association for Computing Machinery.

Buechley, L., & Eisenberg, M. (2008). The LilyPad Arduino: Toward wearable engineering for everyone. *IEEE Pervasive Computing*, *7*(2), 12–15. doi:10.1109/MPRV.2008.38

Bull, G., Hodges, C., Mouza, C., Grant, M., Archambault, L., Borup, J., Ferdig, R.E., & Schmidt-Crawford, D. A. (2019). Conceptual dilution. *Contemporary Issues in Technology and Teacher Education, 19*(2). https://citejournal.org/volume-19/issue-2-19/editorial/editorial-conceptual-dilution

Bunyamin, M. A. H., Talib, C. A., Ahmad, N. J., & Surif, J. (2020). Current teaching practice of physics teachers and implications for integrated STEM education. *Universal Journal of Educational Research.*, *8*(5A), 18–28. doi:10.13189/ujer.2020.081903

Burrows, A. C., Borowczak, M., Myers, A., Schwortz, A. C., & McKim, C. (2021). Integrated STEM for teacher professional learning and development: "I need time for practice." *Education Sciences*, *11*(21), 1–23. doi:10.3390/educsci11010021

Burrows, A. C., Swarts, G. P., Hutchison, L., Katzmann, J. M., Thompson, R., Freeman, L., Schanke, A., Kilty, T., & Reynolds, T. (2021). Finding spaces: Teacher education technology competencies (TETCs). *Education in Science*, *11*(11), 733. doi:10.3390/educsci11110733

Büssing, A. G., Schleper, M., & Menzel, S. (2019). Emotions and pre-service teachers' motivation to teach the context of returning wolves. *Environmental Education Research*, *25*(8), 1174–1189. doi:10.1080/13504622.2018.1487034

Bybee, R. W. (2013). *Case for STEM education: Challenges and opportunities*. National Science Teachers Association.

Cakir, M. (2008). Constructivist approaches to learning in science and their implications for science pedagogy: A literature review. *International Journal of Environmental and Science Education*, *3*(4), 193–206.

Calabrese Barton, A., & Tan, E. (2019). Designing for rightful presence in STEM: The role of making present practices. *Journal of the Learning Sciences*, *28*(4-5), 616–658. doi:10.1080/10508406.2019.1591411

Çalik, M., Ebenezer, J., Özsevgeç, T., Küçük, Z., & Artun, H. (2015). Improving science student teachers' self-perceptions of fluency with innovative technologies and scientific inquiry abilities. *Journal of Science Education and Technology*, *24*(4), 448–460. doi:10.100710956-014-9529-1

Campbell, T., & Abd-Hamid, N. H. (2013). Technology use in science instruction (TUSI): Aligning the integration of technology in science instruction in ways supportive of science education reform. *Journal of Science Education and Technology*, *22*(4), 572–588. doi:10.100710956-012-9415-7

Capobianco, B. M., DeLisi, J., & Radloff, J. (2018). Characterizing elementary teachers' enactment of high-leverage practices through engineering design-based science instruction. *Science Education*, *102*(2), 342–376. doi:10.1002ce.21325

Carey, C. C., & Gougis, R. D. (2017). Simulation modeling of lakes in undergraduate and graduate classrooms increases comprehension of climate change concepts and experience with computational tools. *Journal of Science Education and Technology*, *1*(26), 1–11. doi:10.100710956-016-9644-2

Carpenter, S. K., Cepeda, N. J., Rohrer, D., Kang, S. H. K., & Pashler, H. (2012). Using spacing to enhance diverse forms of learning: Review of recent research and implications for instruction. *Educational Psychology Review*, *24*(3), 369–378. doi:10.100710648-012-9205-z

Carrabba, C., & Farmer, A. (2018). The impact of project-based learning and direct instruction on the motivation and engagement of middle school students. *Language Teaching and Educational Research*, *1*(2), 163–174.

Carrigan, J., Bodzin, A. M., Hammond, T. C., Rutzmoser, S., Popejoy, K., Farina, W., Hanson, I., Salter, S., Anastasio, D., Kangas, S., Holland, B., & Sahagian, D. (2019). Investigating urban trees. Exploring the impact of trees around our school with geospatial technologies. *Science Teacher (Normal, Ill.)*, *86*(8), 27–35.

Cartee, J. (2021). Strategic empathy in virtual learning and instruction: A contemplative essay about teacher-student rapport during times of crisis. *Journal of Instructional Research*, 12-19.

Cartwright, J. (2016). Technology: Smartphone science. *Nature*, *531*(7596), 669–671. doi:10.1038/nj7596-669a PMID:27035011

Chang, C. C., & Chen, Y. (2020). Using mastery learning theory to develop task-centered hands-on STEM learning of Arduino-based educational robotics: Psychomotor performance and perception by a convergent parallel mixed method. *Interactive Learning Environments*, 1–16.

Chang, C. J., Liu, C. C., Wen, C. T., Tseng, L. W., Chang, H. Y., Chang, M. H., Fan Chiang, S.-H., Hwang, F.-K., & Yang, C.-W. (2020). The impact of light-weight inquiry with computer simulations on science learning in classrooms. *Computers & Education*, *146*, 103770. doi:10.1016/j.compedu.2019.103770

Cheng, C. H., Bråten, I., Yang, F. Y., & Brandmo, C. (2021). Investigating structural relationships among upper-secondary school students' beliefs about knowledge, justification for knowing, and Internet-specific justification in the domain of science. *Journal of Research in Science Teaching*, *58*(7), 980–1009. doi:10.1002/tea.21689

Chen, J., & Bogachenko, T. (2022). Online community building in distance education: The case of social presence in the blackboard discussion board versus multimodal voicethread interaction. *Journal of Educational Technology & Society*, 62–75.

Chin, P., Munby, H., Hutchinson, N. L., Taylor, J., & Clark, F. (2004). Where's the science? Understanding the form and function of workplace science. In E. Scanlon, P. Murphy, J. Thomas, & E. Whitelegg (Eds.), *Reconsidering science learning* (pp. 118–134). Routledge Falmer. doi:10.4324/9780203464021_chapter_2.6

Christian, K. B., Kelly, A. M., & Bugallo, M. F. (2021). NGSS-based teacher professional development to implement engineering practices in STEM instruction. *International Journal of STEM Education*, *8*(1), 1–18. doi:10.118640594-021-00284-1

Chu, W. W., Ong, E. T., Ayop, S. K., Azmi, M. S. M., Abdullah, A. S., Abd Karim, N. S., & Tho, S. W. (2021). The innovative use of smartphone for sound STEM practical kit: A pilot implementation for secondary classroom. *Research in Science & Technological Education*, 1–23. Advance online publication. doi:10.1080/02635143.2021.1978963

City, E. A., Elmore, R. F., Fiarman, S. E., Teital, L., & Lachman, A. (2009). *Instructional rounds in education: A network approach to improving teaching and learning*. Harvard Education Press.

Clark, R. C., & Mayer, R. E. (2008). Learning by viewing versus learning by doing: Evidence-based guidelines for principled learning environments. *Performance Improvement*, *47*(9), 5–13. doi:10.1002/pfi.20028

Cleary, T. J., Gregory, A., Kitsantas, A., Slemp, J., & Panish, D. (2020). Implementation of self- regulated learning-focused interventions in schools: The intersection of sociocultural factors and social validity principles. In G. A. D. Liem & D. M.McInerney (Eds.), Promoting motivation and learning in contexts: Sociocultural perspectives on educational interventions (pp. 143-172). Information Age Publishing.

Cleary, T. J. (2018). *The self-regulated learning guide: Teaching students to think in the language of strategies*. Routledge. doi:10.4324/9781315693378

Cleary, T. J., Velardi, B., & Schnaidman, B. (2017). Effects of the Self-Regulation Empowerment Program (SREP) on middle school students' strategic skills, self-efficacy, and mathematics achievement. *Journal of School Psychology*, *64*, 28–42. doi:10.1016/j.jsp.2017.04.004 PMID:28735606

Cleary, T. J., & Zimmerman, B. J. (2004). Self-regulation empowerment program: A school-based program to enhance self-regulated and self-motivated cycles of student learning. *Psychology in the Schools*, *41*(5), 537–550. doi:10.1002/pits.10177

Cleary, T. J., & Zimmerman, B. J. (2006). Teachers' perceived usefulness of strategy microanalytic assessment information. *Psychology in the Schools*, *43*(2), 149–155. doi:10.1002/pits.20141

Clement, J. (2022, July 27). *Regional distribution of desktop traffic to Reddit.com as of May 2022 by country*. Statista. https://www.statista.com/statistics/325144/reddit-global-active-user-distribution/

Clement, J. (1985). Misconceptions in graphing. In L. Streefland (Ed.), *Proceedings of the Ninth Conference of the International Group for the Psychology of Mathematics Education* (pp. 369-375). IGPME.

Cole, M. (2005). Cross-cultural and historical perspectives on the developmental consequences of education. *Human Development*, *48*(4), 195–216. doi:10.1159/000086855

Comarú, M. W., Lopes, R. M., Braga, L. A. M., Batista Mota, F., & Galvão, C. (2021). A bibliometric and descriptive analysis of inclusive education in science education. *Studies in Science Education*, *57*(2), 241–263. doi:10.1080/03057267.2021.1897930

Condliffe, B., Quint, J., Visher, M. G., Bangser, M. R., Drohojowska, S., Saco, L., & Nelson, E. (2017). *Project Based Learning: A Literature Review*. MDRC.

Connor, D. (2012). Does Dis/ability Now Sit at the Table (s) of Social Justice and Multicultural Education? A Descriptive Survey of Three Recent Anthologies. *Disability Studies Quarterly*, *32*(3). Advance online publication. doi:10.18061/dsq.v32i3.1770

Council of State Science Supervisors. (2020). *Supporting students' science learning during COVID-19 school closures*. https://stemteachingtools.org/assets/landscapes/Guidance-for-Supporting-Science-Learning-During-COVID-19-School-Closures.pdf

Craig, T. T., & Marshall, J. (2019). Effect of project-based learning on high school students' state-mandated, standardized math and science exam performance. *Journal of Research in Science Teaching*, *56*(10), 1461–1488. doi:10.1002/tea.21582

Creswell, J. W., Hanson, W. E., Clark Plano, V. L., & Morales, A. (2007). Qualitative research designs: Selection and implementation. *The Counseling Psychologist*, *35*(2), 236–264. doi:10.1177/0011000006287390

Crompton, H. (2013). A historical overview of mobile learning: Toward learner-centered education Z.L. Berge. In L. Y. Muilenburg (Ed.), *Handbook of mobile learning* (pp. 3–14). Routledge.

Dailey, D., Jackson, N., Cotabish, A., & Trumble, J. (2018). STEMulate engineering academy: Engaging students and teachers in engineering practices. *Roeper Review*, *40*(2), 97–107. doi:10.1080/02783193.2018.1434709

Dale, D. A., Sutter, J., & Kloster, D. (2019). Asking real-world questions with inquiry-based labs. *The Physics Teacher*, *57*(8), 547–550. doi:10.1119/1.5131122

Daniel, K., McConnell, M., Schuchardt, A., & Peffer, M. E. (2022). Challenges facing interdisciplinary researchers: Findings from a professional development workshop. *PLoS One*, *17*(4), e0267234. doi:10.1371/journal.pone.0267234 PMID:35439277

Darling-Hammond, L., Flook, L., Cook-Harvey, C., Barron, B., & Osher, D. (2020). Implications for the educational practice of the science of learning and development. *Applied Developmental Science*, *24*(2), 97–140. doi:10.1080/10888691.2018.1537791

Daud, A., Daud, A., Aljohani, N., Abbas, R., Lytras, M., Abbas, F., & Alowibdi, J. (2017). Predicting student performance using advanced learning analytics. *26th International Conference on World Wide Web Companion*, 415-421. 10.1145/3041021.3054164

Daugherty, M. K., & Carter, V. (2017). The nature of interdisciplinary STEM education. In M. J. de Vries (Ed.), *Handbook of technology education* (pp. 159–171). Springer.

Davis, M. (2019, April 23). Harvard Business Review, MBA lessons guide principals' ed-tech leadership. *Education Week*. https://www.edweek.org/technology/harvard-business-review-mba-lessons-guide-principals-ed-tech-leadership/2019/04

Dayan, E., & Tsybulsky, D. (2021). *Science teachers' perceptions regarding digital curation as a personalized learning activity that promotes professional learning*. In NARST2021 Annual Meeting, Orlando, FL.

De Vivo, K. (2022, Jan. 24). A new research base for rigorous project-based learning. *Phi Delta Kappan*. Retrieved from: https://kappanonline.org/research-project-based-learning-de-vivo/

Dean, B. (2022, January 5). *How many people use Twitter in 2022?* [New Twitter stats]. Backlinko. https://backlinko.com/twitter-users

Decker, A., & McGill, M. (2019). A systematic review exploring the differences in reported data for pre-college educational activities for computer science, engineering, and other STEM disciplines. *Education Sciences*, *9*(2), 69. doi:10.3390/educsci9020069

Denzin, N. K., Lincoln, Y. S., & Giardina, M. D. (2006). Disciplining qualitative research. *International Journal of Qualitative Studies in Education: QSE*, *19*(6), 769–782. doi:10.1080/09518390600975990

Deschaine, M. E., & Sharma, S. A. (2015). The Five Cs of Digital Curation: Supporting Twenty-First-Century Teaching and Learning. *InSight: A Journal of Scholarly Teaching*, *10*, 19-24.

Design-Based Research Collective. (2003). Design-based research: An emerging paradigm for educational inquiry. *Educational Researcher*, *32*(1), 5–8. doi:10.3102/0013189X032001005

Deutscher, R., Holthuis, N., Maldonado, S., Pecheone, R. L., Schultz, S. E., & Wei, R. C. (2021). *Learning Through Performance Project-Based Learning as a Lever for Engaging the Next-Generation Science Standards*. Retrieved from: https://scienceeducation.stanford.edu/sites/g/files/sbiybj8661/f/ltp_science_report_2.8.21_final.pdf

Dexter, S., & Richardson, J. W. (2020). What does technology integration research tell us about the leadership of technology? *Journal of Research on Technology in Education*, *52*(1), 17–36. doi:10.1080/15391523.2019.1668316

DiBiase, D., Corbin, T., Fox, T., Francica, J., Green, K., Jackson, J., Jeffress, G., Jones, B., Jones, B., Mennis, J., Schuckman, K., & Smith, C. (2010). The new geospatial technology competency model: Bringing workforce needs into focus. *URISA Journal*, *22*(2), 55–72.

Díez, L. F., Valencia, A., & Bermúdez, J. (2017). Agent-based model for the analysis of technological acceptance of mobile learning. *IEEE Latin America Transactions*, *15*(6), 1121–1127. doi:10.1109/TLA.2017.7932700

Dori, Y. J., & Belcher, J. (2005). How does technology-enabled active learning affect undergraduate students' understanding of electromagnetism concepts? *Journal of the Learning Sciences*, *14*(2), 243–279. doi:10.120715327809jls1402_3

Dori, Y. J., & Herscovitz, O. (2005). Case-based long-term professional development of science teachers. *International Journal of Science Education*, *27*(12), 1413–1446. doi:10.1080/09500690500102946

Duda, M., & Wilson, B. (2015). *Using implementation science to close the policy to practice gap* (A Literate Nation White Paper). https://www.wilsonlanguage.com/wp-content/uploads/2016/06/Implementation_Science_White_Paper_for_Literate_Nation.pdf

Duijzer, C., Van den Heuvel-Panheuizen, M., Veldhuis, M., Doorman, M., & Leseman, P. (2019b). Embodied learning environments for graphing motion: A systematic literature review. *Educational Psychology Review*, *31*(3), 597–629. doi:10.100710648-019-09471-7

Duijzer, C., Van den Heuvel-Panhuizen, M., Veldhius, M., & Doorman, M. (2019a). Supporting primary school students' reasoning about motion graphs through physical experiences. *ZDM Mathematics Education*, *51*(6), 899–913. doi:10.100711858-019-01072-6

Dunaetz, D. R. (2022). When technology does more harm than good: Technostress in missionary contexts. *Journal of the Evangelical Missiological Society*, *2*(1), 112–128. doi:10.31219/osf.io/w7b32

Duncan-Howell, J. (2010). Teachers making connections: Online communities as a source of professional learning. *British Journal of Educational Technology*, *41*(2), 324–340. doi:10.1111/j.1467-8535.2009.00953.x

Duncan, J., Punch, R., & Croce, N. (2021). Supporting primary and secondary teachers to deliver inclusive education. *Australian Journal of Teacher Education (Online)*, *46*(4), 92–107. doi:10.14221/ajte.2021v46n4.6

Dunleavy, M. (2014). Design principles for augmented reality learning. *TechTrends*, *58*(1), 28–34. doi:10.100711528-013-0717-2

Dunlosky, J., Rawson, K. A., Marsh, E. J., Nathan, M. J., & Willingham, D. T. (2013). Improving students' learning with effective learning techniques: Promising directions from cognitive and educational psychology. *Psychological Science in the Public Interest*, *14*(1), 4–58. doi:10.1177/1529100612453266 PMID:26173288

Edelbroek, H., Mijnders, M., & Post, G. (2018). *Interdisciplinary learning activities*. Amsterdam University Press.

Educreations, Inc. (2022). *Teach anything to anyone from anywhere*. https://www.educreations.com

EdWeek. (2020). *Map: Coronavirus and school closures in 2019-2020*. https://www.edweek.org/leadership/map-coronavirus-and-school-closures-in-2019-2020/2020/03

Efstathiou, C., Hovardas, T., Xenofontos, N., Zacharia, Z. C., deJong, T., Anjewierden, A., & van Riesen, S. A. N. (2018). Providing guidance in virtual lab experimentation: The case of an experiment design tool. *Educational Technology Research and Development*, *66*(3), 767–791. doi:10.100711423-018-9576-z

Ellamil, M., Berson, J., Wong, J., Buckley, L., & Margulies, D. (2016). One in the dance: Musical correlates of group synchrony in a club environment. *PLoS One*, *11*(10), e0164783. doi:10.1371/journal.pone.0164783 PMID:27764167

Elmore, R. A. (2022). *Teaching presence in a fully online asynchronous undergraduate mathematics course and its impact on social and cognitive presence* [Doctoral dissertation, Michigan State University]. ProQuest Information & Learning.

Emdin, C. (2016). *For White folks who teach in the hood... and the rest of y'all too: Reality pedagogy and urban education.* Beacon Press.

Emdin, C., Adjapong, E., & Levy, I. P. (2021, July). On Science Genius and Cultural Agnosia: Reality Pedagogy and/as Hip-Hop Rooted Cultural Teaching in STEM Education. *The Educational Forum*, *85*(4), 391–405. doi:10.1080/00131725.2021.1957636

Enzingmüller, C., & Prechtl, H. (2021). Constructing graphs in biology class: Secondary biology teachers' beliefs, motivation, and self-reported practices. *International Journal of Science and Mathematics Education*, *19*(1), 1–19. doi:10.100710763-019-09975-2

Epstein, D. (2019). *Range: Why Generalists Triumph in a Specialized World.* Penguin.

Ericsson, K. A., & Charness, N. (1994). Expert performance: Its structure and acquisition. *The American Psychologist*, *49*(8), 725–747. doi:10.1037/0003-066X.49.8.725

Estes, J. S., & Dailey-Hebert, A. (2018). Modeling technology integration in teacher preparation programs. In *Handbook of Research on Pedagogical Models for Next-Generation Teaching and Learning* (pp. 82–97). IGI Global. doi:10.4018/978-1-5225-3873-8.ch005

Evans-Amalu, K., & Claravall, E. B. (2021). Inclusive online teaching and digital learning: Lessons learned in the time of pandemic and beyond. *Journal of Curriculum Studies Research*, *3*(1), i–iii. doi:10.46303/jcsr.2021.4

Faulconer, E., & Gruss, A. (2018). A review to weigh the pros and cons of online, remote and distance science laboratory experiences. *International Review of Research in Open and Distributed Learning*, 155-168.

Ferguson, R. (2012). Learning analytics: Drivers, developments and challenges. *International Journal of Technology Enhanced Learning*, *4*(5/6), 304–317. doi:10.1504/IJTEL.2012.051816

Fernández-Batanero, J. M., Román-Graván, P., Reyes-Rebollo, M. M., & Montenegro-Rueda, M. (2021). Impact of educational technology on teacher stress and anxiety: A literature review. *International Journal of Environmental Research and Public Health*, *18*(2), 548. doi:10.3390/ijerph18020548 PMID:33440738

Fidai, A., Barroso, L. R., Capraro, M. M., & Capraro, R. M. (2020, October). Effects of Engineering Design Process on Science and Mathematics. In 2020 IEEE Frontiers in Education Conference (FIE) (pp. 1-4). IEEE. doi:10.1109/FIE44824.2020.9274167

Fine, C. G. M. (2022). Translanguaging Interpretive Power in Formative Assessment Co-Design: A Catalyst for Science Teacher Agentive Shifts. *Journal of Language, Identity, and Education*, *21*(3), 191–211. doi:10.1080/15348458.2022.2058858

Fitch, E. F., Hulgin, K. M., & Coomer, M. N. (2021). How "special needs" vouchers deceive the public and silence the right to inclusive education. *International Journal of Qualitative Studies in Education*, 1–21. doi:10.1080/09518398.2021.1957174

Fixsen, D., Blase, K., Metz, A., & Van Dyke, M. (2015). Implementation science. International Encyclopedia of the Social and Behavioral Sciences, 11, 695-702. doi:10.1016/B978-0-08-097086-8.10548-3

Flemming, S. A., & Johnston, C. R. (2020). The Nature of a Problem, Problem Diagnosis, and Engineering Design. *Proceedings of the Canadian Engineering Education Association (CEEA).* 10.24908/pceea.vi0.14156

Fleur, D. S., Bredeweg, B., & van den Bos, W. (2021). Metacognition: Ideas and insights from neuro- and educational sciences. *Nature: Science of Learning, 6*(13), 13. doi:10.1038/s41539-021-00089-5

Fontanilla, H. S. (2016). *Comparison of beginning teachers' and experienced teachers' readiness to integrate technology as measured by TPACK scores* [Doctoral dissertation]. Brandman University.

Forkosh Baruch, A., & Gadot, R. (2021). Social curation experience: Towards authentic learning in preservice teacher training. *Technology, Knowledge and Learning*, *26*(1), 105–122. doi:10.100710758-020-09449-3

Foulger, T. S., Wetzel, K., & Buss, R. R. (2019). Moving toward a technology infusion approach: Considerations for teacher preparation programs. *Journal of Digital Learning in Teacher Education, 35*(2), 79–91. doi:10.1080/21532974.2019.1568325

Fowler-Amato, M., & Warrington, A. (2017). Teachers as designers: Social design experiments as vehicles for developing antideficit English education. *Literacy Research: Theory, Method, and Practice*, *66*(1), 358–372. doi:10.1177/2381336917719054

France, B. (2018). Modeling in technology education: A route to technological literacy. In M. J. de Vries (Ed.), *Handbook of technology education* (pp. 1–11). Springer. doi:10.1007/978-3-319-44687-5_45

Francis, G. (2020). Medicine: Art or science? *Lancet*, *395*(10217), 24–25. doi:10.1016/S0140-6736(19)33145-9 PMID:31982047

Gadot, R., & Tsybulsky, D. (in press). Digital curation as a pedagogical approach for promoting critical thinking. *Journal of Science Education and Technology*.

Gao, F., Li, L., & Sun, Y. (2020). A systematic review of mobile game-based learning in STEM education. *Educational Technology Research and Development*, *68*(4), 1791–1827. doi:10.100711423-020-09787-0

Gao, S., Wang, J., & Zhong, Z. (2018). Influence of science instruction reform on academic performance of eighth grade students in Chinese Inner-Mongolia Autonomous Region. *Compare: A Journal of Comparative Education*, *48*(6), 879–895. doi:10.1080/03057925.2017.1365285

Gao, X., Li, P., Shen, J., & Sun, H. (2020). Reviewing assessment of student learning in interdisciplinary STEM education. *International Journal of STEM Education*, *7*(1), 24. doi:10.118640594-020-00225-4

Genc, M. (2015). The project-based learning approach in environmental education. *International Research in Geographical and Environmental Education*, *24*(2), 105–117. doi:10.1080/10382046.2014.993169

Georgiou, Y., & Ioannou, A. (2019). Embodied learning in a digital world: A systematic review of empirical research in K-12 education. In P. Díaz, A. Ioannou, K. K. Bhagat, & J. M. Spector (Eds.), *Learning in a digital world* (pp. 155–177). Springer. doi:10.1007/978-981-13-8265-9_8

Giles, M., & Willis, J. (2019). Technology proficiency self-evaluation: What are they reporting? In *Society for Information Technology & Teacher Education International Conference* (pp. 1831-1834). Association for the Advancement of Computing in Education (AACE).

Glimpse. (2019, May 14). *Glimpse K12 analysis of school spending shows that two-thirds of software license purchases go unused* [Press release]. https://www.glimpsek12.com/blog-posts/glimpse-k12-analysis-of-school-spending-shows-that-two-thirds-of-software-license-purchases-go-unused

Glowatz, M., & O'Brien, O. (2017). Academic engagement and technology: Revisiting the technological, pedagogical and content knowledge framework (TPACK) in higher education (HE)—The academics' perspectives. *IAFOR Journal of Education*, *5*, 133–159.

Goldberg, F. M., & Anderson, J. H. (1989). Student difficulties with graphical representations of negative velocity. *The Physics Teacher*, *27*(4), 254–260. doi:10.1119/1.2342748

González, M. Á., Martín, M. E., Llamas, C., Martínez, Ó., Vegas, J., Herguedas, M., & Hernández, C. (2017). Teaching and learning physics with smartphones. IGI Global.

Goodchild, M. F., & Janelle, D. G. (2010). Toward critical spatial thinking in the social sciences and humanities. *GeoJournal*, *75*(1), 3–13. doi:10.100710708-010-9340-3 PMID:20454588

Goradia, T. (2018). Role of educational technologies utilizing the TPACK framework and 21st century pedagogies: Academics' perspectives. *IAFOR Journal of Education*, *6*(3), 43–61. doi:10.22492/ije.6.3.03

Gordon, N., Brayshaw, M., & Grey, S. (2013). Maximizing gain for minimal pain: Utilizing natural game mechanics. *Innovation in Teaching and Learning in Information and Computer Sciences*, *12*(1), 27–38.

Graesser, A. C., Lu, S., Jackson, G. T., Mitchell, H., Ventura, M., Olney, A., & Louwerse, M. M. (2004). AutoTutor: A tutor with dialogue in natural language. *Behavior Research Methods, Instruments, & Computers*, *36*(2), 180–193. doi:10.3758/BF03195563 PMID:15354683

Greene, M. D., & Jones, W. M. (2020). Analyzing contextual levels and applications of technological pedagogical content knowledge (TPACK) in E nglish as a second language s ubject area. *Journal of Educational Technology & Society*, *23*(4), 75–88.

Greenhow, C., & Lewin, C. (2016). Social media and education: Reconceptualizing the boundaries of formal and informal learning. *Learning, Media and Technology*, *41*(1), 6–30. doi:10.1080/17439884.2015.1064954

Grigal, M., Neubart, D. A., Moon, S. M., & Graham, S. (2003). Self-determination for students with disabilities: Views of parents and teachers. *Exceptional Children*, *70*(1), 97–112. doi:10.1177/001440290307000106

Grubbs, M., & Strimel, G. (2015). Engineering design: The great integrator. *Journal of STEM teacher. Education*, *50*(1), 8.

Guspatni, G. (2021). Student-generated PowerPoint animations: A study of student teachers' conceptions of molecular motions through their expressed models. *Chemistry Education Research and Practice*, *22*(2), 312–327. doi:10.1039/D0RP00229A

Gutiérrez, K. D., Cortes, K., Cortez, A., DiGiacomo, D., Higgs, J., Johnson, P., Ramón Lizárraga, J., Mendoza, E., Tien, J., & Vakil, S. (2017). Replacing representation with imagination: Finding ingenuity in everyday practices. *Review of Research in Education*, *41*(1), 30–60. doi:10.3102/0091732X16687523

Gutiérrez, K. D., & Jurow, A. S. (2016). Social design experiments: Toward equity by design. *Journal of the Learning Sciences*, *25*(4), 565–598. doi:10.1080/10508406.2016.1204548

Guzey, S. S., Caskurlu, S., & Kozan, K. (2020). Integrated STEM pedagogies and student learning. In C. C. Johnson, M. J. Mohr-Schroeder, T. J. Moore, & L. D. English (Eds.), *Handbook of research on STEM education* (pp. 65–75). Taylor & Francis. doi:10.4324/9780429021381-8

Hallström, J., & Schönborn, K. J. (2019). Models and modelling for authentic STEM education: Reinforcing the argument. *International Journal of STEM Education*, *6*(1), 22. doi:10.118640594-019-0178-z

Hamilton, L., & Gross, B. (2021). *How has the pandemic affected students' social-emotional well-being? A review of the evidence to date.* University of Washington, Bothell. Center on Reinventing Public Education (CRPE).

Hammond, T. C., Bodzin, A., Anastasio, D., Holland, B., Popejoy, K., & Sahagian, D. (2019). Shoulder-to-shoulder: Teacher professional development and curriculum design and development for geospatial technology integration with science and social studies teachers. *Contemporary Issues in Technology & Teacher Education*, *19*(2). https://www.citejournal.org/volume-19/issue-2-19/current-practice/shoulder-to-shoulder-teacher-professional-development-and-curriculum-design-and-development-for-geospatial-technology-integration-with-science-and-social-studies-teachers/

Hammond, T. C., Bodzin, A., Anastasio, D., Holland, B., Popejoy, K., Sahagian, D., Rutzmoser, S., Carrigan, J., & Farina, W. (2018). "You know you can do this, right?": Developing geospatial technological pedagogical content knowledge (GS-TPACK) and enhancing teachers' cartographic behaviors with Socio-Environmental Science Investigations (SESI). *Cartography and Geographic Information Science*, *45*(4), 305–318. doi:10.1080/15230406.2017.1419440

Handal, B., Campbell, C., Cavanagh, M., & Dave, K. (2014, January). Appraising mobile maths apps: The TPACK model. In *Australian Computers in Education Conference* (pp. 251-269). Australian Council for Computers in Education.

Hansen, M. (2014). Characteristics of schools successful in STEM: Evidence from two states' longitudinal data. *The Journal of Educational Research*, *107*(5), 374–391. doi:10.1080/00220671.2013.823364

Harris, P. (2001). Going mobile. Learning circuits, *ASTD Online Magazine.* http://www.learningcircuits.org/2001/jul2001/harris.html

Hattie, J., & Timperley, H. (2007). The power of feedback. *Review of Educational Research*, *77*(1), 81–112. doi:10.3102/003465430298487

Hechter, R. P. (2012). Pre-service teachers' maturing perceptions of a TPACK-framed signature pedagogy in science education. *Computers in the Schools*, *29*(1-2), 53–69. doi:10.1080/07380569.2012.657999

Hechter, R. P., & Vermette, L. (2013). Technology integration in K-12 science classrooms: An analysis of barriers and implications. *Themes in Science and Technology Education*, *6*(2), 73–90.

Hellgren, S., & Lindberg, S. (2017). Motivating students with authentic science experiences: Changes in motivation for school science. *Research in Science & Technological Education*, *35*(4), 409–426. doi:10.1080/02635143.2017.1322572

Helppolainen, S., & Aksela, M. (2020). Science teachers' ICT use from a viewpoint of technological pedagogical content knowledge (TPCK). *Teknologia Kemian Opetuksessa*, *1*(1), 2–2.

Henderson, K. J., Coppens, E. R., & Burns, S. (2021). Addressing barriers to implementing problem-based learning. *AANA Journal*, *89*(2). PMID:33832571

Herman, J. (2001). Medicine: The science and the art. *Medical Humanities*, *27*(1), 42–46. doi:10.1136/mh.27.1.42 PMID:23670550

Hermans, R., Tondeur, J., van Braak, J., & Valcke, M. (2008). The impact of primary school teachers' educational beliefs on the classroom use of computers. *Computers & Education*, *51*(4), 1499–1509. doi:10.1016/j.compedu.2008.02.001

Hestenes, D. (1987). Toward a modeling theory of physics instruction. *American Journal of Physics*, *55*(5), 440–454. doi:10.1119/1.15129

Hjalmarson, M. A., Holincheck, N. B., Courtney, K., & Galanti, T. M. (2020). Learning models and modeling across the STEM disciplines. In C. C. Johnson, M. J. Mohr-Schroeder, T. J. Moore, & L. D. English (Eds.), Handbook of research on STEM education (pp. 223–233). Taylor & Francis.

Hobbs, R. (2017). *Create to learn: Introduction to digital literacy*. John Wiley & Sons.

Holincheck, N., & Galanti, T. M. (2022). Are you a STEM teacher?: Exploring PK-12 teachers' conceptions of STEM education. *Journal of STEM Education: Innovations and Research*, *23*(2), 23–29.

Holthuis, N., Deutscher, R., Schultz, S. E., & Jamshidi, A. (2018). The New NGSS Classroom: A Curriculum Framework for Project-Based Science Learning. *American Educator*, *42*(2), 23–27.

Höttecke, D., & Allchin, D. (2020). Reconceptualizing nature-of-science education in the age of social media. *Science Education*, *104*(4), 641–666. doi:10.1002ce.21575

Hsu, P.-S., & Sharma, P. (2008). A case study of enabling factors in the technology integration change process. *Journal of Educational Technology & Society*, *11*, 213–228.

Hsu, T., Chang, S., & Hung, Y. (2018). How to learn and how to teach computational thinking: Suggestions based on a review of the literature. *Computers & Education*, *126*, 296–310. doi:10.1016/j.compedu.2018.07.004

Hughes, J., Luo, W., Kwok, O., & Loyd, L. (2008). Teacher-student support, effortful engagement, and achievement: A 3-year longitudinal study. *Journal of Educational Psychology*, *100*(1), 1–14. doi:10.1037/0022-0663.100.1.1 PMID:19578558

Hurwitz, L. B., & Schmitt, K. L. (2020). Can children benefit from early internet exposure? Short- and long-term links between internet use, digital skill, and academic performance. *Computers & Education*, *146*, 1–11.

Huynh, N. T., & Sharpe, B. (2013). An assessment instrument to measure geospatial thinking expertise. *The Journal of Geography*, *112*(1), 3–17. doi:10.1080/00221341.2012.682227

Hyde, J. M. (2021). Exploring Hubble Constant Data in an Introductory Course. *The Physics Teacher*, *59*(3), 159–161. doi:10.1119/10.0003654

Ibourk, A., & Kendrick, M. (2021). *Elementary students' explanation of variation of traits and teacher's feedback using an online embedded assessment tool*. doi:10.1080/09500693.2021.1903618

Illeris, K. (2016). *Learning, Development, and Education: From Learning Theory to Education and Practice, the Selected Works of Knud Illeris*. Routledge.

Ionnidis, K., Treder, M. S., Chamberlain, S. R., Kraly, F., Redden, S. A., Stein, D. J., & Grant, J. E. (2018). Problematic internet use as an age-related multifaceted problem: Evidence from a two-site survey. *Addictive Behaviors*, *81*, 157–166.

Jackson, C., Tank, K. M., Appelgate, M. H., Jurgenson, K., Delaney, A., & Erden, C. (2020). History of integrated STEM curriculum. In C. C. Johnson, M. J. Mohr-Schroeder, T. J. Moore, & L. D. English (Eds.), *Handbook of research on STEM education* (pp. 169–183). Taylor & Francis. doi:10.4324/9780429021381-17

Jacobs, K. G., Kugler, J., Chi, J., Stuart, E., Merrell, S. B., & Rassbach, C. (2020). A mixed methods approach to understanding curricular impact of a capstone course on the self-efficacy of fourth-year medical students. *Cureus*, *12*(8). Advance online publication. doi:10.7759/cureus.9537 PMID:32905172

Jacques, L. (2020, April). Using the engineering design process to teach computational thinking. *Proceedings of Society for Information Technology & Teacher Education International Conference*, 33-37.

Jacques, L., & Howle, H. (2020). *Computational thinking high school course*. Cyber Innovation Center. Retrieved from https://cyber.org/

Jaggars, S. S., & Xu, D. (2016). How do online course design features influence student performance? *Computers & Education, 95*, 270–284. doi:10.1016/j.compedu.2016.01.014

Jalilifar, A., White, P., & Malekizadeh, N. (2017). Exploring nominalization in scientific textbooks: A cross-disciplinary study of hard and soft sciences. *International Journal of English Studies, 17*(2), 1–20. doi:10.6018/ijes/2017/2/272781

Jang, S. J., & Chen, K. C. (2010). From PCK to TPACK: Developing a transformative model for pre-service science teachers. *Journal of Science Education and Technology, 19*(6), 553–564. doi:10.100710956-010-9222-y

Jang, S. J., & Tsai, M. F. (2013). Exploring the TPACK of Taiwanese secondary school science teachers using a new contextualized TPACK model. *Australasian Journal of Educational Technology, 29*(4). Advance online publication. doi:10.14742/ajet.282

Jimoyiannis, A. (2010). Developing a technological pedagogical content knowledge framework for science education: Implications of a teacher trainers' preparation program. *Proceedings of Informing Science & IT Education Conference, 597*, 607. doi:10.28945/1277

Johnson-Glenberg, M. C. (2018). Immersive VR and education: Embodied design principles that include gesture and hand controls. *Frontiers in Robotics and AI: Hypothesis and Theory, 5*(81), 1–19. doi:10.3389/frobt.2018.00081 PMID:33500960

Johnson-Glenberg, M. C. (2019). The necessary nine: Design principles for embodied VR and active stem education. In P. Díaz, A. Ioannou, K. K. Bhagat, & J. M. Spector (Eds.), *Learning in a digital world: Perspective on interactive technologies for formal and informal education* (pp. 83–112). Springer Singapore. doi:10.1007/978-981-13-8265-9_5

Johnson-Glenberg, M. C., Birchfield, D. A., Tolentino, L., & Koziupa, T. (2014). Collaborative embodied learning in mixed reality motion-capture environments: Two science studies. *Journal of Educational Psychology, 106*(1), 86–104. doi:10.1037/a0034008

Johnson-Glenberg, M. C., Birchfield, D. A., & Usyal, S. (2009). SMALLLab: Virtual geology studies using embodied learning with motion, sound, and graphics. *Educational Media International, 46*(4), 267–280. doi:10.1080/09523980903387555

Johnson-Glenberg, M. C., Birchfield, D., Megowan-Romanowicz, M. C., & Snow, E. L. (2015). If the gear fits, spin it! Embodied education and in-game assessments. *International Journal of Gaming and Computer Based Simulation, 7*, 40–65.

Johnson-Glenberg, M. C., Birchfield, D., Savvides, P., & Megowan-Romanowicz, C. (2011). Semi-virtual embodied learning-real world STEM assessment. In L. Annetta & S. C. Bronack (Eds.), *Serious educational game assessment* (pp. 241–257). Brill Sense. doi:10.1007/978-94-6091-329-7_15

Johnson-Glenberg, M. C., & Megowan-Romanowicz, C. (2017). Embodied science and mixed reality: How gesture and motion capture affect physics education. *Cognitive Research: Principles and Implications, 2*(4), 1–28. doi:10.118641235-017-0060 9 PMID:28603770

Johnson, R. B., & Onwuegbuzie, A. J. (2004). Mixed methods research: A research paradigm whose time has come. *Educational Researcher, 33*(7), 14–26. doi:10.3102/0013189X033007014

Jokhan, A., Sharma, B., & Singh, S. (2018). Early warning system as a predictor for student performance in higher education blended courses. *Studies in Higher Education.*

Kaldenberg, E. R., Watt, S. J., & Therrien, W. J. (2015). Reading instruction in science for students with learning disabilities: A meta-analysis. *Learning Disability Quarterly, 38*(3), 160–173. doi:10.1177/0731948714550204

Kali, Y., McKenney, S., & Sagy, O. (2015). Teachers as designers of technology enhanced learning. *Instructional Science, 43*(2), 173–179. doi:10.100711251-014-9343-4

Kapor, M. (1991). *Internet Quotation Appendix*. Harvard Berkman Klein Center. https://cyber.harvard.edu/archived_content/people/reagle/inet-quotations-19990709.html

Keates, S., & Clarkson, J. (2003). Design exclusion. *Inclusive Design*, 88-102.

Kelley, T. R., Knowles, J. G., Holland, J. D., & Han, J. (2020). Increasing high school teachers self-efficacy for integrated STEM instruction through a collaborative community of practice. *International Journal of STEM Education*, *7*(1), 1–13. doi:10.118640594-020-00211-w

Kelly, G. J. (2018). Developing epistemic aims and supports for engaging students in scientific practices. *Science & Education*, *27*(3-4), 245–246. doi:10.100711191-018-9974-y

Kelly, N., & Antonio, A. (2016). Teacher peer support in social network sites. *Teaching and Teacher Education*, *56*, 138–149. doi:10.1016/j.tate.2016.02.007

Kennedy, M. (2016). Parsing the practice of teaching. *Journal of Teacher Education*, *67*(1), 6–17. doi:10.1177/0022487115614617

Kent, A. M., & Giles, R. M. (2017). Preservice teachers' technology self-efficacy. *SRATE Journal*, *26*(1), 9–20.

Kenward, K., Hipp, S., & Baker, M. (2011, March). Seamless technology integration for pre-service teachers. In *Society for Information Technology & Teacher Education International Conference* (pp. 2542-2544). Association for the Advancement of Computing in Education (AACE).

Kersaint, G., Ritzhaupt, A. D., & Liu, F. (2013). Technology to enhance mathematics and science instruction: Changes in teacher perceptions after participating in a yearlong professional development program. *Journal of Computers in Mathematics and Science Teaching*, *33*, 73–101.

Khairani, A. Z. (2017). Assessing urban and rural teachers' competencies in STEM integrated education in Malaysia. *MATEC Web of Conferences, 87*, 1-5.

Kilty, T. J. (2019). *Technology integration and science education: An instructional design and technology perspective* (Publication No. 13882883) [Doctoral dissertation, University of Wyoming]. ProQuest Dissertations Publishing.

Kilty, T. J., & Burrows, A. C. (2021). Secondary science preservice teachers: Technology integration in methods and residency. *Journal of Science Teacher Education*, *32*(5), 578–600. doi:10.1080/1046560X.2021.1907514

Kim, M. C., Hannafin, M. J., & Bryan, L. A. (2007). Technology-enhanced inquiry tools in science education: An emerging pedagogical framework for classroom practice. *Science Education*, *91*(6), 1010–1030. doi:10.1002ce.20219

Kirscher, P. A., Sweller, J., & Clark, R. E. (2006). Why minimal guidance during instruction does not work: An analysis of the failure of constructivist, discovery, problem-based, experiential, and inquiry-based teaching. *Educational Psychologist*, *41*(2), 75–86.

Kita, S., Alibali, M. W., & Chu, M. (2017). How do gestures influence thinking and speaking? The gesture-for-conceptualization hypothesis. *Psychological Review*, *124*(3), 245–266. doi:10.1037/rev0000059 PMID:28240923

Kitsantas, A., Baylor, A. L., & Hiller, S. E. (2019). Intelligent system support to optimize performance: Augmenting cognitive capacity and supporting self-regulation of critical thinking skills in decision making. *Cognitive Systems Research*, *58*, 387–397. doi:10.1016/j.cogsys.2019.09.003

Kitsantas, A., & Dabbagh, N. (2010). *Learning to Learn with Integrative LearningTechnologies (ILT): A Practical Guide for Academic Success*. Information Age Publishing.

Kobulnicky, H., & Dale, D. (2016). A community mentoring model for STEM undergraduate research experiences. *Journal of College Science Teaching*, *45*(6), 17–23. doi:10.2505/4/jcst16_045_06_17

Koehler, M. J., & Mishra, P. (2009). What is technological pedagogical content knowledge? *Contemporary Issues in Technology & Teacher Education*, *9*(1), 60–70.

Koehler, M. J., Mishra, P., & Cain, W. (2013). What is technological pedagogical content knowledge (TPACK)? *Journal of Education*, *193*(3), 13–19. doi:10.1177/002205741319300303

Koehler, M., & Mishra, P. (2009). What is technological pedagogical content knowledge (TPACK)? *Contemporary Issues in Technology & Teacher Education*, *9*(1), 60–70.

Kohen, Z., Keller, N., & Koichu, B. (2020). Metacognitive processes in online collaborative problem solving forums: mathematics teachers' dual roles. *Proceedings of the International Commission on Mathematical Instruction (ICMI 25)*, 484-491.

Koomen, M., Kahn, S., Atchison, C. L., & Wild, T. A. (Eds.). (2018). *Towards inclusion of all learners through science teacher education*. Brill Publishers. doi:10.1163/9789004368422

Korucu-Kis, S., & Ozmen, K. S. (2019). Exherent and inherent value beliefs about technology: Missing pieces in the puzzle of technology integration? *International Journal of Educational Technology*, *6*(1), 1–11.

Kotter, J. (1995, May-June). Leading change: Why transformation efforts fail. *Harvard Business Review*. https://hbr.org/1995/05/leading-change-why-transformation-efforts-fail-2

Krahenbuhl, K. S. (2021). *Building Effective Learning Environments: A Framework for Merging the Best of Old and New Practices*. Routledge.

Krajcik, J. S., & Blumenfeld, P. C. (2006). Project-based learning. In R. K. Sawyer (Ed.), *The Cambridge handbook of the learning sciences* (pp. 317–333). Cambridge University Press.

Krajcik, J., McNeill, K. L., & Reiser, B. J. (2008). Learning-goals-driven design model: Developing curriculum materials that align with national standards and incorporate project-based pedagogy. *Science Education*, *92*(1), 1–32. doi:10.1002ce.20240

Kramaraski, B., & Kohen, Z. (2017). Promoting preservice teachers' dual self-regulation roles as learners and as teachers: Effects of generic vs. specific prompts. *Metacognition and Learning*, *12*(2), 157–191. doi:10.100711409-016-9164-8

Krauss, J., & Prottsman, K. (2018). *Computational thinking and coding for every student: The teacher's getting started guide*. Corwin Press.

Kulkarni, S., Nusbaum, E., & Boda, P. A. (2021). DisCrit at the margins of teacher education: Informing curriculum, visibilization, and disciplinary integration. *Race, Ethnicity and Education*, *24*(5), 654–670. doi:10.1080/13613324.2021.1918404

Kümmel, E., Moskaliuk, J., Cress, U., & Kimmerle, J. (2020). Digital learning environments in higher education: A literature review of the role of individual vs. social settings for measuring learning outcomes. *Education Sciences*, *10*(3), 78. doi:10.3390/educsci10030078

Kuykendall, K. (2022, March 31). LearnPlatform Launches Certification for Ed Tech Providers Meeting ESSA Evidence and Student Data, Privacy Standards. *THE Journal*. https://thejournal.com/articles/2022/03/31/learnplatform-launches-certification-for-ed-tech-providers-meeting-essa-evidence-and-data-standards.aspx

Kwan, M. (2012, Jan. 15). Sunday snippet: Scott Adams (Dilbert). *Beyond the Rhetoric*. Retrieved from: https://btr.michaelkwan.com/2012/01/15/sunday-snippet-scott-adams-dilbert/

Ladson-Billings, G. (2018). The social funding of race: The role of schooling. *Peabody Journal of Education*, *93*(1), 90–105. doi:10.1080/0161956X.2017.1403182

Lang, D., & Savageau, J. A. (2017). *Starting at the end: Measuring learning using retrospective pre-post evaluations. Center for Health Policy and Research (CHPR)*. Publications.

Langen, T. A., Mourad, T., Grant, B. W., Gram, W. K., Abraham, B. J., Fernandez, D. S., Carroll, M., Nuding, A., Balch, J. K., Rodriguez, J., & Hampton, S. E. (2014). Using large public datasets in the undergraduate ecology classroom. *Frontiers in Ecology and the Environment*, *12*(6), 362–363. doi:10.1890/1540-9295-12.6.362

Lee, C., Miller, E., & Januszyk, R. (2015). *Next generation science standards (NGSS) for all students*. NSTA Press.

Lee, L.-K., Cheung, S. K. S., & Kwok, L.-F. (2020). Learning analytics: Current trends and innovative practices. *Journal of Computers in Education*, *7*(1), 1–6. doi:10.100740692-020-00155-8

Lee, S. Y. (2021). Rethinking teacher agency: Cybernetics, action research, and the process-oriented rationality. *Journal of Curriculum Studies*, *53*(6), 821–840. doi:10.1080/00220272.2020.1858451

Lemley, J. B., Schumacher, G., & Vesey, W. (2014). What learning environments best address 21st-century students' perceived needs at the secondary level of instruction. *NASSP Bulletin*, *98*(2), 101–125. doi:10.1177/0192636514528748

Li, H. (2013). *Educational Change towards PBL (Problem Based Learning): An Organizational Perspective*. River Publishers.

Lindgren, R., & Johnson-Glenberg, M. (2013). Emboldened by embodiment: Six precepts for research on embodied learning and mixed reality. *Educational Researcher*, *42*(8), 445–452. doi:10.3102/0013189X13511661

Lindgren, R., & Moshell, J. M. (2011). *Supporting children's learning with body-based metaphors in a mixed reality environment* [Paper presentation]. *10th International Conference on Interaction Design and Children*, Ann Arbor, MI, United States.

Lindgren, R., Tscholl, M., Wang, S., & Johnson, E. (2016). Enhancing learning and engagement through embodied interaction within a mixed reality simulation. *Computers & Education*, *95*, 174–183. doi:10.1016/j.compedu.2016.01.001

Lindsay, L. (2016). Transformation of teacher practice using mobile technology with one-to-one classes: M-learning pedagogical approaches. *British Journal of Educational Technology*, *47*(5), 883–892. doi:10.1111/bjet.12265

Lin, K., Wu, Y., Hsu, Y., & Williams, P. J. (2021). Effects of infusing the engineering design process into STEM project-based learning to develop preservice technology teachers' engineering design thinking. *International Journal of STEM Education*, *8*(1), 1. doi:10.118640594-020-00258-9

Lin, T. C., Tsai, C. C., Chai, C. S., & Lee, M. H. (2013). Identifying science teachers' perceptions of technological pedagogical and content knowledge (TPACK). *Journal of Science Education and Technology*, *22*(3), 325–336. doi:10.100710956-012-9396-6

Liu, C., Bano, M., Zowghi, D., & Kearney, M. (2021). Analysing user reviews of inquiry-based learning apps in science education. *Computers & Education*, *164*, 104–119. doi:10.1016/j.compedu.2020.104119

Lloyd, P. (2000). Storytelling and the development of discourse in the engineering design process. *Design Studies*, *21*(4), 357–373. doi:10.1016/S0142-694X(00)00007-7

Lockwood, J., & Mooney, A. (2017). Computational thinking in education: Where does it fit? A systematic literary review. *International Journal of Computer Sciences and Engineering Systems*, *2*(1), 41–60.

Lopez, R. E., & Hamed, K. (2004). Student interpretations of 2-D and 3-D renderings of the substorm current wedge. *Journal of Atmospheric and Solar-Terrestrial Physics*, *66*(15-16), 1509–1517. doi:10.1016/j.jastp.2004.03.030

Loughran, J. (2019). Pedagogical reasoning: The foundation of the professional knowledge of teaching. *Teachers and Teaching*, *25*(5), 523–535. doi:10.1080/13540602.2019.1633294

Lowenthal, P., & Dunlap, J. (2011). Investigating students' perceptions of various instructional strategies to establish social presence. In *Annual Meeting of the American Educational Research Association.* Google Scholar.

Lowenthal, P., & Dunlap, J. (2020). Social presence and online discussions: A mixed method investigation. *Distance Education*, 490-514.

Ludlow, B. (2015). Virtual reality: Emerging applications and future directions. *Rural Special Education Quarterly*, *34*(3), 3–10. doi:10.1177/875687051503400302

Luetzenburg, G., Kroon, A., & Bjørk, A. A. (2021). Evaluation of the Apple iPhone 12 Pro LiDAR for an application in geosciences. *Scientific Reports*, *11*(1), 1–9. doi:10.103841598-021-01763-9 PMID:34782692

Lungu, B., & Lungu, M. (2021). Exploring the effects on student learning and engagement of COVID-19: An innovative and interdisciplinary approach. *Journal of Microbiology & Biology Education*, *22*(1), ev22i1.2429. Advance online publication. doi:10.1128/jmbe.v22i1.2429 PMID:33884068

Macia, M., & García, I. (2016). Informal online communities and networks as a source of teacher professional development: A review. *Teaching and Teacher Education*, *55*, 291–307. doi:10.1016/j.tate.2016.01.021

Maddux, J. E., & Kleiman, E. M. (2020). 31 Self-Efficacy. The Oxford Handbook of Positive Psychology, 443.

Maddux, J. E. (2002). Self-efficacy: The power of believing you can. In S. Lopez & C. Snyder (Eds.), *The Oxford handbook of positive psychology* (2nd ed., pp. 335–344). Oxford University Press.

Madsen, A. M., Larson, A. M., Loschky, L. C., & Sanjay Rebello, N. (2012). Differences in visual attention between those who correctly and incorrectly answer physics problems. *Physical Review Special Topics. Physics Education Research*, *8*(1), 010122. doi:10.1103/PhysRevSTPER.8.010122

Maeng, J. L., Whitworth, B. A., Bell, R. L., & Sterling, D. R. (2020). The effect of professional development on elementary science teachers' understanding, confidence, and classroom implementation of reform-based science instruction. *Science Education*, *104*(2), 326–353. doi:10.1002ce.21562

Makgato, M. (2007). Factors associated with poor performance of learners in mathematics and physical science in secondary schools in Soshanguve, South Africa. *Africa Education Review, 4*(1), 89-103.

Markula, A., & Aksela, M. (2022). The key characteristics of project-based learning: How teachers implement projects in K-12 science education. *Disciplinary and Interdisciplinary Science Education Research, 4*(2).

Massachusetts Department of Education. (2006). Massachusetts science and technology/engineering curriculum framework. Author.

Massey, A., Zhang, W., & Amar, A. (2021). A comparison of non-traditional online and traditional wet-lab experiences in human anatomy and physiology: An innovative approach for pre-licensure nursing education. *Nurse Education Today*, *107*, 105149. doi:10.1016/j.nedt.2021.105149 PMID:34592576

Mathias, B., Andrä, C., Schwager, A., Macedonia, M., & von Kriegstein, K. (2022). Twelve- and fourteen-year-old school children differentially benefit from sensorimotor- and multisensory-enriched vocabulary training. *Educational Psychology Review*, *34*(3), 1739–1770. doi:10.100710648-021-09648-z

Mayer, R. E. (2020). *Multimedia Learning* (3rd ed.). Cambridge University Press.

Mazzolini, M., & Maddison, S. (2007). When to jump in: The role of the instructor in online discussion forums. *Computers & Education*, *49*(2), 193–213. doi:10.1016/j.compedu.2005.06.011

McCarter, S. A. (2019). Adolescence. In E. D. Hutchison (Ed.), *Dimensions of Human Behavior: The Changing Life Course* (pp. 189–229). Sage Publications.

McDermott, L. C., Rosenquist, M. L., & van Zee, E. H. (1987). Student difficulties in connecting graphs and physics: Examples from kinematics. *American Journal of Physics*, *55*(6), 503–513. doi:10.1119/1.15104

McElhaney, K. W., Chang, H.-Y., Chiu, K. L., & Linn, M. (2015). Evidence for effective uses of dynamic visualizations in science curriculum materials. *Studies in Science Education*, *51*(1), 49–85. doi:10.1080/03057267.2014.984506

McKinsey Global Institute. (2017). *Jobs Lost, jobs gained: Workforce transitions in a time of automation*. McKinsey & Company.

McKnight, K., O'Malley, K., Ruzic, R., Horsley, M. K., Franey, J. J., & Bassett, K. (2016). Teaching in a digital age: How educators use technology to improve student learning. *Journal of Research on Technology in Education*, *48*(3), 194–211.

Mehalik, M. M., Doppelt, Y., & Schuun, C. D. (2008). Middle-school science through design-based learning versus scripted inquiry: Better overall science concept learning and equity gap reduction. *Journal of Engineering Education*, *97*(1), 71–85. doi:10.1002/j.2168-9830.2008.tb00955.x

Menon, D., & Sadler, T. D. (2016). Preservice elementary teachers' science self-efficacy beliefs and science content knowledge. *Journal of Science Teacher Education*, *27*(6), 649–673. doi:10.100710972-016-9479-y

Menon, D., & Sadler, T. D. (2018). Sources of science teaching self-efficacy for preservice elementary teachers in science content courses. *International Journal of Science and Mathematics Education*, *16*(5), 835–855. doi:10.100710763-017-9813-7

Mentzer, G. A., Czerniak, C. M., & Lisa, B. (2017). An examination of teacher understanding of project-based science as a result of participating in an extended professional development program: Implications for implementation. *School Science and Mathematics*, *117*(1–2), 76–86. doi:10.1111sm.12208

Mentzer, N. (2011). High school engineering and technology education integration through design challenges. *Journal of STEM Teacher Education*, *48*(2), 103–136. doi:10.30707/JSTE48.2Mentzer

Mentzer, N., Huffman, T., & Thayer, H. (2014). High school student modeling in the engineering design process. *International Journal of Technology and Design Education*, *24*(3), 293–316. doi:10.100710798-013-9260-x

Metz, A., & Louison, L. (2018). *The Hexagon tool: Exploring context*. National Implementation Research Network.

Microsoft. (2017). *Why Europe's girls aren't studying STEM*. https://ncmedia.azureedge.net/ncmedia/2017/02/Microsoft_girls_in_STEM_final-Whitepaper.pdf

Mihailidis, P., & Cohen, J. N. (2013). Exploring curation as a core competency in digital and media literacy education. *Journal of Interactive Media in Education*, *2013*(1), 1–19. doi:10.5334/2013-02

Millar, R., & Osborne, J. F. (Eds.). (1998). *Beyond 2000: Science education for the future*. King's College London.

Millis, F. C., Butler, H., Wallace, P., Graesser, A., & Halpern, D. (2011). Operation ARIES! A serious game for teaching scientific inquiry. In Serious Games and Edutainment Applications (pp. 169–195). Springer London.

Ministry of Industry and Technology, The Southeastern Anatolia Project Regional Development Administration. (2019). *GAP green.* http://www.gapgreen.org

Mishra, P., & Koehler, M. J. (2006). Technological pedagogical content knowledge: A framework for teacher knowledge. *Teachers College Record, 108*(6), 1017–1054. doi:10.1111/j.1467-9620.2006.00684.x

Monteiro, M., & Martí, A. (2022). Resource letter MDS-1: Mobile devices and sensors for physics teaching. *American Journal of Physics, 90*(5), 328–343. doi:10.1119/5.0073317

Moore, T. J., Glancy, A. W., Tank, K. M., Kersten, J. A., Smith, K. A., & Stohlmann, M. S. (2014b). A framework for quality K-12 engineering education: Research and development. *Journal of Pre-College Engineering Education Research, 4*(1), Article 2. . doi:10.7771/2157-9288.1069

Moore, T. J., Johnston, A. C., & Glancy, A. W. (2020). STEM integration: A synthesis of conceptual frameworks and definitions. In C. C. Johnson, M. J. Mohr-Schroeder, T. J. Moore, & L. D. English (Eds.), *Handbook of research on STEM education* (pp. 3–16). Taylor & Francis. doi:10.4324/9780429021381-2

Moore, T. J., Stohlmann, M. S., Wang, H.-H., Tank, K. M., Glancy, A. W., & Roehrig, G. H. (2014a). Implementation and integration of engineering in K-12 STEM education. In J. Strobel, S. Purzer, & M. Cardella (Eds.), *Engineering in precollege settings: Research into practice* (pp. 35–60). Sense Publishers. doi:10.2307/j.ctt6wq7bh.7

Moore, T. J., Tank, K. M., Glancy, A. W., & Kersten, J. A. (2013, January). A framework for implementing engineering standards in K-12. Paper to be presented at the *International Conference of Association for Science Teacher Education.*

Mueller, J., Wood, E., Willoughby, T., Ross, C., & Specht, J. (2008). Identifying discriminating variables between teachers who fully integrate computers and teachers with limited integration. *Computers & Education, 51*(4), 1523–1537. doi:10.1016/j.compedu.2008.02.003

Mungan, C. E. (2021). Using desmos to understand the difference between phase and group velocity. *The Physics Teacher, 59*(1), 30–33. doi:10.1119/10.0003012

Murgatroyd, S. (2010). 'Wicked problems' and the work of the school. *European Journal of Education, 45*(2), 259–279. doi:10.1111/j.1465-3435.2010.01428.x

Mutambara, D., & Bayaga, A. (2021). Determinants of mobile learning acceptance for STEM education in rural areas. *Computers & Education, 160*, 104010. doi:10.1016/j.compedu.2020.104010

Naraian, S. (2010). General, special and… inclusive: Refiguring professional identities in a collaboratively taught classroom. *Teaching and Teacher Education, 26*(8), 1677–1686. doi:10.1016/j.tate.2010.06.020

Naraian, S. (2017). *Teaching for inclusion: Eight principles for effective and equitable practice.* Teachers College Press.

Naraian, S. (2021). Making inclusion matter: Critical disability studies and teacher education. *Journal of Curriculum Studies, 53*(3), 298–313. doi:10.1080/00220272.2021.1882579

Nariman, N., & Chrispeels, J. (2016). PBL in the era of reform standards: Challenges and benefits perceived by teachers in one elementary school. *The Interdisciplinary Journal of Problem-Based Learning, 10*(1). Advance online publication. doi:10.7771/1541-5015.1521

National Academies of Sciences, Engineering, and Medicine. (2017). *Information technology and the U.S. workforce: Where are we and where do we go from here?* National Academies Press.

National Academies of Sciences, Engineering, and Medicine. (2019). *Science and engineering for grades 6-12: Investigations and design at the center.* The National Academies Press.

National Academies of Sciences, Engineering, and Medicine. (2020). *Teaching K-12 science and engineering during a crisis*. National Academies Press.

National Academy of Engineering & National Research Council. (2009). *Engineering in K-12 education understanding the status and improving the prospects.* Washington, DC: National Academies Press.

National Aeronautics and Space Administration. (2011). *Engineering design process*. Retrieved from https://www.nasa.gov/audience/foreducators/best/edp.html

National Center for Educational Statistics (NCES). (2013). *The condition of education 2013*. Washington, DC: National Center for Educational Statistics. Retrieved from https://nces.ed.gov/pubs2013/2013037.pdf

National Governors Association Center for Best Practices & Council of Chief State School Officers. (2010). *Common core state standards for mathematics*. Authors.

National Implementation Research Network. (n.d.). *Topic 3: Exploration stage*. https://nirn.fpg.unc.edu/module-4/topic-3-exploration

National Research Council (NRC). (2010). Report of a workshop on the scope and nature of computational thinking. Committee for the Workshops on Computational Thinking, Computer Science and Telecommunications Board, Division on Engineering and Physical Sciences. Washington, DC: National Academies Press.

National Research Council (NRC). (2011). Report of a workshop on the pedagogical aspects of computational thinking. Committee for the Workshops on Computational Thinking, Computer Science and Telecommunications Board, Division on Engineering and Physical Sciences. Washington, DC: National Academies Press.

National Research Council. (1991). *Mathematical sciences, technology, and economic competitiveness*. National Academies Press. doi:10.17226/1786

National Research Council. (2006). *Learning to think spatially: GIS as a support system in K-12 education*. National Academies Press.

National Research Council. (2010). *New worlds, new horizons in astronomy and astrophysics*. National Research Council.

National Research Council. (2011). *Expanding underrepresented minority participation: America's science and technology talent at the crossroads*. National Academies Press.

National Research Council. (2012). *A framework for K-12 science education: Practices, crosscutting concepts, and core ideas*. National Academies Press.

National Research Council. (2012a). *A framework for K-12 science education*. National Academies Press.

National Research Council. (2012b). *Discipline-based education research: Understanding and improving learning in undergraduate science and engineering*. National Academies Press.

National Research Council. (2013). *High magnetic field science and its application in the United States: Current status and future directions*. National Academies Press.

National Research Council. (2013). Next generation science standards: For states, by states. Washington, DC: The National Academies Press. doi:10.17226/18290

National Science Board. (2015). *Revisiting the STEM workforce*. National Science Foundation.

National Science Foundation. (2017). *NSF's 10 big ideas*. https://www.nsf.gov/news/special_reports/big_ideas/

National Society of Professional Engineers. (2022). *Peer Review Legislation*. Retrieved from: https://www.nspe.org/resources/issues-and-advocacy/professional-policies-and-position-statements/peer-review-legislation

Nawawi, N., & Dafrita, I. E. (2022). STEM Teaching Materials Integrated with Arduino Science Journal For Biology Prospective Teachers. *Jurnal Pendidikan Sains Universitas Muhammadiyah Semarang*, *10*(1), 36–44. doi:10.26714/jps.10.1.2022.36-44

Ng, O.-L., Shi, L., & Ting, F. (2020). Exploring differences in primary students' geometry learning outcomes in two technology-enhanced environments: Dynamic geometry and 3D printing. *International Journal of STEM Education*, *7*(1), 50. doi:10.118640594-020-00244-1

NGSS Lead States. (2013). *Next Generation Science Standards. For States, By States*. The National Academies Press.

NGSS Lead States. (2013). *Next Generation Science Standards: For states, by states*. National Academies Press.

NGSS Lead States. (2013). *Next generation science standards: For states, by states*. The National Academies Press.

NGSS Lead States. (2013). *Next Generation Science Standards: For States, By States*. The National Academies Press.

NGSS Lead States. (2013). *Next Generation Science Standards: For States, By States*. Washington, DC: The National Academies Press. Retrieved from: https://www.nextgenscience.org/

Noman, M., & Amrita, K. (2014). Differentiated assessment: a new paradigm in assessment practices for diverse learners. *International Journal of Education and Applied Sciences*, 167-174.

Norris, C. M., Morris, J. E., & Lummis, G. W. (2018). Preservice teachers' self-efficacy to teach primary science based on 'science learner' typology. *International Journal of Science Education*, *40*(18), 2292–2308. doi:10.1080/09500693.2018.1528645

Nugent, G., Barker, B., Lester, H., Grandgenett, N., & Valentine, D. (2019). Wearable textiles to support student STEM learning and attitudes. *Journal of Science Education and Technology*, *28*(5), 470–479. doi:10.100710956-019-09779-7

Nusbaum, E. A. (2013). Vulnerable to exclusion: The place for segregated education within conceptions of inclusion. *International Journal of Inclusive Education*, *17*(12), 1295–1311. doi:10.1080/13603116.2013.826292

O'Brien, D. (2021). A guide for incorporating e-teaching of physics in a post-COVID world. *American Journal of Physics*, *89*(4), 403–412. doi:10.1119/10.0002437

Ohler, J. (2013). *Digital storytelling in the classroom: New media pathways to literacy, learning, and creativity*. Corwin Press. doi:10.4135/9781452277479

Oladokun, V., Adebanjo, A., & Charles-Owaba, O. (2008). Predicting students' academic performance using artificial neural network: A case study of an engineering course. *Pac J Sci Tech*, 72-9.

Oyarzun, B., Barreto, D., & Conklin, S. (2018). Instructor Social Presence Effects on Learner Social Presence, Achievement, and Satisfaction. *TechTrends*, *62*(6), 625–634. doi:10.100711528-018-0299-0

Page, M. J., McKenzie, J. E., Bossuyt, P. M., Boutron, I., Hoffmann, T. C., Mulrow, C. D., Shamseer, L., Tetzlaff, J. M., Akl, E. A., Brennan, S. E., Chou, R., Glanville, J., Grimshaw, J. M., Hróbjartsson, A., Lalu, M. M., Li, T., Loder, E. W., Mayo-Wilson, E., McDonald, S., ... Moher, D. (2021). The PRISMA 2020 statement: An updated guideline for reporting systematic reviews. *Systematic Reviews*, *10*(1), 89. Advance online publication. doi:10.118613643-021-01626-4 PMID:33781348

Panda, S. C. (2006). Medicine: Science or art? *Mens Sana Monographs*, *4*(1), 127–138. doi:10.4103/0973-1229.27610 PMID:22013337

Pang, V. O., Alvarado, J. L., Preciado, J. R., & Schleicher, A. R. (2021). Culturally relevant education: Think local within a holistic orientation. *Multicultural Perspectives*, *23*(1), 3–16. doi:10.1080/15210960.2021.1877546

Paris, S., & Turner, J. (1994). *Student motivation, cognition, and learning: Essays in honor of Wilbert J. McKeachie*. Academic Press.

Park, Y. S., & Green, J. (2019). Bringing computational thinking into science education. *Korea Science*, *40*(4), 340–352. doi:10.5467/JKESS.2019.40.4.340

Patrick, F. (2019). Second life and virtual learning: An educational alternative for neurodiverse students in college. *College Student Journal*, 33–41.

Paulus, M., Villegas, S. G., & Howze-Owens, H. (2020). Professional learning communities: Bridging the technology integration gap through effective professional development. *Peabody Journal of Education*, *95*(2), 193–202. doi:10.1080/0161956X.2020.1745610

Penuel, W. R., & Reiser, B. J. (2018). Designing NGSS-aligned curriculum materials. *Committee to Revise America's Lab Report*, 1-51.

Penuel, W. R., & Gallagher, D. J. (2017). *Creating research-practice partnerships in education*. Harvard Education Press.

Penuel, W., Fishman, B., Yamaguchi, R., & Gallagher, L. (2007). What makes professional development effective? Strategies that foster curriculum implementation. *American Educational Research Journal*, *44*(4), 921–995. doi:10.3102/0002831207308221

Peppler, K., & Glosson, D. (2013). Stitching circuits: Learning about circuitry through e-textile materials. *Journal of Science Education and Technology*, *22*(5), 751–763. doi:10.100710956-012-9428-2

Perdana, R., Apriani, A. N., Richardo, R., Rochaendi, E., & Kusuma, C. (2021). Elementary students' attitudes towards STEM and 21st-century skills. *International Journal of Evaluation and Research in Education*, *10*(3), 1080–1088. doi:10.11591/ijere.v10i3.21389

Perignat, E., & Katz-Buonincontro, J. (2019). STEAM in practice and research: An integrative literature review. *Thinking Skills and Creativity*, *31*, 31–43. doi:10.1016/j.tsc.2018.10.002

Peters-Burton, E. E., Cleary, T. J., & Kitsantas, A. (2018). Computational thinking in the context of science and engineering practices: A self-regulated learning approach. In D. Sampson, D. Ifenthaler, J.M. Spector & P. Isaias (Eds.), Digital technologies: Sustainable innovations for improving teaching and learning (pp. 223-240). Springer. doi:10.1007/978-3-319-73417-0_13

Peters-Burton, E. E., & Kysar-Mattietti, G. (2017). Building student self-awareness of learning to enhance diversity in the sciences. In D. D. Liston & R. Rahimi (Eds.), *Promoting social justice through the scholarship of teaching and learning* (pp. 270–296). Indiana University Press. doi:10.2307/j.ctt2005zc2.18

Peters, E. E., & Kitsantas, A. (2010). The effect of nature of science metacognitive prompts on science students' content and nature of science knowledge, metacognition, and self-regulatory efficacy. *Journal of School Science and Math*, *110*(8), 382–396. doi:10.1111/j.1949-8594.2010.00050.x

Pew Research Center. (2019). *Mobile connectivity in emerging economies*. https://www.pewresearch.org/internet/2019/03/07/mobile-connectivity-in-emerging-economies/

Pew Research Center. (2020). *Parenting children in the age of screens.* https://www.pewresearch.org/internet/2020/07/28/parenting-children-in-the-age-of-screens/

Pew Research Center. (2021). *Mobile fact sheet.* https://www.pewresearch.org/internet/fact-sheet/mobile/

Phillips, M. (2017). Processes of practice and identity shaping teachers' TPACK enactment in a community of practice. *Education and Information Technologies*, *22*(4), 1771–1796. doi:10.100710639-016-9512-y

Pierson, A. E., Clark, D. B., & Brady, C. E. (2021). Scientific modeling and translanguaging: A multilingual and multimodal approach to support science learning and engagement. *Science Education*, *105*(4), 776–813. doi:10.1002ce.21622

Pinou, T., Pitzrick, M. S., Rahaman, A., & Young, M. A. (2021). Letter to the editor. *Journal of Microbiology & Biology Education*, *22*(1), e00096-21. Advance online publication. doi:10.1128/jmbe.v22i1.2199 PMID:33953820

Pires, F., Masanet, M. J., & Scolari, C. A. (2021). What are teens doing with YouTube? Practices, uses and metaphors of the most popular audio-visual platform. *Information Communication and Society*, *24*(9), 1175–1191. doi:10.1080/1369118X.2019.1672766

Ponners, P., & Asim, S. (2016). Beyond the classroom walls: Technology infusion advancing science education. *The Delta Kappa Gamma Bulletin*, *83*(1), 61–67. https://www.proquest.com/docview/1822382441

Power, M. (2008). The emergence of a blended online learning environment. *Journal of Online Learning and Teaching*, *4*(4), 503–514.

Pritchard, A. (2018). *Ways of Learning: Learning Theories for the Classroom* (4th ed.). Routledge.

Psycharis, S. (2020). Editorial: A new era with STEM education? *Hellenic Journal of STEM Education*, *1*(2), 43–44. doi:10.51724/hjstemed.v1i2.14

Psycharis, S., & Kalovrektis, K. (2022, March). Assessment and integrated STEAM in engineering education. In *2022 IEEE Global Engineering Education Conference (EDUCON)* (pp. 695-703). IEEE. 10.1109/EDUCON52537.2022.9766654

Pugach, M. C., Matewos, A. M., & Gomez-Najarro, J. (2021). Disability and the meaning of social justice in teacher education research: A precarious guest at the table? *Journal of Teacher Education*, *72*(2), 237–250. doi:10.1177/0022487120929623

Putri, A. A. (2021). The effectiveness of Khan Academy as a science learning support to improve student's mastery of skills: Literature review. *Journal of Environmental and Science Education*, *1*(2), 52–56. doi:10.15294/jese.v1i2.50370

Qiu, K., Buechley, L., Baafi, E., & Dubow, W. (2013). A curriculum for teaching computer science through computational textiles. In *Proceedings of the 12th International Conference on Interaction Design and Children* (pp. 20-27). ACM. 10.1145/2485760.2485787

Rainey, K., Dancy, M., Mickelson, R., Stearns, E., & Moller, S. (2019). A descriptive study of race and gender differences in how instructional style and perceived professor care influence decisions to major in STEM. *International Journal of STEM Education*, *1*(6), 6. doi:10.118640594-019-0159-2

Raza, S. H., Sharma, B. N., & Chaudhary, K. (2021). A new pair of watchful eyes for students in online courses. *Frontiers in Applied Mathematics and Statistics*, *7*, 620080. doi:10.3389/fams.2021.620080

Reddish, E. F., Saul, J. M., & Steinberg, R. N. (1997). On the effectiveness of active-engagement microcomputer-based laboratories. *American Journal of Physics*, *65*(1), 45–54. doi:10.1119/1.18498

Reeves, P., Ng, S. L., Harris, M., & Phelan, S. K. (2022). The exclusionary effects of inclusion today:(re) production of disability in inclusive education settings. *Disability & Society*, *37*(4), 612–637. doi:10.1080/09687599.2020.1828042

Reid, D. K., & Knight, M. G. (2006). Disability justifies exclusion of minority students: A critical history grounded in disability studies. *Educational Researcher*, *35*(6), 18–23. doi:10.3102/0013189X035006018

Reiser, R. A. (2012). What field did you say you were in? Defining and naming our field. In R. A. Reiser & J. V. Dempsey (Eds.), *Trends and Issues in Instructional Design and Technology* (3rd ed., pp. 1–7). Pearson.

Resnick, I., Kastens, K. A., & Shipley, T. F. (2018). How students reason about visualizations from large professionally collected data sets: A study of students approaching the threshold of data proficiency. *Journal of Geoscience Education*, *66*(1), 55–76. doi:10.1080/10899995.2018.1411724

Richardson, J. C., Maeda, Y., Lv, J., & Caskurlu, S. (2017). Social presence in relation to students' satisfaction and learning in the online environment: A meta-analysis. *Computers in Human Behavior*, *71*, 402–417. doi:10.1016/j.chb.2017.02.001

Ritzhaupt, A. D., Rehman, M., Wilson, M. L., & Ruggles, K. (2022). Exploring the factors associated with undergraduate students' online learning anxiety: Development of the online learner anxiety scale. *Online Learning, 26*(3), 111–131. https://doi-org.libproxy.calbaptist.edu/10.24059/olj.v26i3.2810

Roche, J., Bell, L., Galvão, C., Golumbic, Y. N., Kloetzer, L., Knoben, N., Laakso, M., Lorke, J., Mannion, G., Massetti, L., Mauchline, A., Pata, K., Ruck, A., Taraba, P., & Winter, S. (2020). Citizen science, education, and learning: Challenges and opportunities. *Frontiers in Sociology*, *5*, 613–814. doi:10.3389/fsoc.2020.613814 PMID:33869532

Roediger, H. L. III, & Butler, A. C. (2011). The critical role of retrieval practice in long-term retention. *Trends in Cognitive Sciences*, *15*(1), 20–27.

Roessingh, H., & Chambers, W. (2011). Project-based learning and pedagogy in teacher preparation: Staking out the theoretical mid-ground. *International Journal on Teaching and Learning in Higher Education*, *23*(1), 60–71.

Rotherham, A. J., & Willingham, D. (2009). 21st century skills: The challenges ahead. *Educational Leadership*, *67*(1), 16–21.

Rotherham, A. J., & Willingham, D. (2009). 21st century. *Educational Leadership*, *67*(1), 16–21. PMID:19670694

Rubacha, M., Rattan, A., & Hosselet, S. (2011). A review of electronic laboratory notebooks available in the market today. *Journal of Laboratory Automation*, *16*(1), 90–98. doi:10.1016/j.jala.2009.01.002 PMID:21609689

Sadler, T. D., Amirshokoohi, A., Kazempour, M., & Allspaw, K. M. (2006). Socioscience and ethics in science classrooms: Teacher perspectives and strategies. *Journal of Research in Science Teaching*, *43*(4), 353–376. doi:10.1002/tea.20142

Sadoski, M., & Paivio, A. (2013). *Imagery and text: A dual coding theory of reading and writing*. Routledge. doi:10.4324/9780203801932

Salar, R. (2021). An arduino-based experiment designed to investigate gas pressure. *The Physics Teacher*, *59*(3), 202–204. doi:10.1119/10.0003665

Salend, S. J. (1998). Using an activities-based approach to teach science to students with disabilities. *Intervention in School and Clinic*, *34*(2), 67–72. doi:10.1177/105345129803400201

Sancho-Gil, J. J. (2020). Digital technology as a trigger for learning promises and realities. *Digital Education Review*, *37*, 195–207.

Sarıçam, U. (2020). 5. sınıf açık kaynak kodlu yazılımlar içeren stem aktivite planları [5th grade- open source software integrated STEM activities]. In M. Yıldırım & G.Y. Uz (Eds.), Açık kaynak kodlu yazılımlar içeren STEM aktiviteleri: Ortaokul Fen Bilimleri Öğretim Programı Kazanımlarına Uygun [Open source software integrated STEM activities for middle school science curriculum objectives] (pp. 15-32). Anı yayıncılık.

Sarı, U., Çelik, H., Pektaş, H. M., & Yalçın, S. (2022). Effects of STEM-focused Arduino practical activities on problem-solving and entrepreneurship skills. *Australasian Journal of Educational Technology*, *38*(3), 135–149. doi:10.14742/ajet.7293

Sat, M., & Cagiltay, K. (2022). The design of educational wearable technologies with teachers: Issues, challenges, and suggestions. *2022 International Conference on Advanced Learning Technologies (ICALT)*, 201-205. 10.1109/ICALT55010.2022.00066

Savec, V. F. (2020). The opportunities and challenges for ICT in science education. *Teknologia kemian opetuksessa, 1*(1), 1-1.

Schibeci, R. A., & Hickey, R. (2000). Is it natural or processed? Elementary school teachers and conceptions about materials. *Journal of Research in Science Teaching: The Official Journal of the National Association for Research in Science Teaching*, *37*(10), 1154–1170. doi:10.1002/1098-2736(200012)37:10<1154::AID-TEA7>3.0.CO;2-6

Schmidt-Crawford, D. A., Lindstrom, D., & Thompson, A. D. (2018). Addressing the "why" for integrating technology in teacher preparation. *Journal of Digital Learning in Teacher Education*, *34*(3), 132–133. doi:10.1080/21532974.2018.1465783

Schraw, G., Crippen, K. J., & Hartley, K. (2006). Promoting self-regulation in science education: Metacognition as part of a broader perspective on learning. *Research in Science Education*, *36*(1-2), 111–139. doi:10.100711165-005-3917-8

Schriver, M., & Czerniak, C. M. (1999). A comparison of middle and junior high science teachers' levels of efficacy and knowledge of developmentally appropriate curriculum and instruction. *Journal of Science Teacher Education*, *10*(1), 21–42. doi:10.1023/A:1009472629345

Schulze, S., & Bosman, A. (2018). Learning style preferences and mathematics achievement of secondary school learners. *South African Journal of Education*, *38*(1), 1–8. doi:10.15700aje.v37n1a1276

Schwortz, A. C. (2021). *Physics and Astronomy Mentoring and Professional Development: Equity and Empowerment Across K-20 STEM Spaces* (Publication No. 2621664057) [Doctoral dissertation, University of Wyoming]. ProQuest Dissertations Publishing. https://www.proquest.com/docview/2621664057

Schwortz, A. C., & Burrows, A. C. (2021). Authentic science experiences with STEM datasets: Post-secondary results and potential gender influences. *Research in Science & Technological Education*, *39*(3), 347–367. doi:10.1080/02635143.2020.1761783

Science Interactive. (2022). *Digital Lessons Hands On Experiments*. Retrieved from Science Interactive: https://www.scienceinteractive.com

Semali, L. M., Hristova, A., & Owiny, S. A. (2015). Integrating Ubunifu, informal science, and community innovations in science classrooms in East Africa. *Cultural Studies of Science Education*, *10*(4), 865–889. doi:10.100711422-014-9640-x

Semken, S. (2005). Sense of place and place-based introductory geoscience teaching for American Indian and Alaska Native undergraduates. *Journal of Geoscience Education*, *53*(2), 149–157. doi:10.5408/1089-9995-53.2.149

Semken, S., Ward, E. G., Moosavi, S., & Chinn, P. W. U. (2017). Placed-based education in geoscience: Theory, research, practice, and assessment. *Journal of Geoscience Education*, *65*(4), 542–562. doi:10.5408/17-276.1

Semmens, K., Fosbenner, L., Maxfield, K., Carr, R. H., Ramage, J., O'Leary, J., Larouche, C., Hayward, J., & Henderson, M. (2020). Teaching Earth systems connections to middle school students in an informal learning environment with Science on a Sphere®, personal relevance and arts-based activities. *Journal of Geoscience Education*, *69*(1), 71–84. doi:10.1080/10899995.2020.1829396

Seneviratne, K., Hamid, J. A., Khatibi, A., Azam, F., & Sudasinghe, S. (2019). Multi-faceted professional development designs for science teachers' self-efficacy for inquiry-based teaching: A critical review. *Universal Journal of Educational Research*, *7*(7), 1595–1611. doi:10.13189/ujer.2019.070714

Senger, K., Betlem, P., Birchall, T., Buckley, S. J., Coakley, B., Eide, C. H., Flaig, P. P., Forien, M., Galland, O., Gonzaga, L. Jr, Jensen, M., Kurz, T., Lecomte, I., Mair, K., Malm, R. H., Mulrooney, M., Naumann, N., Nordmo, I., Nolde, N., ... Smyrak-Sikora, A. (2020). Using digital outcrops to make the high Arctic more accessible through the Svalbox database. *Journal of Geoscience Education*, *69*(2), 123–137. doi:10.1080/10899995.2020.1813865

Shafie, H., Majid, F. A., & Ismail, I. S. (2019). Technological pedagogical content knowledge (TPACK) in teaching 21st century skills in the 21st century classroom. *Asian Journal of University Education*, *15*(3), 24–33. doi:10.24191/ajue.v15i3.7818

Shahali, E. H. M., Halim, L., Rasul, M. S., Osman, K., & Zulkifeli, M. A. (2016). STEM learning through engineering design: Impact on middle secondary students' interest towards STEM. *Eurasia Journal of Mathematics, Science and Technology Education*, *13*(5), 1189–1211.

Shea, M. V., & Sandoval, J. (2020). Using historical and political understanding to design for equity in science education. *Science Education*, *104*(1), 27–49. doi:10.1002ce.21555

Sheffield, R., Blackley, S., & Moro, P. (2018). A professional learning model supporting teachers to integrate digital technologies. *Issues in Educational Research*, *28*, 487–510. http://www.iier.org.au/iier28/sheffield.pdf

Shirky, C. (2010). *Cognitive Surplus: How Technology Makes Consumers into Collaborators*. Penguin.

Shulman, L. S. (1986). Those who understand: Knowledge growth in teaching. *Educational Researcher*, *15*(2), 4–14. doi:10.3102/0013189X015002004

Shute, V. J., Sun, C., & Asbell-Clarke, J. (2017). Demystifying computational thinking. *Educational Research Review*, *22*, 142–158. doi:10.1016/j.edurev.2017.09.003

Siemens, G. (2005). Connectivism: A learning theory for the digital age. *International Journal of Instructional Technology & Distance Learning*, *2*(1), 3–10.

Silva, E. (2009). Measuring skills for 21st-century learning. *Phi Delta Kappan*, *90*(9), 630–634. doi:10.1177/003172170909000905

Siverling, E. A., Suazo-Flores, E., Mathis, C. A., & Moore, T. J. (2019). Students' use of STEM content in design justifications during engineering design-based STEM integration. *School Science and Mathematics*, *119*(8), 457–474. doi:10.1111sm.12373

Smith, A. (2018). Greek optics. In A. Jones & L. Taub (Eds.), *The Cambridge history of science* (pp. 413–427). Cambridge University Press. doi:10.1017/9780511980145.023

Sobel, D. (2004). *Place-Based Education: Connecting Classrooms and Communities*. Orion Society.

Song, M. (2017). Teaching integrated STEM in Korea: Structure of teacher competence. *LUMAT-B: International Journal on Math. Science and Technology Education*, *2*(4), 61–72.

Speer, J. E., Lyon, M., & Johnson, J. (2021). Gains and losses in virtual mentorship: A descriptive case study of undergraduate mentees and graduate mentors in stem research during the covid-19 pandemic. *CBE Life Sciences Education*, *20*(2), ar14. doi:10.1187/cbe.20-06-0128 PMID:33734867

Spuck, T. (2014). Putting the "authenticity" into science learning. In T. Spuck, L. Jenkins, T. Rust, & R. Dou (Eds.), Best practices in STEM education: Innovative approaches from Einstein Fellow alumni (Vol. 27, pp. 107-139). Peter Lang.

Srivastava, S., Rao, D. S., & Nandakumar, H. (2016). Novel optical gyroscope: Proof of principle demonstration and future scope. *Scientific Reports*, *6*(34634). Advance online publication. doi:10.1038rep34634 PMID:27694987

Stapleton, S. R. (2021). Teacher participatory action research (TPAR): A methodological framework for political teacher research. *Action Research*, *19*(2), 161–178. doi:10.1177/1476750317751033

Starkey, L. (2010). Teachers' pedagogical reasoning and action in the digital age. *Teachers and Teaching*, *16*(2), 233–244. doi:10.1080/13540600903478433

Stasiewicz, C. G. (2022). Interdisciplinary Education Equips People to Face Unique Challenges. *Journal of National Collegiate Honors Council*, 89-91.

Statista. (2022). *Number of smartphone subscriptions worldwide from 2016 to 2027*. https://www.statista.com/statistics/330695/number-of-smartphone-users-worldwide/

Stehle, S. M., & Peters-Burton, E. E. (2019). Developing student 21 st Century skills in selected exemplary inclusive STEM high schools. *International Journal of STEM Education*, *6*(1), 1–15. doi:10.118640594-019-0192-1

Strubbe, L., & McKagan, S. (2020, 25 March). *I suddenly have to move my lab course online! What should I do?* PhysPort. https://www.physport.org/recommendations/Entry.cfm?ID=119927

Struck, W., & Yerrick, R. (2010). The effect of data acquisition-probeware and digital video analysis on accurate graphical representation of kinetics in a high school physics class. *Journal of Science Education and Technology*, *19*(2), 199–211. doi:10.100710956-009-9194-y

Struyf, A., De Loof, H., Boeve-de Pauw, J., & Van Petegem, P. (2019). Students' engagement in different STEM learning environments: Integrated STEM education as promising practice? *International Journal of Science Education*, *41*(10), 1387–1407. doi:10.1080/09500693.2019.1607983

Suárez, E. (2020). "Estoy Explorando Science": Emergent bilingual students problematizing electrical phenomena through translanguaging. *Science Education*, *104*(5), 791–826. doi:10.1002ce.21588

Sullivan, A. L., & Thorius, K. A. K. (2010). Considering intersections of difference among students identified as disabled and expanding conceptualizations of multicultural education. *Race, Gender, & Class*, •••, 93–109.

Sung, E., & Mayer, R. (2012). Computers in Human Behavior Five facets of social presence in online distance education. *Computers in Human Behavior*, 1783–1747.

Svihla, V., & Reeve, R. (Eds.). (2016). *Design as scholarship: Case studies from the learning sciences*. Routledge. doi:10.4324/9781315709550

Tal, T., & Dallashe, S. (2021). Engaging the hands, heads, and hearts in a medical simulation informal learning environment. *Journal of Research in Science Teaching*, *58*(6), 759–789. doi:10.1002/tea.21677

Tam, A. C. F. (2015). Exploring teachers' beliefs about teacher learning in professional learning communities and their influence on collegial activities in two departments. *Compare: A Journal of Comparative Education*, *45*(3), 422–444. doi:10.1080/03057925.2013.872025

Tanak, A. (2020). Designing TPACK-based course for preparing student teachers to teach science with technological pedagogical content knowledge. *Kasetsart Journal of Social Sciences*, *41*(1), 53–59.

Tanguma, J., Underwood, D., & Mayo, N. (2004). A validation study of the technology and teaching efficacy scale. *Southeast Educational Research Association Annual Conference*, Dallas, TX.

Tank, K. M., Moore, T. J., Pettis, C., & Gajdzik, E. (2017). *PictureSTEM designing paper baskets.* PictureSTEM @ 2017 Purdue University Research Foundation. http://picturestem.org/wp-content/uploads/2017/07/PictureSTEM-Designing-Paper-Baskets_July-2017.pdf

Teach Engineering. (n.d.). *Engineering design process.* University of Colorado Boulder. Retrieved from: https://www.teachengineering.org/k12engineering/designprocess

The Association for Computing Machinery, Code.org, Computer Science Teachers Association, Cyber Innovation Center, & National Math and Science Initiative. (2016). *K-12 computer science framework.* http://www.k12cs.org

The Cornell Lab of Ornithology. (n.d.). *Young birder blogs.* eBird. Retrieved from https://ebird.org/about/resources/for-young-birders/young-birder-blogs

Thomas, G. (2013). A review of thinking and research about inclusive education policy, with suggestions for a new kind of inclusive thinking. *British Educational Research Journal*, *39*(3), 473–490. doi:10.1080/01411926.2011.652070

Thomas, J. W. (2000). *A review of research on project-based learning.* Autodesk Foundation.

Thompson, A. E. & Sorbet, S. (2020). The impact of self-efficacy on leveraging technology in the classroom. *Leveraging Technology to Improve School Safety and Student Wellbeing*, 244-261. doi:10.4018/978-1-7998-1766-6

Thornton, R. K. (1987). Tools for scientific thinking-microcomputer-based laboratories for physics teaching. *Physics Education*, *22*(4), 230–238. doi:10.1088/0031-9120/22/4/005

Thornton, R. K., & Sokoloff, D. R. (1990). Learning motion concepts using real-time microcomputer-based laboratory tools. *American Journal of Physics*, *58*(9), 858–867. doi:10.1119/1.16350

Thuy, N. T. T., Bien, N. V., & Quy, D. X. (2020). Fostering teachers' competence of the integrated STEM education. *Jurnal Penelitian dan Pembelajaran IPA*, *2*(6), 116.

Tosun, C. (2022). Bibliometric and Content Analyses of Articles Related to Science Education for Special Education Students. *International Journal of Disability Development and Education*, *69*(1), 1–18. doi:10.1080/1034912X.2021.2016659

Tour, E. (2017). Teachers' self-initiated professional learning through personal learning networks. *Technology, Pedagogy and Education*, *26*(2), 179–192. doi:10.1080/1475939X.2016.1196236

Trautmann, N. M., & MaKinster, J. G. (2010). Flexibly adaptive professional development in support of teaching science with geospatial technology. *Journal of Science Teacher Education*, *21*(3), 351–370. doi:10.100710972-009-9181-4

Traxler, J. (2005). Mobile learning: It's here, but what is it. *Interaction*, *9*(1), 1–12.

Trevathan, M., & Willis, J. (2019, March). Teacher education, experience, training, and course factors in STEM implementation: A look at engineering. In *Society for Information Technology & Teacher Education International Conference* (pp. 2070-2075). Association for the Advancement of Computing in Education (AACE).

Tricot, A., & Sweller, J. (2014). Domain-specific knowledge and why teaching generic skills does not work. *Educational Psychology Review*, *26*, 265–283.

Trowbridge, D. E., & McDermott, L. C. (1980). Investigation of student understanding of the concept of velocity in one dimension. *American Journal of Physics*, *48*(12), 1020–1028. doi:10.1119/1.12298

Trundle, K. C., Atwood, R. K., & Christopher, J. E. (2002). Preservice elementary teachers' conceptions of moon phases before and after instruction. *Journal of Research in Science Teaching: The Official Journal of the National Association for Research in Science Teaching*, *39*(7), 633–658. doi:10.1002/tea.10039

Trust, T. (2012). Professional learning networks designed for teacher learning. *Journal of Digital Learning in Teacher Education*, *28*(4), 133–138. doi:10.1080/21532974.2012.10784693

Trust, T. (2016). New model of teacher learning in an online network. *Journal of Research on Technology in Education*, *48*(4), 290–305. doi:10.1080/15391523.2016.1215169

Trust, T., & Prestridge, S. (2021). The interplay of five elements of influence on educators' PLN actions. *Teaching and Teacher Education*, *97*, 97. doi:10.1016/j.tate.2020.103195

Tschannen-Moran, M., & Hoy, A. W. (2007). The differential antecedents of self-efficacy beliefs of novice and experienced teachers. *Teaching and Teacher Education*, *23*(6), 944–956. doi:10.1016/j.tate.2006.05.003

Tseng, A. S. (2017). *Reimagining the role of technology in education: 2017 national education technology plan update.* Office of Educational Technology.

Tsybulsky, D. (2019). Transformations and emerging implementations of scientific practices in the digital age. *Proceedings of the International History, Philosophy, and Science Teaching Conference*, 532-541.

Tsybulsky, D., & Levin, I. (2016). SAMR framework for study technology integration in science education. New Perspectives in Science Education Conference Proceedings of the 5th Edition. Florence, Italy: Libreriauniversitaria.it.

Tsybulsky, D. (2020). Digital curation for promoting personalized learning: A study of secondary-school science students' learning experiences. *Journal of Research on Technology in Education*, *52*(3), 429–440. doi:10.1080/15391523.2020.1728447

Tsybulsky, D., & Levin, I. (2017). Inquiry-based science education and the digital research triad. In *Digital Tools and Solutions for Inquiry-Based STEM Learning* (pp. 140–165). IGI Global. doi:10.4018/978-1-5225-2525-7.ch006

Tsybulsky, D., & Levin, I. (2019). Science teachers' worldviews in the age of the digital revolution: Structural and content analysis. *Teaching and Teacher Education*, *86*, 102921. doi:10.1016/j.tate.2019.102921

Turiman, P., Omar, J., Daud, A. M., & Osman, K. (2012). Fostering 21st century skills through scientific literacy and science process skills. *Procedia: Social and Behavioral Sciences*, *59*, 110–116. doi:10.1016/j.sbspro.2012.09.253

Turkle, S. (2017). *Alone together: Why We Expect More from Technology and Less from Each Other*. Basic Books.

Vakil, S. (2020). "I've Always Been Scared That Someday I'm Going to Sell Out": Exploring the relationship between political identity and learning in computer science education. *Cognition and Instruction*, *38*(2), 87–115. doi:10.1080/07370008.2020.1730374

Vakil, S., & McKinney de Royston, M. (2022). Youth as philosophers of technology. *Mind, Culture, and Activity*, 1–20. doi:10.1080/10749039.2022.2066134

Valtonen, T., Sointu, E., Kukkonen, J., Kontkanen, S., Lambert, M. C., & Mäkitalo-Siegl, K. (2017). TPACK updated to measure pre-service teachers' twenty-first century skills. *Australasian Journal of Educational Technology*, *33*(3). Advance online publication. doi:10.14742/ajet.3518

Vernon, A. (1999). The dialectics of multiple identities and the disabled people's movement. *Disability & Society*, *14*(3), 385–398. doi:10.1080/09687599926217

Vieyra, R., & Vieyra, C. (2022). Immersive learning experiences in augmented reality (AR): Visualizing and interacting with magnetic fields. In S. Macrine & J. Fugate (Eds.), Movement matters: How embodied cognition informs teaching and learning. MIT Press.

Vieyra, R. (2018). Teaching students metacognition through discipline-based research and technology. In T. Spuck, J. Leigh, T. Rust, & R. Dou (Eds.), *Best practices in STEM education: Innovative approaches from Einstein Fellow alumni* (2nd ed., pp. 503–528). Peter Lang, Inc.

Vieyra, R., & Vieyra, C. (2014). Analyzing forces on amusement park rides with mobile devices. *The Physics Teacher*, *52*(3), 149–151. doi:10.1119/1.4865516

Vieyra, R., Vieyra, C., Marti, A., Monteiro, M., & Jeanjacquot, P. (2015). Turn your smartphone into a science laboratory. *Science Teacher (Normal, Ill.)*, *82*(9), 32–40. doi:10.2505/4/tst15_082_09_32

Vieyra, R., Vieyra, C., Pendrill, A.-M., & Xu, B. (2020). Gamified physics challenges for teachers and the public. *Physics Education*, *55*(4), 1–7. doi:10.1088/1361-6552/ab8779

Villa, R., & Thousand, J. (2016). The Inclusive Education Checklist: A Self-Assessment of Best Practices. National Professional Resources, Inc./Dude Publishing.

Villanueva, M. G., Taylor, J., Therrien, W., & Hand, B. (2012). Science education for students with special needs. *Studies in Science Education*, *48*(2), 187–215. doi:10.1080/14703297.2012.737117

Vinluan, M. V. (2021). *(Holks) Home-Lab Kits Utilizing Arduino Science Journal App: Enhancing scientific attitudes in a new normal grade 7 classroom of biñan secondary school of applied academics*. Retrieved from https://www.depedbinan.com/media/research/SDO-RES_001.433_V786_2021.pdf

Voogt, J., Erstad, O., Dede, C., & Mishra, P. (2013). Challenges to learning and schooling in the digital networked world of the 21st century. *Journal of Computer Assisted Learning*, *29*(5), 403–413. doi:10.1111/jcal.12029

Wackenhut, A. F., & Gillette, M. B. (2022). Covid-19 and (re)learning teaching: Never let a crisis go to waste. *Högre Utbildning*, *12*(1), 52. doi:10.23865/hu.v12.3562

Walker, R. X. (2017). *Assessing teachers' technological, pedagogical, and content knowledge in elementary schools* [Doctoral dissertation]. Wilmington University.

Walkington, C., Nathan, M. J., Hunnicutt, J., Washington, J., & Holcomb-Webb, K. (2022). Learning geometry through collaborative, embodied explorations with augmented reality holograms. In C. Chinn, E. Tan, C. Chan, & Y. Kali (Eds.) *Proceedings of the 16th International Conference of the Learning Sciences – ICLS 2022* (pp. 1992-1993). International Society of the Learning Sciences.

Wang, L., Ertmer, P. A., & Newby, T. J. (2004). Increasing preservice teachers' self-efficacy beliefs for technology integration. *Journal of Research on Technology in Education*, *36*(3), 231–250. doi:10.1080/15391523.2004.10782414

Wang, Y. M. (2006). Stand-alone computer courses in teachers' IT training. *EDUCAUSE Quarterly*, *29*(3), 8.

Ware, L. (2001). Writing, identity, and the other: Dare we do disability studies? *Journal of Teacher Education*, *52*(2), 107–123. doi:10.1177/0022487101052002003

Warr, M., Mishra, P., & Scragg, B. (2019, March). Beyond TPACK: Expanding technology and teacher education to systems and culture. In *Society for Information Technology & Teacher Education International Conference* (pp. 2558-2562). Association for the Advancement of Computing in Education (AACE).

Warren, C. A. (2021). *About centering possibility in Black education*. Teachers College Press.

Watson, J. A., & Pecchioni, L. L. (2011). Digital natives and digital media in the college classroom: Assignment design and impacts on student learning. *Educational Media International*, *48*(4), 307–320. doi:10.1080/09523987.2011.632278

Wehmeyer, M. L., Agran, M., & Hughes, C. A. (2000). National survey of teachers' promotion of self-determination and student-directed learning. *The Journal of Special Education*, *34*(2), 58–68. doi:10.1177/002246690003400201

Weinsten, Y., & Sumeracki, M. (2019). *Understanding How We Learn: A Visual Guide*. Routledge.

Weintrop, D., Beheshti, E., Horn, M., Orton, K., Jona, K., Trouille, L., & Wilensky, U. (2016). Defining computational thinking for mathematics and science classrooms. *Journal of Science Education and Technology*, *25*(1), 127–147. doi:10.100710956-015-9581-5

Wells, M., Hestenes, D., & Swackhamer, G. (1995). A modeling method for high school physics instruction. *The Physics Teacher*, *63*(7), 606–619.

Wemyss, T., & van Kampen, P. (2013). Categorization of first-year university students' interpretations of numerical linear distance-time graphs. *Physical Review Special Topics. Physics Education Research*, *9*(1), 010107. doi:10.1103/PhysRevSTPER.9.010107

Wenbo, D., Weina, L., Xipei, G., Xinyi, L., Wenshuang, G., & Weiping, H. (2022). Effects of regulatory focus on online learning engagement of high school students: The mediating role of self-efficacy and academic emotions. *Journal of Computer Assisted Learning*, 707–718.

Wen, C. T., Chang, C. J., Chang, M. H., Chiang, S. F., Liu, C. C., Hwang, F. K., & Tsai, C. C. (2018). The learning analytics of model-based learning facilitated by a problem-solving simulation game. *Instructional Science*, *46*(6), 847–867. doi:10.100711251-018-9461-5

Wen, C. T., Liu, C. C., Chang, H. Y., Chang, C. J., Chang, M. H., Chiang, S. F., ... Hwang, F. K. (2020). Students' guided inquiry with simulation and its relation to school science achievement and scientific literacy. *Computers & Education*, *149*, 103830. doi:10.1016/j.compedu.2020.103830

Whitaker, S. C. (2019). *Middle School Teachers' Perceptions of Project-based Learning as It Impacts First Year Implementation* [Doctoral dissertation]. University of South Carolina.

Wilcox, B. R., & Lewandowski, H. J. (2016). Open-ended versus guided laboratory activities: Impact on students' beliefs about experimental physics. *Physical Review. Physics Education Research*, *12*(2), 020132. doi:10.1103/PhysRevPhysEducRes.12.020132

Williams-Diehm, K. L., Miller, C. R., Sinclair, T. E., & Wronowski, M. L. (2018). Technology-based employability curriculum and culturally diverse learners with disabilities. *Journal of Special Education Technology*, *33*(3), 159–170. doi:10.1177/0162643417749933

Willingham, D. T. (2021). *Why don't Students Like School? A Cognitive Scientist Answers Questions About How the Mind Works and What it Means for the Classroom*. Jossey-Bass.

Willis, J. M. (2015). Examining technology and teaching efficacy of preservice teacher candidates: A deliberate course design model. *Current Issues in Education (Tempe, Ariz.)*, *18*(3).

Willis, J., Weiser, B., & Kirkwood, D. (2014). Bridging the gap: Meeting the needs of early childhood students by integrating technology and environmental education. *International Journal of Early Childhood Environmental Education*, *2*(1), 140–155.

Wilson, M. (2002). Six views of embodied cognition. *Psychonomic Bulletin & Review*, *9*(4), 625–636. doi:10.3758/BF03196322 PMID:12613670

Wilson, T. D., Houston, C. E., Etling, K. M., & Brekke, N. (1996). A new look at anchoring effects: Basic anchoring and its antecedents. *Journal of Experimental Psychology*, *125*(4), 387.

Wing, J. M. (2006). Computational thinking. *Communications of the ACM*, *49*(3), 33–35. doi:10.1145/1118178.1118215

Wofford, J. (2009). K-16 computationally rich science education: A ten-year review of the Journal of Science Education and Technology (1998-2008). *Journal of Science Education and Technology*, *18*(1), 29–36. doi:10.100710956-008-9127-1

Wolf, M. (2018). *Reader, Come Home: The Reading Brain in the Digital World*. Harper-Collins.

Wu, M., Pathak, P. H., & Mohapatra, P. (2015). Monitoring building door events using barometer sensor in smartphones. *Proceedings of the 2015 ACM International Joint Conference on Pervasive and Ubiquitous Computing*, 319-323. 10.1145/2750858.2804257

Xu, Q., Zheng, R., & Hranilovic, S. (2015). IDyLL: Indoor localization using inertial and light sensors on smartphones. *Proceedings of the 2015 ACM International Joint Conference on Pervasive and Ubiquitous Computing*, 307-318. 10.1145/2750858.2807540

Yamada, M. (2009). The role of social presence in learner-centered communicative language learning using synchronous computer-mediated communication: Experimental study. *Experimental Study. Computers & Education*, *52*(4), 820–833. doi:10.1016/j.compedu.2008.12.007

Yang, D., Swanson, S. R., Chittoori, B. B., & Baek, Y. (2018, June). Board 70: Work in Progress: Integrating Computational Thinking in STEM Education through a Project-based Learning Approach. *2018 ASEE Annual Conference & Exposition*. 10.18260/1-2--30091

Yoon, S., & Wang, J. (2014). Making the invisible visible in science museums through augmented reality devices. *TechTrends*, *58*(1), 49–55. doi:10.100711528-013-0720-7

Yu, K., Wu, P., & Fan, S. (2019). Structural relationships among high school students' scientific knowledge, critical thinking, engineering design process, and design product. *International Journal of Science and Mathematics Education*. Advance online publication. doi:10.100710763-019-10007-2

Zarske, M. S., Cooper, L., & Carlson, D. (2009). Design step 2: Research the problem. In *Teach Engineering*. Regents of the University of Colorado. Retrieved from: https://www.teachengineering.org/activities/view/cub_creative_activity2

Zarzar, L., Kalow, J., Xinping, H., Walish, J., & Swager, T. (2017). Optical visualization and quantification of enzyme activity using dynamic droplet lenses. *Proceedings of the National Academy of Sciences of the United States of America*, *114*(15), 3821–3825. doi:10.1073/pnas.1618807114 PMID:28348236

Zeidler, D. L., & Nichols, B. H. (2009). Socioscientific Issues: Theory and Practice. *Journal of Elementary Science Education*, *21*(2), 49–58. doi:10.1007/BF03173684

Zhao, Y., & Bryant, F. L. (2006). Can teacher technology integration training alone lead to high levels of technology integration? A qualitative look at teachers' technology integration after state mandated technology training. *Electronic Journal for the Integration of Technology in Education*, *5*(1), 53–62.

Zimmerman, B. J. (2000). Attaining self-regulation: A social-cognitive perspective. In M. Boekaerts, P. Pintrich, & M. Zeidner (Eds.), *Handbook of self-regulation* (pp. 13–39). Academic Press. doi:10.1016/B978-012109890-2/50031-7

Zohar, A., & Dori, Y. J. (2003). Higher order thinking skills and low achieving students: Are they mutually exclusive? *Journal of the Learning Sciences*, *12*(2), 145–183. doi:10.1207/S15327809JLS1202_1

Zyngier, D. (2011). (Re)conceptualizing risk: Left numb and unengaged and lost in a no-man's lands or what (seems to) work for at-risk students. *International Journal of Inclusive Education*, *15*(2), 211–231. doi:10.1080/13603110902781427

About the Contributors

Sumreen Asim is a coffee enthusiast. Currently teaches STEM courses to both undergraduate and graduate students at Indiana University Southeast. Over the last two decades she has focused her work to share, love and grow with fellow educators. She enjoys helping others implement inquiry-based instructional approaches. Her research interests are in collaborative efforts STEM education, culturally-responsive teaching, and educational technology.

Joshua Ellis is an Associate Professor of STEM Education in the Department of Teaching and Learning and Affiliated Faculty in the School of Universal Computing, Construction, and Engineering Education at Florida International University. He serves as the President of the Southeastern Association for Science Teacher Education and the incoming Co-Editor of Contemporary Issues in Technology and Teacher Education - Science. Dr. Ellis engages in research that facilitates the development of pre-service and in-service teachers' professional practice in online, blended, and face-to-face learning environments. He explores the role of technology-mediated learning experiences for both K-12 and adult learners in STEM education environments and uses research to inform the design of new, innovative pedagogical strategies.

David Slykhuis is a nineteen year veteran in higher education with much of his career focused on educational technology. He is currently the Dean of the College of Education and Human Services at Valdosta State University. He is the Chair of the National Technology Leadership Summit, co-Chair of the American Association of Colleges of Teacher Education (AACTE) standing committee on Innovation and Technology, and a member of the Steering Committee of the EdTech Genome Project sponsored by the EdTech Evidence Exchange. David is a past President of the Society for Information Technology and Teacher Education (SITE). He is one of the four researchers who developed the Teacher Educator Technology Competencies.

Jason Trumble is an Associate Professor of Education at the University of Central Arkansas where he teaches preservice and in-service teachers to effectively integrate technology, content, and pedagogy and think critically about their practice. He is also the program coordinator for the UCA education specialist degree in digital-age teaching and learning. Dr. Trumble's research focuses on the intersection of teaching and technology as well as STEM learning, assessment practices, and emerging technologies. He is an ISTE Certified Educator. He is the President of the Arkansas Instructional Innovation Association, and he serves as the chair of the Teacher Education Council at SITE.

* * *

Curby Alexander is an Associate Professor of Professional Practice in the TCU College of Education. He teaches courses on the foundations of education, instructional methods, digital communication and collaboration, and outdoor education. His current scholarship is focused on technology implementation in large lecture-based courses, the intersection of gamification and motivation, and cybersecurity education and workforce development.

Salma Ali is a Ph.D. candidate in Curriculum and Instruction emphasizes Teaching and Technology in the College of Education and Human Development at Texas A&M University. Her research focus is on teacher education & teacher preparation programs. She specifically examines the various technology-based frameworks that help teachers (various disciplines including Social sciences/studies) to integrate emerging educational technologies with content and pedagogy in K12 and higher education settings. Salma also researches teachers' lives and teacher education and preparation in high-needs areas.

Evrim Baran is an associate professor of educational technology and human-computer education at Iowa State University. Her research examines effective strategies for the design of mobile, online, and flexible learning environments in teacher education, engineering education, and STEM learning contexts. Dr. Baran projects were funded by national and international agencies (e.g., NSF, European Commission, German Research Foundation, and the Scientific and Technological Research Council of Turkey) to implement technology solutions in the educational contexts of different countries, including the USA, Canada, Germany, and Turkey. She published numerous journal articles, book chapters, and conference proceedings on technology and teacher education, online teaching, and STEM education. Her scholarship, service, and teaching have been recognized by numerous awards at international organizations, including AERA, SITE, AECT. Dr. Baran teaches courses on instructional design, program evaluation, and online teaching/learning.

Phillip Boda's work leverages the affordances of relationship-building with students, their local communities, and teachers to explore questions around the overlapping nexuses of historically marginalized identities (i.e., race, class, gender, disability, and native language). In this way, he draws on Intersectionality, Philosophies of Liberation, and Epistemic Disobedience to challenge sustained colonialities of power in education. His work has explored how researchers analyze student-, teacher-, and school-level data that, when combined with Design-based Research methods, can change the landscape of learning environments by design to respond to those students erased from STEM, Science, and broader educational praxes that sustain marginality. Pluralizing these approaches, Dr. Boda asks: Who decides when justice is met? How do researchers shift beyond comfortable proximities seeded in their epistemic commitments? Why do oppressive colonial logics of subjugation around achievement, affect, and disciplinary hegemonies remain prevalent even among socially just inquiries in education?

Alec M. Bodzin is Professor in the Teaching, Learning, and Technology program and is a core faculty member of the Lehigh Environmental Initiative. He joined the Lehigh faculty in 1999. He received his Ph.D. in Science Education from North Carolina State University as a National Science Foundation Fellow in Instructional Technology in Science Education. Dr. Bodzin is the Principal Investigator of Socio-environmental Science Investigations (SESI) and Environmental Literacy and Inquiry (ELI), an inquiry-based secondary school curriculum that uses geospatial technologies including Web GIS to investigate environmental issues in the areas of energy, climate change, and land use change. Dr. Bodzin's research interests involve the design of immersive virtual reality learning environments, engagement and learning, learning with spatial thinking tools including GIS and Google Earth, learning design, design and implementation of inquiry-based science and environmental curriculum, learning technologies, game-based learning, and environmental literacy and education for sustainable development. He has published over seventy articles in peer-reviewed science, technology, and environmental education journals. Dr. Bodzin teaches science methods, environmental education, and curriculum and instructional design courses at Lehigh University.

Kristen Brown is a doctoral student in science education at Texas Christian University. Her research interests include relationships between universities and in-service science teachers, professional development for science educators, and integrating geospatial technology into secondary science curriculum. She serves as a graduate assistant on a collaborative NSF grant with TCU, Lehigh University, and Washington State University Tri-Cities. Kristen previously taught middle school and high school science and served in administrative roles including assistant principal, instructional coach, and curriculum director in a suburban school district in Texas. She was also an adjunct instructor for undergraduate education courses and supervised student teachers across a variety of school districts.

Andrea Burrows Borowczak is a Professor of Science Education and the Director of the School of Teacher Education at the University of Central Florida. She conducts research at the intersection of science education, engineering education, and computer science education, in partnership settings. She engages pre-service and in-service STEM teachers in real-world experiences. Dr. Burrows Borowczak utilizes quantitative, qualitative, and action research methodologies as a means of gaining greater insight into integrated STEM education. In addition, she is an expert in conflict management and professional development for K-12 teachers. She has received over 10 million in grant funding from NSF and other agencies, and she publishes regularly.

Sedef Canbazoglu Bilici is an Associate Professor at the department of science education at Gazi University, Turkey. She worked as a visiting scholar at the University of Minnesota, STEM Education Center in 2011-2012. She received her Ph.D. in science education at Gazi University, Turkey in 2012. Her dissertation research focused on pre-service teachers' TPACK and their self-efficacy beliefs toward TPACK. Her research interests include the development of technology-enriched activities for developing science teachers' TPACK, and STEM education. Dr. Canbazoglu Bilici has worked on TPACK and science teachers' competencies as a project coordinator and a researcher in projects supported by the European Union and the Scientific and Technological Research Council of Turkey.

Timothy J. Cleary (PhD) is Professor in the Graduate School of Applied and Professional Psychology (GSAPP) at Rutgers University. His primary research interests include the development and application of self-regulated learning (SRL) and motivation assessment and intervention practices across academic, athletic, medical, and clinical contexts. He has published over 60 peer-review journal articles and book chapters along with several books addressing SRL-related principles and practices. Dr. Cleary has received extensive extramural grant funding as PI and co-PI throughout his career, with the total funding over $13 million. His work has been supported by the National Science Foundation (NSF), Institute of Education Sciences (IES), Department of Education's Fund for the Improvement of Postsecondary

Education, Spencer Foundation, and others. Dr. Cleary currently teaches doctoral courses in learning theory, academic interventions, and statistics.

Daniel Dale joined UW in 2001 as a professor in the Department of Physics & Astronomy. His research specialty is multi-wavelength studies of nearby galaxies.

Efrat Dayan is a Ph.D. student in the Faculty of Education in Science and Technology at the Technion- Israel Institute of Technology. She graduated from the Weizmann Institute of Science in 2012 with an M.Sc. degree in Science Education, at the Rothschild-Weizmann Program. Efrat is the Haredi District Science Inspector in the Ministry of Education. Before that, she had been mentoring science teachers parallel to teaching biology in high school during a long period of time. Her research focuses on the digital curation in science education.

Jonah B. Firestone is an Associate Professor of Science Education, WSU. He researches change in teachers' beliefs and practices in STEM and the use of innovative technology in science learning. He brings expertise in researching change in critical thinking and cognition. His research focuses on the use of emerging technologies in Virtual Reality (VR), Augmented Reality (AR), and neurocognition in decision making, attitude formation, and problem-solving in students and under-represented population and the use of these technologies in the development of Serious Educational Games (SEG) and simulations to foster learning.

Rivka Gadot is a Post-Doctoral Fellow at the Faculty of Education in Science and Technology at the Technion. Her multidisciplinary academic background includes Computer Science, Economics, Business Administration and Education. Rivka is a faculty member of the Computer Science Department at Lev Academic Center. Her Ph.D. dissertation from Tel-Aviv University focused on social curation as a learning activity. Her main research interests include social media, social curation, and data mining.

Michelle Giles, PhD, is an Assistant Professor and Program Coordinator of Instructional Design and Technology at University of Houston-Clear Lake. Since 2001, she has instructed instructional design/technology, web design, and multimedia courses at both the graduate/undergraduate level. Her research interests include technology integration in the EC-12 classroom. Specifically, improving in-service teachers and preservice teacher candidates' technology proficiency to support effective technology integration in the classroom. She holds a doctorate degree in Educational Technology from the University of North Texas.

Thomas C. Hammond is Associate Professor of Teaching, Learning, and Technology in Lehigh University's College of Education. Prior to earning his doctorate in Instructional Technology at the University of Virginia, he taught middle and high school social studies and computer science. His research addresses teachers' use of innovative instructional technologies, particularly the integration of geospatial tools into social studies and science instruction.

Allen Hays is an instructor and director of the Delta STEM Education Center at Arkansas State University. He received his MSE in Educational Leadership through Arkansas State University and is currently working on his EdD in Educational Technology at Boise State University. His work includes teacher professional development, school and community outreach, and grant writing for school districts throughout Eastern Arkansas. Mr. Hays started at Arkansas State University in 2017 as an instructor in Teacher Education. Before starting his current position, he taught secondary science at a high school. He strives to incorporate project-based learning, making, STEM, and other areas of educational technology into all the courses he teaches. His research interests are Makerspace, STEM, artificial technology, and virtual learning.

James Hernandez is a Ph.D. candidate in the College of Education and Human Development at the University of Texas. His research interest is technology and gamification in school. He also worked for Mark Twain Dual language Academy, Dan Antonio Independent School District.

Barbara Hickman has worked in site, district and state level leadership positions including as an associate superintendent for curriculum, instruction, and human resources, and as a superintendent. She is currently an assistant professor and program coordinator for Educational Leadership at the University of Wyoming. Her research interests center on large scale systems, processes, and policies, with one area of focus on the application of constitutional law in schools and the other on the implementation of systemic initiatives, including leadership for STEM and educational technology.

Heather Howle is an experienced STEM educator with a demonstrated history in program development, project management, instructional design, curriculum development, policy, and leadership. In recent years, her work has focused on training in-service teachers in computational thinking skills and cyber education. Currently, she is an adjunct instructor at LSU-Shreveport in STEM education.

Lorraine Jacques is an Assistant Professor of Computer Science at the University of Tampa. Her background includes ten years' MS/HS teaching experience in math and computer science, curriculum development/research in computational thinking and STEAM, disciplinary literacy research/education, and a PhD in Learning Sciences. Her current research focuses on computer science education, especially with under-represented groups, and informal learning environments for computational thinking.

Mina C. Johnson-Glenberg is a Research Professor at Arizona State University in the Psychology department. She is also an entrepreneur and President of the award-winning learning technology company called Embodied Game, LLC. At her university lab and in the spinout company, she leads a team that creates, researches, and disseminates innovative virtual and mixed reality (XR) educational content for 4th graders through life-long learners. Several of the NSF-funded STEM games can be downloaded for free at . She has a varied background as a screenwriter, experimental psychologist, and game designer. Her lab has been at the forefront of creating "embodied" and kinesthetically-active games using sensors and motion capture technologies for over a decade and a half. Dr. Johnson-Glenberg presents nationally and internationally. Some recent publications include design guidelines for learning VR at Frontiers in Robotics and AI and, a rubric that teachers can use to score the quality of a VR lesson which can be found under blog/tools on the Embodied Games website.

Natalie A. Johnson-Leslie is an associate professor of secondary education at Arkansas State University (A-State). She received her doctoral degrees from Iowa State University in the areas of Educational Leadership and Policy Studies as well as Curriculum and Instructional Technology. She joined the Teacher Education faculty at A-State in the fall of 2004. She is a former high school science teacher who integrates technology in all her courses. In her current teaching roles, her classes include, integrating technology in the elementary classroom, introduction to technology, and tools for teaching technology at the graduate level. Her research interests lie in the field of educational leadership; technology integration in the curriculum; assessing and evaluating student behavior; motivating gifted and talented students, and surviving school violence. Her work including book chapters has been published in; IGI-Global, the Journal of Educational Technology Systems (JETS), International Journal of Reflection (IJR), the Journal of European Teacher Education Network (JETEN), and the International Journal of Learning (IJL). She served on the review board of the Journal of College Student Development for four years. Currently, she is an associate editor for the Journal of Technology and Teacher Education (JTATE).

Anastasia Kitsantas, Ph.D., is Professor of Educational Psychology in the College of Education and Human Development at George Mason University. Her research interests focus on the development of Self-Regulated Learning (SRL) and learner motivational beliefs across diverse areas of functioning including academic learning, athletics, and health. She has also studied the role of learning technologies in supporting student SRL. This work has resulted in more than 100 publications, many of which are directed toward the training of self-regulation.

Kevin S. Krahenbuhl is Associate Professor of Education and Program Director for the Assessment Learning and Student Success EdD Program at Middle Tennessee State University. He is author of four books, dozens of peer reviewed articles and presentations. His work focuses on application of cognition to instructional practice.

Doug Leeson is a doctoral student in Lehigh's Teaching, Learning, and Technology program.

Danielle J. Malone is a graduate research assistant in the College of Education at Washington State University, Tri-Cities. Her research interests include exploring the relationship between various social structures and student agency as well as the affordances of student agency within science classrooms and identity formation.

Rebekah S. Marsh earned her master's degree in Biology from the University of Nebraska at Omaha in 1999 and her doctorate in Biochemistry from Iowa State University in 2005. She completed two post-docs in Immunology (Kansas State University) and Human Nutrition/Biochemistry (The Ohio State University). She is a trained protein biochemist, experienced in biochemical, cell, genetic and molecular techniques. Dr. Marsh started her teaching experience while pursuing her undergraduate degree and continued through her master's and doctoral degrees. During her doctorate began volunteering and continues to do each academic year with PlantingScience – an outreach program where scientists share their passion with young scientists on student-centered research projects. She has been teaching online since 2009 and has taught a variety of college-level science courses. She thoroughly enjoys teaching and helping others in their educational journey, especially as it relates to science education.

Colleen Megowan-Romanowicz is a Senior Fellow, American Modeling Teachers Association, responsible for research and development. Specializing in research on STEM teaching and learning (particularly in the physical sciences and astronomy/astrophysics), often in the context of Modeling Instruction. special interests: modeling, computational modeling, distributed and embodied cognition, augmented and virtual learning environments, representations and inscriptions.

Judith Morrison is a professor of science education at Washington State University Tri-Cities where she researches STEM teaching and learning and the integration of technology in K-12 classrooms.

Adam Myers is an Associate Professor in the Department of Physics & Astronomy at the University of Wyoming in Laramie, Wyoming. His research focus is studying the distribution of quasars across cosmic history. Dr. Myers' specific interests include; constraining cosmology, and models of quasar activity and galaxy formation, by analyzing clustering within large surveys of quasars and galaxies; using proximate pairs of quasars and AGN to constrain models of quasar fueling and galaxy evolution; statistical lensing (cosmic magnification) as a trace of dark matter; and the development of rigorous probabilistic techniques to classify and characterize quasars in large imaging surveys.

Dan O'Brien recently completed a PhD in physics at Georgetown University. He specializes in microelectromechanical systems and biosensors, but has authored publications in soft matter physics, microscale device engineering, materials science, and physics education technology. Dan now works in the climate field, modeling the effects of energy policies on emissions, public health, and the economy. Most importantly, Dan enjoys getting out of lab to hike long trails, watch baseball, and give dogs belly rubs.

Erin Peters-Burton is the Donna R. and David E. Sterling Endowed Professor in Science Education at George Mason University and is the Director of the Center for Social Equity through Science Education. She was a public secondary science and mathematics teacher for 15 years and joined Mason's faculty in 2007. Her research interests involve interactions between teachers and students during self-regulated learning processes in science and engineering practices, particularly data practices and computational thinking. She has been PI on several National Science Foundation Grants and is an active member of the National Association for Research on Science Teaching and the Association for Science Teacher Educators, where she was awarded Outstanding Science Teacher Educator in 2016.

Kate Popejoy is a researcher in science education and technology, whose current research includes promoting spatial thinking and reasoning with GIS-integrated curricula. She received her PhD in Science Education from the University of British Columbia in 2007. Dr. Popejoy has an extensive background in science education and science teacher preparation, which included 16 years in academia. She has been an Investigator on three National Science Foundation grants, and serves as an editorial review board member for a number of professional journals. In addition to her current grant-funded research, she also is the Executive Director for the Association for Science Teacher Education (ASTE).

Peter Rich is an Associate Professor of Instructional Psychology and Technology at Brigham Young University in Provo, Utah, USA. He researches Computational Thinking in K-8, teaching elementary teachers and kids to code, play with Lego Robotics, and do fun engineering projects. He also teaches courses in instructional design and development and educational psychology.

Jim Rost is Assistant Professor of Education and Methodologist for the Assessment Learning and Student Success EdD Program at Middle Tennessee State University.

Andria Schwortz is a recent PhD graduate in Physics from the University of Wyoming, studying issues in equity and empowerment in post-secondary physics and astronomy students, and STEM teacher professional development. Dr. Schwortz is also a tenured Full Professor of Physics at Quinsigamond Community College in Worcester, MA.

Dina Tsybulsky is an Assistant Professor and Head of the Biology Education research group in the Faculty of Education in Science and Technology at the Technion Israel Institute of Technology. Her research interests include approaches to biology teaching and learning; nature of science and scientific practices; biology teacher education and professional development. Dr. Tsybulsky has served as the co-chair of the History, Philosophy, Sociology and the Nature of Science research strand and as a member of the Outstanding Doctoral Research Award Committee at the NARST Association. In Israel, Dr. Tsybulsky serves as a member of several committees and as a co-chair of the Israeli Biology Education Research Forum. Currently, she has been a member of the Taking Citizen Science to School research center, a PI of the Smart BioEd Lab, and a PI of several research projects funded by the EuroTech Universities Alliance, the Israeli Science Foundation and the Israeli Ministry of Education.

Jeremy Dean Unruh received his Doctorate of Education in Curriculum and Instruction at the University of Houston-Clear Lake with a specialization in STEM education and a focus on educational technology and instructional design in 2019. Teaching Biology courses at San Jacinto Community College since 2010 has shown him that learning is not simply a transmission process between a teacher and student but a dynamic relationship that involves the commitment of both parties. He then applied this kind of student-centered learning at the University of Houston-Clear Lake in 2021 for the doctorate learning and cognition course, teaching the interplay of biology and learning in the human mind. He has also co-founded Limitless Learning Solutions in 2008. He feels that his role as an educator has evolved from being a book of knowledge from which students can read to a pen with which students can write their book.

Rebecca Vieyra is a physics education researcher and prior high school physics teacher. Until late 2021, she managed the Inter-American Teacher Education Network, a $2.5M project of the Organization of American States to network Ministries of Education and education leaders to improve the quality of STEM teaching in the Americas. Previously, she served as the K-12 Program Manager at the American Association of Physics Teachers. In this capacity, she managed collaborative grants valued at $8.5M total from NASA, NSF, and private foundations. She collaborates with her husband, Chrystian Vieyra, to develop Vieyra Software mobile sensor apps for science education, used by 2M+ individuals worldwide.

Chrystian Vieyra is an Android and iOS app developer with a specialty in visualizing sensor data. Born in Celaya, Mexico, he immigrated to the United States to complete his B.S. in Computer Science from Western Illinois University. Chrystian resides in Washington, DC.

Molly Weinburgh is the Andrew Chair of Mathematics & Science Education and Director of the Andrews Institute of Mathematics & Science Education at Texas Christian University. Her honors include the Chancellor's Distinguished Achievement as a Creative Teachers and Scholar, Piper Professorship, Fellow of the American Association for the Advancement of Science, and 2020 Award for Liftetime Outstanding Service to ASTE. Her scholarship focuses on academic language acquisition and conceptual understanding in science by emerging multilingual students. She is the co-author of Supporting Emergent Multilingual Students using the 5R Instructional Model: The Hybrid Language of Inquiry-based Science (NSTA Press) and co-author on a chapter outlining the 5R Instructional Model in the Handbook of Educational Theories (IAP, 2013). She is also co-editor of Science Teacher Preparation in Content-Based Second Language Acquisition and Becoming and Being: Women's Experience in Leadership in K-16 Science Education Communities. She has been the PI on two NSF, one FIPSE, one CPB, 31 Eisenhower/Teacher Quality, five Texas Higher Education Coordinating Board, and nine foundation grants for a total of about $14M in external funding. She has severed as evaluator or educational consultant on four NSF grants.

Jana M. Willis has worked in STEM related areas for over 30 years. Dr. Willis has a Ph.D. from Texas A&M University in Educational Psychology with an emphasis in Educational Technology. She is a Professor in the College of Education Instructional Design and Technology program at the University of Houston-Clear Lake and Chair of the Literacy, Library, and Learning Technologies Department. In her university role she designs, develops, and implements online/hybrid courses. Research includes online course development, project/problem-based learning, K-12 technology integration, teacher development, games in education, self-efficacy, and digital storytelling. Her interest in games in education has led to involvement in the UHCL Game and Simulation Initiative, the Annual Ed Game Expo, and involvement in the recently approved BA and MS Programs in Serious Games and Simulations. Scholarly works include international, national, and regional refereed/non-refereed journal articles, edited book chapters, and professional conference presentations. Willis' service spans beyond teaching to community engagement and research, earning her the 2016 UHCL President's Distinguished Service Award. She is part of UHCL's Million Dollar Club for collaborative efforts developing interdisciplinary projects with over $3 million awarded.

Hatice Yildiz Durak is a researcher in the Department of Educational Science, Instructional Technology at the Eregli Faculty of Education, Konya Necmettin Erbakan University, Turkey. She received his MSc degree in Computer Education and Instructional Technologies from Hacettepe University in 2011. She received her PhD in Educational Science, Department of Computer Education and Instructional Technology from Gazi University, Turkey in 2016. Her dissertation research focused on teaching programming to gifted primary school students. Her research interests are related to instructional technology, programming training in K12, flipped learning, e-learning, teacher training, integration of ICT into the learning, and computer science education.

Effat Zeidan Hatoum has a background in Nanotechnology, a highly interdisciplinary field and is currently serving as a distant learning educator at California Baptist University. She is currently directing and teaching the general education courses and two programs, BA in Interdisciplinary Studies and MA in Interdisciplinary Studies in an Online environment and is serving a traditional and non-traditional student community. Her focus has been on improving the quality of online science and math education through the introduction of practical learning tools and motivational learning strategies. During the last year, this effort has been focused on ensuring the emotional well-being of students who have been facing challenges due to the pandemic year, as a precursor for effective online learning.

Index

K

L

M

O

P

R

S

T

V

Recommended Reference Books

IGI Global's reference books are available in three unique pricing formats: Print Only, E-Book Only, or Print + E-Book.

Shipping fees may apply.

www.igi-global.com

ISBN: 9781522589648
EISBN: 9781522589655
© 2021; 156 pp.
List Price: US$ **155**

ISBN: 9781799840756
EISBN: 9781799840763
© 2021; 282 pp.
List Price: US$ **185**

ISBN: 9781799855989
EISBN: 9781799856009
© 2021; 382 pp.
List Price: US$ **195**

ISBN: 9781799864745
EISBN: 9781799864752
© 2021; 271 pp.
List Price: US$ **195**

ISBN: 9781799875710
EISBN: 9781799875734
© 2021; 252 pp.
List Price: US$ **195**

ISBN: 9781799857709
EISBN: 9781799857716
© 2021; 378 pp.
List Price: US$ **195**

Do you want to stay current on the latest research trends, product announcements, news, and special offers?
Join IGI Global's mailing list to receive customized recommendations, exclusive discounts, and more.
Sign up at: **www.igi-global.com/newsletters.**

Publisher of Timely, Peer-Reviewed Inclusive Research Since 1988

IGI Global
PUBLISHER of TIMELY KNOWLEDGE

Ensure Quality Research is Introduced to the Academic Community

Become an Evaluator for IGI Global Authored Book Projects

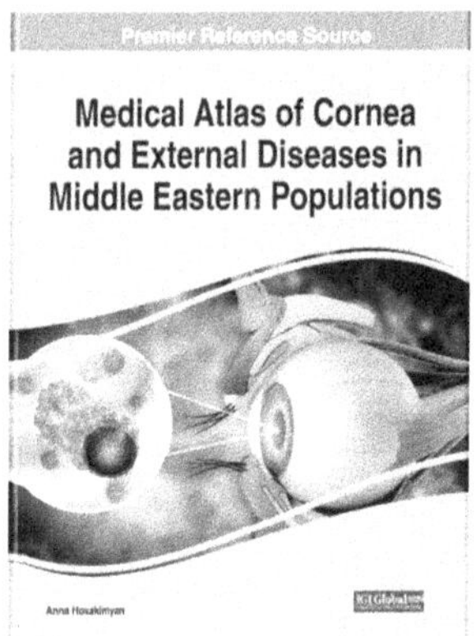

The overall success of an authored book project is dependent on quality and timely manuscript evaluations.

Applications and Inquiries may be sent to:
development@igi-global.com

Applicants must have a doctorate (or equivalent degree) as well as publishing, research, and reviewing experience. Authored Book Evaluators are appointed for one-year terms and are expected to complete at least three evaluations per term. Upon successful completion of this term, evaluators can be considered for an additional term.

If you have a colleague that may be interested in this opportunity, we encourage you to share this information with them.

www.ingramcontent.com/pod-product-compliance
Lightning Source LLC
LaVergne TN
LVHW060425300926
849162LV00006B/1528
9781668455852